ENABLING TECHNOLOGIES FOR MOBILE SERVICES

ENABLING TECHNOLOGIES FOR MOBILE SERVICES

The MobiLife Book

Editor

Mika Klemettinen
Nokia, Finland

BICENTENNIAL 1807 WILEY 2007 BICENTENNIAL

John Wiley & Sons, Ltd

Telephone (+44) 1243 779777

Email (for orders and customer service enquiries): cs-books@wiley.co.uk
Visit our Home Page on www.wiley.com

Other Wiley Editorial Offices

John Wiley & Sons Inc., 111 River Street, Hoboken, NJ 07030, USA

Jossey-Bass, 989 Market Street, San Francisco, CA 94103-1741, USA

Wiley-VCH Verlag GmbH, Boschstr. 12, D-69469 Weinheim, Germany

John Wiley & Sons Australia Ltd, 42 McDougall Street, Milton, Queensland 4064, Australia

John Wiley & Sons (Asia) Pte Ltd, 2 Clementi Loop #02-01, Jin Xing Distripark, Singapore 129809

John Wiley & Sons Canada Ltd, 22 Worcester Road, Etobicoke, Ontario, Canada M9W 1L1

Wiley also publishes its books in a variety of electronic formats. Some content that appears in print may not be available in electronic books.

Library of Congress Cataloging-in-Publication Data

Enabling technologies for mobile services : the MobiLife book / editor, Mika Klemettinen.
p. cm.
Includes bibliographical references and index.
ISBN 978-0-470-51290-6
1. Mobile communication systems–Technological innovations. 2. Wireless communication systems–Technological innovations. I. Klemettinen, Mika.
TK6570.M6E53 2007
621.3845′6–dc22

2007014669

British Library Cataloguing in Publication Data

A catalogue record for this book is available from the British Library

ISBN 978-0-470-51290-6 (HB)

Typeset in 10/12pt Times by Aptara, New Delhi, India
Printed and bound in Great Britain by Antony Rowe Ltd, Chippenham, England.
This book is printed on acid-free paper responsibly manufactured from sustainable forestry in which at least two trees are planted for each one used for paper production.

Contents

Preface

In September 2004, IST Integrated Project MobiLife for mobile applications and services research was launched with a large consortium comprising manufacturers, operators and solution providers, SMEs and academia. The strategic goal of MobiLife was to bring advances in mobile applications and services within the reach of users in their everyday life by innovating and deploying new applications and services based on the evolving capabilities of the 3G systems and beyond. The research challenge of MobiLife was to address the multi-dimensional diversity in end-user devices, available networks, interaction modes, applications and services. To deal with this complexity and to reach its strategic goal, MobiLife researched with a user-centric approach context-awareness, privacy and trust, adaptation, semantic interoperability, and their embodiment in novel services and applications matching key use scenarios of everyday life. This book collects together main findings from the joint research that was concluded at the end of 2006.

The book is organised into 11 chapters and an Appendix.

- Chapter 1 gives an introduction to the book.
- Chapter 2 describes the User-Centred Design process behind the book approach.
- Chapter 3 gives an overview of a complete and coherent Mobile Services Architecture meeting the requirements of the new Mobile World.
- Chapter 4 introduces a Context Management Framework that is a key component in the Mobile Services Architecture, as well as other context awareness-related technologies.
- Chapter 5 presents new multimodal and personalisation technologies that are essential in providing natural and adapted user interaction in the novel mobile applications.
- Chapter 6 goes into privacy and trust issues; i.e. how to provide privacy-preserving and secure approaches for mobile applications. Additionally, novel group awareness solutions are introduced.
- Chapter 7 describes a number of reference applications that use the introduced Mobile Services Architecture and its technological components presented in Chapters 4–6.
- Chapter 8 defines both best practices and key learning from user and technical evaluation of the Mobile Service Architecture, the technological components and the reference applications.
- Chapter 9 analyses the marketplace dynamics and business model perspectives related to the new Mobile World.
- Chapter 10 complements the analyses from legal and regulatory perspectives.

- Chapter 11 is a short conclusion to the book.
- The Appendix contains the mobile scenarios used throughout the book.

This book is targeted at all people working with enabling technologies and service architectures either in companies or academia, and students studying applications/services, enabling technologies and service architectures at universities, but also to anyone interested in the general issues surrounding mobile technology and user research.

Also, as an extra resource, additional materials are available on the companion website at http://www.wiley.com/go/klemettinen.

Acknowledgements

Thanks to all of those people who contributed to this book and to the research that made this book possible, with special thanks to all of the Chapter Editors and Section Editors for their efforts as well as to Andy Aftelak for the additional proofreading. Thanks also to the European Commission for the partial funding for the research carried out in the MobiLife project and to the Wiley team for practical assistance.

Finally, thanks to my wife Mimi with her patience and love while finalising this book, especially during the last weeks and days of her pregnancy for our first child, Minni.

Contributors

Book editor	Mika Klemettinen (Nokia)
Chapter 1	
Editor	Mika Klemettinen (Nokia)
Authors	Andy Aftelak (Motorola Ltd) Mika Klemettinen (Nokia) Jukka T Salo (Nokia)
Chapter 2	
Editor	Annakaisa Häyrynen (Elisa)
Authors	Andy Aftelak (Motorola Ltd) Luca Galli (Neos) Annakaisa Häyrynen (Elisa) Ulla Killström (Elisa) Esko Kurvinen (Helsinki University of Technology) Harri Lehmuskallio (Helsinki University of Technology) Mia Lähteenmäki (Nokia) Kevin Mercer (Motorola Ltd) Antti Salovaara (Helsinki University of Technology)
Chapter 3	
Editors	Bernd Mrohs (Fraunhofer FOKUS) Stephan Steglich (Fraunhofer FOKUS)
Authors	Bharat Bhushan (Fraunhofer FOKUS) Mathieu Boussard (Alcatel-CIT) Alexander Domene (Fraunhofer FOKUS) Renata Guarneri (Siemens SpA) Denis Leclerc (Alcatel-CIT) Alessandro Mamelli (HP Italiana) Bernd Mrohs (Fraunhofer FOKUS) Christian del Rosso (Nokia) Christian Räck (Fraunhofer FOKUS)

Alfons Salden (Telematica Instituut)
Jukka T Salo (Nokia)
Stephan Steglich (Fraunhofer FOKUS)

Chapter 4

Editors

Patrik Floréen (University of Helsinki)
Matthias Wagner (DoCoMo Euro-Labs)

Authors

Agathe Battestini (Nokia)
Adrian Flanagan (Nokia)
Patrik Floréen (University of Helsinki)
Stefan Gessler (NEC)
Johan Koolwaaij (Telematica Instituut)
Eemil Lagerspetz (University of Helsinki)
Sian Lun Lau (University of Kassel)
Marko Luther (DoCoMo Euro-Labs)
Miquel Martin (NEC)
Jean Millerat (Motorola SAS)
Bernd Mrohs (Fraunhofer FOKUS)
Petteri Nurmi (University of Helsinki)
Massimo Paolucci (DoCoMo Euro-Labs)
Julien Robinson (Alcatel-CIT)
Jukka Suomela (University of Helsinki)
Claudia Villalonga (NEC)
Matthias Wagner (DoCoMo Euro-Labs)

Chapter 5

Editors

David Bonnefoy (Motorola SAS)
Olaf Drögehorn (University of Kassel)
Ralf Kernchen (University of Surrey)

Authors

David Bonnefoy (Motorola SAS)
Mathieu Boussard (Alcatel-CIT)
Nermin Brgulja (University of Kassel)
Alexander Domene (Fraunhofer FOKUS)
Olaf Drögehorn (University of Kassel)
Giovanni Giuliani (HP Italiana)
Ralf Kernchen (University of Surrey)
Sian Lun Lau (University of Kassel)
Jean Millerat (Motorola SAS)
Bernd Mrohs (Fraunhofer FOKUS)
Petteri Nurmi (University of Helsinki)
Pekka J Ollikainen (Nokia)
Mateusz Radziszewski (BLStream)
Christian Räck (Fraunhofer FOKUS)
Marcin Salacinski (BLStream)
Alfons Salden (Telematica Instituut)
Michael Sutterer (University of Kassel)

Chapter 6

Editors

Göran Schultz (LM Ericsson)
Olivier Coutand (University of Kassel)
Ronald van Eijk (Telematica Instituut)
Johan Hjelm (Ericsson AB)
Silke Holtmanns (Nokia)
Markus Miettinen (Nokia)
Rinaldo Nani (Neos)

Authors

Stefano Campadello (Nokia)
Olivier Coutand (University of Kassel)
Peter Ebben (Telematica Instituut)
Ronald van Eijk (Telematica Instituut)
Johan Hjelm (Ericsson AB)
Silke Holtmanns (Nokia)
Theo Kanter (Ericsson AB)
Sian Lun Lau (University of Kassel)
Miquel Martin (NEC)
Björn Melén (LM Ericsson)
Markus Miettinen (Nokia)
Rinaldo Nani (Neos)
Petteri Nurmi (University of Helsinki)
Mateusz Radziszewski (BLStream)
Marcin Salacinski (BLStream)
Göran Schultz (LM Ericsson)
Esa Turtiainen (LM Ericsson)

Chapter 7

Editor

Dario Melpignano (Neos)

Authors

Péter Boda (Nokia)
Nermin Brgulja (University of Kassel)
Stefan Gessler (NEC)
Giovanni Giuliani (HP Italiana)
Johan Koolwaaij (Telematica Instituut)
Miquel Martin (NEC)
Dario Melpignano (Neos)
Jean Millerat (Motorola SAS)
Rinaldo Nani (Neos)
Petteri Nurmi (University of Helsinki)
Pekka J Ollikainen (Nokia)
Petr Polasek (UNIS)
Mateusz Radziszewski (BLStream)
Marcin Salacinski (BLStream)
Göran Schultz (LM Ericsson)
Michael Sutterer (University of Kassel)
Dari Trendafilov (Nokia)
Libor Ukropec (UNIS)

Chapter 8

Editors
Esko Kurvinen (Helsinki University of Technology)
Renata Guarneri (Siemens SpA)
Jukka T Salo (Nokia)

Authors
Agathe Battestini (Nokia)
Luca Galli (Neos)
Renata Guarneri (Siemens SpA)
Annakaisa Häyrynen (Elisa)
Mika Karlstedt (University of Helsinki/Nokia)
Esko Kurvinen (Helsinki University of Technology)
Harri Lehmuskallio (Helsinki University of Technology)
Kari Lehtinen (Elisa)
Mia Lähteenmäki (Nokia)
Rinaldo Nani (Neos)
Pekka J Ollikainen (Nokia)
Marcin Salacinski (BLStream)
Nicoletta Salis (Telecom Italia)
Jukka T Salo (Nokia)
Antti Salovaara (Helsinki University of Technology)

Chapter 9

Editor
Ulla Killström (Elisa)

Authors
Luca Galli (Neos)
Timber Haaker (Telematica Instituut)
Olli Immonen (Nokia)
Ulla Killström (Elisa)
Mark de Reuver (Telematica Instituut/TU Delft)

Chapter 10

Author
Olli Pitkänen (Helsinki Institute for Information Technology HIIT, Helsinki University of Technology and University of Helsinki)

Chapter 11

Editor
Mika Klemettinen (Nokia)

Authors
Andy Aftelak (Motorola LTD)
Mika Klemettinen (Nokia)
Jukka T Salo (Nokia)

List of Figures

List of Tables

1

Introduction

Edited by Mika Klemettinen (Nokia, Finland)

1.1 Overview

During the past decade, two simultaneous phenomena have changed the role of information and communications technology in the everyday lives of most people living in the developed part of the world: the Internet and mobile telephony.

Mobile phones especially have had a tremendous impact on our life and societies. In many developed countries people are used to being able to contact almost anyone, anywhere, and at any time – being always reachable by others. The ability to stay informed and to respond quickly has become vital in our business and private lives, and has changed how we agree on things to do, schedule them, adapt to changes, and maintain a general feel of each others' situation. Yet, as profound as these capabilities and changes may be, they are only the first stage in an ongoing development towards the mobile knowledge-based society, the new Mobile World.

This Mobile World is expected to fulfil the vision of ubiquitous computing and communications providing access to not only Internet-like or telecommunications-like, but also entirely new kinds of service at any time and anywhere. Through the Mobile World, computing and communications seem to be destined to invade and inhabit, for better or worse, every part of our everyday environment – every place and social situation that people go to and spend time in:

> Traditionally, people used a number of life spaces of different functionality or significance (home, workplace, school, hospitals, to name a few) for specific purposes. Mobile services, however, can provide many of these functionalities transcending spaces. In other words, it enables people to literally study, play, work and shop anywhere they want. [2]

Enabling Technologies for Mobile Services: The MobiLife Book Edited by Mika Klemettinen

Ultimately, the integration of Mobile World technology with wirelessly networked devices embedded in various products and the physical environment will open the door to truly universal and ubiquitous services. This view of the future is also shared by international consortia such as the Wireless World Research Forum (WWRF) in its *Book of Visions* [4].

In order to be able to adjust to the new requirements, industry is facing a major transformation, where telecommunications, information technologies, consumer electronics and media industries are converging. However, the changing game is not only visible in industry. As described above, the use of mobile communication services will be taken for granted in the daily lives of people, and their behaviour may significantly change from the conventional behavioural patterns.

Within this setting, the expected future evolution of mobile and wireless communication technologies will enable a whole new generation of mass-market-scale ubiquitous services and applications. At the same time, the convergence and the changing marketplace dynamics imply also new software, systems and services related requirements and challenges: the creation of services should become easier, faster and cheaper; interoperability of different systems and services from different domains is required; complex and changing business ecosystems should be supported; etc.

The realisation of the Mobile World vision will ultimately depend on its acceptability and value to end-users who are 'increasingly living in a 'mosaic society' with lifestyles characterised by greater mobility, diversity and change' as decribed in an ISTAG report [1]. The same report continues:

> Increasingly many people live at considerable distances from family, friends and feel little connection with community organisations and local and national democratic structures. There is a feeling that face-to-face communication, social gatherings and 'neighbourliness' are in decline which has led a number of commentators to warn that our stock of social capital – the very fabric of our connections with each other – has plummeted, impoverishing our lives and communities.

Indeed, the ultimate success of the services and applications created will depend on how well they match the values of the intended end-users in binding the disconnected, calming the hectic, empowering the weakened, enriching the impoverished, and bringing close the distant aspects of modern life.

How to provide all that added value? People today take part in varying social contexts and play different roles in their everyday life. There is a growing need to manage today's complex lifestyles – this requires facilities and tools to support communication, and to share information and time with others. Future communication environments may give new possibilities to do this, but also new challenges due to increasing heterogeneity of technological environments, user needs and expectations.

In order to be able to overcome these technological and non-technological barriers, certain issues have to be addressed. What will the end-users use their devices and services for? Which new devices and services will be acceptable to end-users, many of whom already feel overwhelmed by the hectic lifestyle? How can desirable service characteristics be facilitated on the basis of new and emerging enabling technologies such as positioning, context sensitivity, and adaptive multimodal interaction? What kinds of technical infrastructures and platforms are needed? How can the services be provisioned and adapted to match the variable needs and profiles of potentially hundreds of millions of end-users? How will different players on the market – network operators, service operators, content and application providers, public

Figure 1.1 Multiple aspects of life in today's society. Copyright © 2007 Nokia Corporation.

authorities, user groups and individual users – interact and co-operate to create and provide the services effectively, timely, and securely while balancing the various interests and values of the different players?

To fulfil the requirements of the different actors, many technology enablers have been introduced by different initiatives and organisations focusing on one of the requirement subsets.

The result is a quite complex mobile landscape where the technology enablers are incomplete, inconsistent, competing and even conflicting with each other. Mobile services were originally implemented as logic installed in the mobile handset and in the mobile network. To improve mobile services, technology enablers both in the mobile handset and the mobile network are required. In addition, to enable the introduction of innovative services, there should also be technology enablers that bridge the mobile network to the Internet and to intranets.

In order to be able to cope with the complex mobile landscape, innovations on the architectures side are needed as well. Existing software architectures are often component-based, while service architecture (or Service Oriented Architecture, SOA) means a new paradigm shift emerged from the growing complexity of distributed software. It means a dynamic architecture – i.e. the structure and behaviour of software is changing at run-time as well as the location where the software is executed. A pervasive computing environment also brings new non-functional requirements related to interoperability, heterogeneity, mobility and adaptability. This means that pervasive services have to be loosely coupled, and service composition will be supported by service architectures that simplify complexity and allow dynamic service compositions; i.e. services will be self-descriptive. Dynamic service compositions and dynamic binding techniques are required to enable dynamic architectures. Distributed self-aware software is required to manage the distributed service communication in a dynamic environment.

Aspects of service architecture are rarely visible to the end-user. Thus, it is only natural that the end-user point of view has gained less attention and developmental work is driven from the technological angle. Preconditions for success, even from a solely technological perspective, are considerable.

This book takes a comprehensive approach to these challenges and provides practical guidelines on building new, innovative applications and services, and gives also learning from a collaborative research project MobiLife where the methods and technologies were applied and utilised [3].

Methodologically, the approach used in this book acknowledges that the user requirements must be learned partially from experiments with novel service and application prototypes illustrating the end-user value and possibilities of new enabling technologies. To implement this, this book assumes a user-centric stance, focusing on the viewpoint of an individual user and his/her everyday life defined by the complex web of relationships and interactions with other people and groups such as family members, friends, relatives, work colleagues, and the various scenes delineating the flow of everyday life such as home, office, car, and various public spaces such as school, medical centre, library, hobby clubs, shopping centre, entertainment facilities, and public spaces in cities and towns. The objective is to recognise and realise novel services and applications that address the true user needs emerging from this complex picture and provide sustained added value and positive experiences to the end-user.

To make these services and applications real, the book likewise investigates key application enablers and technologies deemed crucial for their implementation, keeping in mind qualitative constraints such as the very large number of end-users and their diversity. Technologies for maintaining a 'shared cognition' amongst groups of users – such as modelling and reasoning for contextual awareness, technologies for facilitating and maintaining privacy and trust, and technologies for creating and sharing various kinds of content and media related to everyday life – are key areas covered in the book. The enablers and technologies are embodied in application experience prototypes to provide further opportunities to learn how they can facilitate providing sustained added value to the end-users. This complies closely with the recommendation of ISTAG [1] that 'functional, technical, social and economic requirements of systems, gathered from users and stakeholders, are put at the centre of the development process and are revisited throughout design, implementation, checking, and testing'.

None of the services and applications developed will ever reach the end-users, if they cannot in practice be created and provided by some value network consisting of network operators, service operators, content providers, integrators, and others as may be needed. Therefore, the full lifecycle of service creation, packaging, configuration, provision and support are also addressed by this book, thus complementing the user-centric view with the equally decisive value network view. In the context of marketplace dynamics, the book studies the relevant business model and societal issues, in particular potential legal problems and their solutions.

1.2 Acknowledgements

The following individuals contributed to this chapter: Andy Aftelak (Motorola LTD, UK), Mika Klemettinen (Nokia, Finland), and Jukka T Salo (Nokia, Finland). This chapter also presents the joint vision and motivation behind the MobiLife project [4], where numerous people have contributed to and which has materialised partially in the form of this book.

References

[1] ISTAG: 'Strengthening Competitiveness through Cooperation: European Research in Information and Communication Technologies'. IST document, September 2004. 24 pp. Online: ftp.cordis.europa.eu/pub/ist/docs/strengthening-european-research-in-ict.pdf.

[2] mITF (mobile IT Forum): 'Flying Carpet: Towards the 4th Generation Mobile Communications Systems, version 2.0'. mITF, 2004. Online: www.mitf.org/public_e/archives/Flying_Carpet_Ver200.pdf.

[3] MobiLife project. Online: www.ist-mobilife.org.

[4] Tafazolli R. (ed.): *Technologies for the Wireless Future: Wireless World Research Forum* (WWRF), Volume 2, ISBN 978-0-470-02905-3, Wiley, April 2006. 520 pp.

2

Users, Applications and Services, and User Centricity

Edited by Annakaisa Häyrynen (Elisa, Finland)

In the previous chapter, the idea of the Mobile World, in which the user can access a variety of services at any time and anywhere, using his/her personal wireless devices as well as other resources and embedded devices in the environment, was envisioned.

In this chapter, the current *mobility landcape* is reviewed by examining particular trends that are increasingly making this Mobile World a reality, such as the convergence of fixed and mobile systems, the trend towards intelligent environments, and the increasing number of mobile applications and services (Section 2.1).

Then, a generically applicable *user-centred design (UCD)* approach is introduced to take into account the end-users, the people for whom the mobile applications and services are created. Section 2.2 discusses UCD and user perspectives related to future services with advanced methodologies and processes for user research. UCD is not one single process description but rather a set of potential methods and tools that can quite flexibly be implemented case by case. Section 2.3 gives examples of and explanations on the approach of this book. It describes in detail the selected focus group (families) and their identified needs, the scenario-driven process leading to key user requirements, and the implementation of the 'Big Loop' UCD process described in Section 2.2.

Finally, Section 2.4 is a short conclusion summarising some key conclusions relating to the other chapters in this book and a summary on implementing a user-centred design process in order to create new applications, services and related enabling technologies.

Enabling Technologies for Mobile Services: The MobiLife Book Edited by Mika Klemettinen

2.1 Mobility Landscape

From the user's point of view, the new kind of 'mobility' introduced in Chapter 1 means the ability to access information, services and people in new ways, expanding the current mobile phone and Internet usage trends into new kinds of *seamless service access*.

The popularity of mobile phones and other portable communication devices is a first step towards this kind of mobility. The growth of the mobile market has been remarkable. For example, most countries in Europe have an average of more than 75 mobile phones per 100 people in the population [40]. According to the International Telecommunication Union, the number of mobile phone subscriptions across the world has been higher than the number of traditional fixed/landline subscriptions since 2002 (1.155 billion mobile cellular subscribers, compared to 1.129 billion fixed telephone lines) [26].

However, the concept of 'mobility' means much more than simply accessing web content and services from a portable device. As shown, for example, in the WWRF MultiSphere model [58], it is expected that the user will be able to use an ensemble of wearable and portable devices (forming Body Area or Personal Area Networks), which will be able to determine her current situation (context). Specific contextual information and personal preferences can then be taken into account in tailoring the user's content and services to her particular situation. As the user moves through the world, she will encounter new services and resources in the environment (e.g. services being offered in a WiFi hotspot) and these contextual preferences will be increasingly important. In addition, the user should be able to use applications and services *seamlessly* when mobile, without needing to know any details about the underlying technologies. In general, people will expect that higher bandwidth and more constant connectivity, and thus higher quality of service, will be available in this new Mobile World.

2.1.1 Fixed, Mobile and IP Convergence

A key element of the seamlessness described above is the *convergence of fixed line and mobile phone infrastructures*. There is an increasing trend towards the blending of wireline and wireless networks, especially in the core network; one research report, for example, predicts that 'In the core network, the boundaries between fixed and mobile technologies will be largely dissolved by 2010–2012 ... driven mainly by the universal migration to an all-IP network' [17].

From the user's point of view, the fixed–mobile convergence will mean that the user will be able to use her services seamlessly across different locations, by switching between different networks in an uninterrupted way. Karasti *et al.* describe the benefits to the user in the following terms:

> This convergence enables a customer seamlessly and transparently to hand over call or service usage between a mobile network and a fixed network. For example, when arriving at home, the current call or service experience is transferred on the home network, thus providing more bandwidth and better transmission quality and perhaps being less expensive. Moreover, this would also allow users to have only one number, one handset, one bill, whatever networks they use.' [29]

The fixed–mobile convergence will enable the user to continue using her services no matter which network is available, and wherever possible to upgrade her quality of service to the highest possible level allowed by any of the available networks (not just the network upon which the user started accessing the service).

2.1.2 Mobility and Intelligent Environments

Another trend supporting the new idea of mobility is the emergence of various context-aware technologies and intelligent environments. This trend goes by many names, including ubiquitous computing, pervasive computing, ambient intelligence, and 'the disappearing computer'. The concept of 'intelligent space' is a key to this trend. MIT's Project Oxygen defines intelligent spaces in the following way:

> Space-centered computation embedded in ordinary environments defines *intelligent spaces* populated by cameras, microphones, displays, sound output systems, radar systems, wireless networks, and controls for physical entities such as curtains, lighting, door locks, soda dispensers, toll gates, and automobiles. People interact in intelligent spaces naturally, using speech, gesture, drawing, and movement, without necessarily being aware that computation is present. [48]

2.1.2.1 Sensors and Sensor Networks

In order to tailor the performance of the user's applications and services to her situation, there must be some way for the technologies surrounding the user to know what that situation is. Is the user inside or outside? Walking, driving, or standing still? At home or at work or at school? Certain features of the user's situation, such as her location, can be gathered without the use of sensors (e.g. by using the user's cell-of-origin to approximate her physical location). To detect other information about the user, it is possible to use various kinds of sensor. Sensors can be based on many different enabling technologies (e.g. piezo-materials, VLSI video, optical gyros and MEMS (micro electro-mechanical systems) [19]; bio-sensors measuring, e.g., respiration, blood pressure, skin conductivity and muscle tension [47]). The use of sensors can determine many characteristics of the user's context, including:

- location and position information (for both the user and objects in the environment);
- auditory context;
- visual context;
- physiological context (sometimes used to make inferences emotional context, such as measuring the sweatiness of someone's palms to determine her stress level);
- environmental characteristics (temperature, etc.).

In addition, individual sensors can be combined into sensor networks, which enable the system to make inferences based on a wide array of context information rather than a single data point such as the ambient temperature [4]. The promise of sensor-based (in general, context-aware) computing is that one can dramatically improve the user's interaction with services by, for example, adjusting the interface mechanisms to suit her current situation (using various multimodal interaction techniques, such as pointing, speaking and gesturing) and filtering incoming messages to suit her personal preferences. Furthermore, context information about one user can be shared with other users, for example for presence and awareness services.

2.1.2.2 Intelligent Homes

Another way of understanding the user's current environment is to make that environment 'intelligent': for example, by equipping it with sensors and other embedded devices that can determine the user's position and activities automatically. Much of the research in this area has focused on intelligent work and school environments (offices, university classrooms, and so on). However, one active research topic for the intelligent environments/spaces concept is the *intelligent networked home* [32,43]. Research initiatives about intelligent homes aim to understand how transforming the home into an intelligent environment can make everyday life easier for the people living there.

In a further example, the scenarios from the Microsoft EasyLiving project [7] show the user interacting with a variety of displays and embedded devices in the home, with the following user tasks and behaviours emphasised:

- viewing information on various displays in the home (including large wall screens, etc.) which are chosen automatically based on the direction of the user's gaze;
- controlling the lights, music and other features of the room through interaction with these displays;
- moving information from one display to another based on convenience (e.g. move to a larger display for sharing).

The Aware Home at Georgia Tech [31] senses users' location within the home and helps them find lost objects. The specific user application for this project is to support elderly users by adding intelligence to their surroundings; for example, sensors embedded in the environment could help remote family members keep track of how their elderly relative is doing.

Finally, the Digital Living Network Alliance (DLNA), a consortium of consumer electronics, mobile, PC manufacturers and other stakeholders, posits a future intelligent home in which the user can use her personal and commercial media seamlessly within the home. This calls for interoperability among PCs (including home media servers), consumer electronics devices such as DVD players and set-top boxes, and mobile devices. Of particular interest to this book are the DLNA's ideas about the use of mobile multimedia terminals in the intelligent home environment:

> Unlike devices permanently located at home, mobile devices can easily be used to capture content while away from the [home] network and bring it back. In order to enjoy the content, users need to be able to upload it to a media server on the home network using Wi-Fi® or Bluetooth .... Another unique capability of handheld devices is the potential to use them like remote controls to control other devices in the home. [14]

Similarly, in this book, it is envisioned that the home will be equipped with technologies that allow the user to use her mobile devices as controllers for other devices in the home, and that the mobile devices will form part of a network of resources in the home.

2.1.2.3 Applications and Services

Another trend supporting the new notion of mobility is the steady increase in the number and variety of available services for both fixed and mobile use. Not only is the number of standalone user applications and services increasing, but concepts such as generic web services, the Semantic Web, and service composition (service chaining) mean that independent services can also be combined to offer advanced and personalised functionality to the user [42]. Access to this wide array of applications and services is a key driver for the new notion of mobility. The users (in most cases) will be interested in the utility of the applications and services and not in the mobile technologies underlying them.

There are many ways of categorising digital applications and services. Some researchers have posited that there are four categories: information, communication, transaction and entertainment [57]. For example, communication services include fixed line phone calls, mobile calls, SMS, e-mail, instant messaging, and Internet chat. Other researchers have categorised services according to whether they are pragmatic or amusing. For example, Kruse and Carlsson [35] list the following as examples of services with practical benefits for the mobile user:

- navigation, finding the right place;
- directory services – 'yellow pages' (to find people and places);
- information on transportation;
- restaurant guides, with reviews of restaurants;
- TV guides (to find out what's on tonight);
- buying movie tickets;
- dictionaries (while abroad);
- security services (such as emergency alarms);
- reminders (for instance details about a dentist appointment);

They contrast these with 'amusement services', such as:

- music recordings;
- FM radio;
- TV clips, movie trailers, home movie video clips;
- games, horoscopes, jokes;
- video clips of news, sports highlights;
- shopping tips, advertisements for shops or products I like.

This section gives a brief overview of common applications and services for home, leisure and work use, and discusses the increasingly digital access to public services.

Home/leisure

The number of people using the Internet from home is still climbing in Europe. People use the Internet at home for many different purposes, including news and entertainment browsing, email, e-commerce, gaming, and downloading video clips, music and other media. Various web-based applications and services that support community groups have also emerged, such as Yahoo!Groups, a free service that makes it easy to set up a website (with file sharing and scheduling tools) and e-mail list for a particular group (such as a group of family and friends). Instant messaging services such as Instant Messenger and ICQ are also popular for home use. Finally, the trend of web logging (blogging) is also on the increase. All of these activities have an analogue in the mobile world.

Many mobile applications and services are focused on leisure activities. Mobile operators are offering various presence services, which users can use to locate each other and which are often targeted at groups of friends, such as ImaHima in Japan, which allows mobile users to share their current status (location, mood, activity) with a buddy list of their friends [49]. Other location-based services for leisure include restaurant finders, city guides and shopping services [27]. Mobile multiplayer games, including location-aware games, have also increased in popularity.

Mobile users are increasingly able to capture and share their own multimedia content. People use camera phones to take pictures of everyday occurrences and send these to friends and family [54]. Parallel to the blogging trend is the idea of a 'moblog', in which these images are made available in online forums. With these services comes the idea of the user as a 'prosumer', one who both produces and consumes content.

Work

Increasingly people are working from locations other than their offices, using laptops, PDAs and mobile phones to access their work contacts and information. For example, people can get secure mobile access to corporate Local Area Networks (LANs) and Virtual Private Networks (VPNs). The growing popularity of broadband connections from home has also facilitated this kind of remote working.

Other work-related applications and services include shared calendars, instant messaging with colleagues, collaborative meeting applications such as Microsoft's NetMeeting and a variety of web meeting tools. In addition, the Internet is widely used as a source of information for work purposes. All of these services and applications will also be expected to be accessible while the user is mobile.

Public Services

Information about public services is increasingly available online, including public transportation schedules and maps, medical and healthcare information, directory services ('yellow pages'), and government information. In some cases the service itself is available online; for example, many people can now use the Internet to renew their library books, make appointments with their doctors, and pay their taxes. All of these services are also expected to be available to mobile Internet users.

2.2 Collaborative User-centric Design Process

Section 2.1 painted a landscape for mobility and described a world where people are living with their multiple roles and needs. Before developing architectures, enabling technologies or applications based on these generic trends and notions, one should carefully find out what technologies and solutions to use and how to exploit them in a way that users are willing to adopt. In this section, a user-centred design (UCD) process approach is introduced to take into account different user requirements and to consider user researchers and developers as part of the innovation process.

2.2.1 Introduction to User-centred Design

User-centred design (UCD; also known as human-centred design, ISO 13407 [24]) is an empirical research and product development orientation that utilises end-user or customer information for making better (efficient, usable, enjoyable, etc.) and thus commercially

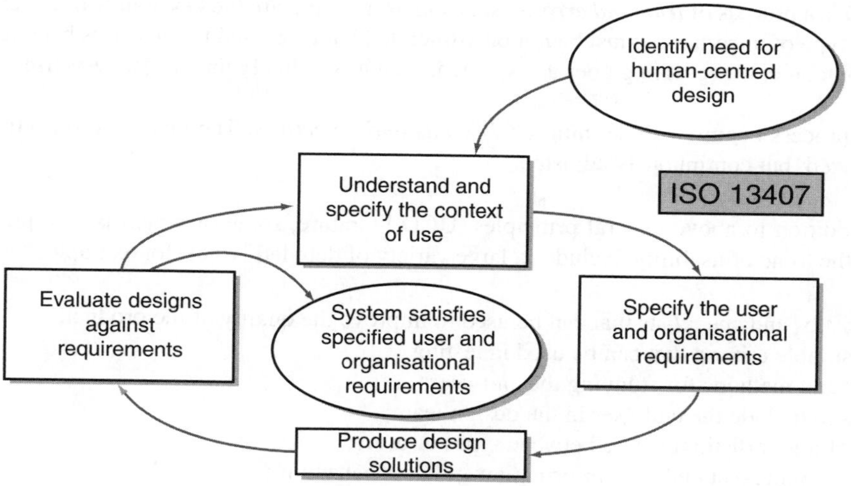

Figure 2.1 ISO13407: user-centred process.

successful products. In practice, this is achieved by involving the end-user in the product development process.

Gould and Lewis [21] list key principles of UCD, dating back to the 1970s:

- '*Early Focus on Users and Tasks*. First, designers must understand who the users will be. This understanding is arrived at in part by directly studying their cognitive, behavioural, anthropometric, and attitudinal characteristics, and, in part, by studying the nature of the work expected to be accomplished.
- *Empirical Measurement*. Second, early in the development process, intended users should actually use simulations and prototypes to carry out real work, and their performance and reactions should be observed, recorded, and analysed.
- *Iterative Design*. Third, when problems are found in user testing, as they will be, they must be fixed. This implies design must be iterative. There must be a cycle of design, test and measure, and redesign, repeated as often as necessary.' [21]

As user centricity is an attribute of the process through which products are developed, it cannot be encapsulated in a single stage that delivers user needs or requirements. Instead, organisational structures for maintaining the end-user focus throughout the process must be created. Rubin [52] outlines some attributes of user-centric organisations:

- UCD is a *phased approach*. This means that there is user input at all crucial points. These milestones are to be specified beforehand and put in the project schedule.
- *Multidisciplinary teams*. User centricity is more likely to take place in teams that have people from multiple disciplines.
- *UCD-concerned management*. Management monitors that user involvement milestones are followed. For decisions that affect the end-user experience, teams are made accountable: 'How was this decision informed by user data?'

- UCD is a process of *trial and error*. Users and user testing are the key source of uncertainty. This type of uncertainty must be not only tolerated but promoted because it is beneficial for the outcome. If everything goes as expected, then it is unlikely that project was user-centric [34].
- The process begins with defining *UCD goals and objectives*. The goals and objectives are not fixed, but continuously adjusted.

In addition to above general principles, UCD literature, some of which is also presented under the topic of usability, includes a large variety of detailed issues, for example:

- heuristics and checklists that can be used to improve the quality of the product;
- measurable criteria that can be used in testing;
- tools and methods for studying the end-user;
- ways to include the end-user in the design team;
- modelling or distinguishing between types or users;
- ways to represent end-user information and user behaviour;
- project planning and management tools for end-user inclusion at key phases;
- organisational (i.e. company-level) policies and approaches for creating and maintaining customer orientation;
- tools for measuring the maturity level of the organisation so that the appropriate UCD approach can be selected.

It is recognised that general guidelines or principles do not always apply to specific problems. User-centricity and usability are highly context-dependent. In practice, processes, requirements, guidelines and lists are interpreted and refined for the purposes of each individual project [53].

2.2.2 UCD Process: the Big Loop

Following the UCD principles presented above, a user-centric applications and services development process [38] consists of four iterations of technical R&D and user research. The key steps in this *Big Loop* (Figure 2.2) are the *technical specification stages* (the technical research and development activities), and the related *prototype evaluation stages* (user research and evaluation activities).

In the following, the process is described in chronological order as a stepwise flow. In reality, the process involves frequent interaction between users, user experts and technical experts (like multimodality, sensors, privacy and trust, contextual related issues, integration experts, architecture experts, ontologists, etc.). Also, the division between design and testing phases does not always make sense, since in order to make prototypes testable, it is typical that some adjustments are needed.

First Round of the Big Loop

At the beginning of the Big Loop, existing scenarios in the area of ubiquitous, context-aware and ambient computing should be collected and analysed from both technological and user perspectives. The idea is that these scenarios best summarise the key benefits of the technology,

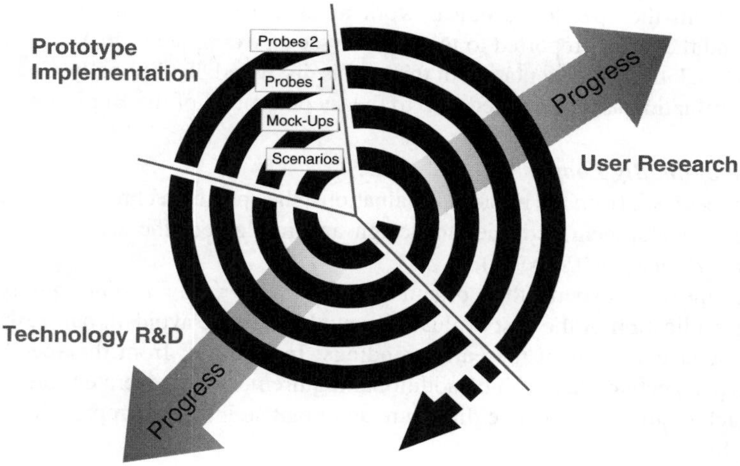

Figure 2.2 Iterative R&D process.

as anticipated by previous research and industrial experience. They also provide a starting point for the development project.

During the analysis process, the specific technical work areas, based on the scenario material, should also build their own scenarios, or rather *use cases*. These work area *internal scenarios* are constructed because the initial scenarios are mostly tools for communication, and, as such, are not detailed enough in describing specific technological solutions.

Using all the scenarios as background materials, it is possible to identify the *key user tasks/behaviours*. These tasks/behaviours, together with jointly created high-level scenarios including these tasks/behaviours (see examples in the Appendix), provide a summarised view of the vision with prioritised functionalities to be used in the applications, services, enabling technologies and architectures development.

In order to verify the vision, one should organise user evaluation of the scenarios (see an example in [13]) and feed the results into the development project.

Sections 2.3.1 and 2.3.2 take a more detailed look at this process with an analysis of the selected focus group, families. Furthermore, the scenarios created for this book (see also the Appendix) and requirements derived are discussed in more detail.

Second Round of the Big Loop

In parallel to the scenario user evaluation, specific technical work areas should edit and map their internal scenarios to the joint high-level scenarios. This *hooking process* ensures that the technical work-area specific internal scenarios that form the basis for technical research and development are relevant from the point of view of the vision encapsulated in the high-level scenarios.

The specific scenarios provide also material for the design and implementation of the *first low-fidelity prototype versions of the applications and services (mock-ups)*.

Mock-ups from the specific technical work areas are used as part of a series of *user evaluations* and the results reported to the project (see an example in [36]).

In Section 2.3.3, the implementation of this step is discussed. Additionally, Chapter 8 covers certain user evaluation aspects related also to the second round of the Big Loop.

Third Round of the Big Loop

Based on the feedback from the mock-up evaluations, the specific technical work areas should continue their development activities to design and implement the *second versions of the applications and services* (Probes 1).

As before, the user experts then organise *expert and/or user evaluations of Probe 1s.* Unnecessary replication of the first evaluation round should be avoided, but fresh viewpoints are collected that complement the earlier findings. In addition, from the specific technical work areas's perspective, there can be additional requirements to generate more enabler-level findings, which requires a reflective, literature and practice informed approach (Section 2.3.3 and Chapter 8).

Fourth Round of the Big Loop

The final iteration of the process has to be carefully considered. In typical 'industrial' user evaluation cases the challenge is to go 'out-and-wild' and really to trial the novel applications and services with users. User trials in real usage environments, as part of everyday life, is a big challenge in many ways. User trials are time-consuming, a major challenge being the process of *getting the application to the stage that it can be given to non-expert users.* The application finalisation can take several months depending on integration challenges on platforms of various kinds (e.g. mobile devices on which the application or service should run). To accelerate this stage, multidisciplinary collaboration is needed; the user experts can assist the technical developers in defining how the trial application/service should be designed to be useful for the trial.

Then, the final round of user evaluations should be real user trials in true contexts of use with *prototypes that are functional to the extent needed to study them properly with the end users.* Special attention should be given to the social dimension of the applications and to enabler-level technologies.

In general, before the actual user evaluations, it is likely that at least one round of *heuristic/expert evaluation* is needed for each application. The purpose of these expert evaluations is to fix most obvious usability problems, but also to inform planning of the final evaluations, especially related to the environment and time span of the user tests.

Aspects related to the fourth and final round of the Big Loop are covered in Section 2.3.3 (implementation aspects). Additionally, Section 2.4 gives general remarks on the process and Chapter 8 describes the user evaluation process and learning with the technologies and applications presented in this book, having an emphasis on the trials.

2.3 UCD in Action

This section gives more details of the user-centred design approach. Section 2.3.1 starts with an important step for any application or service development project, identification of the focus groups (i.e. potential customers), and their needs. In this example case, the focus group

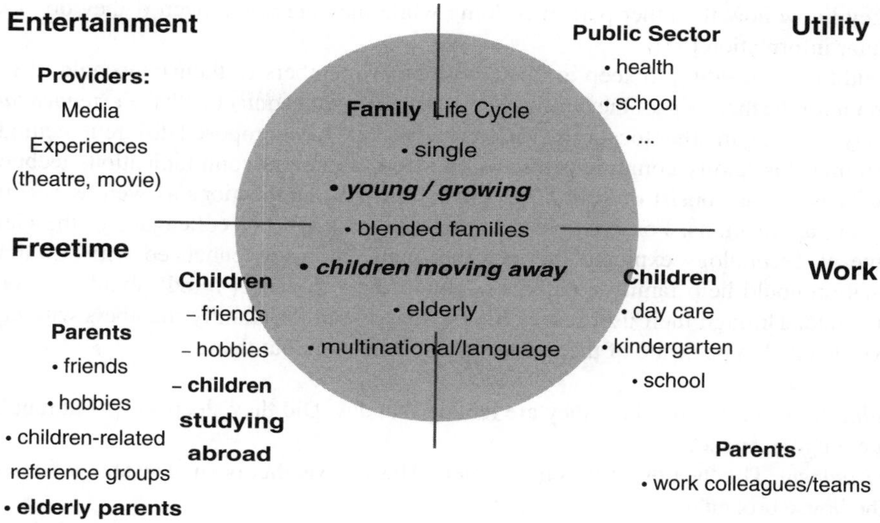

Figure 2.3 Types of family, all trying to balance the demands of family, work, freetime, entertainment, etc.

is *families*. Section 2.3.2 tackles the UCD steps from scenarios to requirements, and Section 2.3.3 covers the steps after the requirements phase.

2.3.1 Identifying the Focus Group and Its Needs

In this section, some of the most important aspects of modern family life, including communication among family members inside and outside the home, the organisation of time and tasks, and the need for work–life balance, are identified and introduced. Especially considered are the needs of *people in a family who are juggling multiple roles*, including their roles in the family, their roles at work and school, and their other social roles. Figure 2.3 shows some of the different types of family, such as families with young children, blended families (e.g. families that are formed when parents who already have children from other marriages come together), and older families (with older children). Figure 2.3 also shows some of the factors that families are trying to balance in their busy lives, such as work, free time and entertainment.

These aspects, together with the mobility aspects discussed in Section 2.1, have given a basis for the technical development described later on in this book (from Chapter 3 to Chapter 6) and also for the reference applications introduced in Chapter 7.

2.3.1.1 Communicating with Family Members at Different Locations

Family members need to be able to *contact other members of the family when they are all in different locations*, such as school and work. Daily communication is important in part so that the family can coordinate its actions; for more on this aspect, see Section 2.3.1.3 below. On the less pragmatic side, family members also like to have '*intimacy at a distance*' – a feeling that they know how the other person is doing while they are apart, even if they don't share particular information [51].

In addition to needing to keep in touch with other members of their household, it is also important for the members of the family to be able to maintain contact with the *extended family*, who may be living in other locations. Various researchers have proposed different methods for maintaining this family communication using information and communications technology (ICT). These ideas consist of both *direct communication* applications as well as *peripheral awareness* applications. For example, Elizabeth Mynatt and her colleagues at the Georgia Institute of Technology explored the idea that family portraits enhanced with 'awareness' information could help family members maintain their 'peace of mind' about each other's current state. Through their user research [45], they found that family members with elderly relatives wanted to be aware of the following information about them:

- *Health*. In general terms, how they are feeling that day. Did they sleep well? Eat regularly? Get enough exercise?
- *Environment*. The 'health' of the environment. Has the weather been pleasant? Is something in the house broken?
- *Relationships*. Interaction with other people is important to one's emotional well-being. This category includes a range of social interactions, whether in person, on the phone, or through written correspondence.
- *Activity*. The general level of physical activity can be a good indicator of the caliber of a person's day in both extremes. A low level of activity may indicate declining health, while a high level of activity may indicate the onset of incessant wandering behaviour.
- *Events*. The occurrence of special events is an indication of the richness and variety in a person's life. This category includes activities, both planned and unplanned, as well as special outings.

While this particular study, along with others such as [56], focused on the needs of older people and their relatives, one can hypothesise that similar information might also be of interest to any family members who are apart. The ASTRA project [41] identified the costs and benefits in Table 2.1 associated with the use of this kind of awareness system.

These are important factors to keep in mind as one designs new types of communication and awareness functions for an application or a system. These are also among the key principles behind the *Mobile Services Architecture* described in Chapter 3.

2.3.1.2 Communicating with Family Members at Home

In addition to staying in touch with each other when they are not together, families also need ways to *communicate within the home*. In addition to simple face-to-face interaction, families leave each other *notes and reminders*, for example. Several papers talk about the special features of 'home' as the main location for interaction among the family members. For example, Debby Hindus and her colleagues [22] describe the following features of the home as a location for family communication:

- *Households are displays*. Households are stages upon which household members imprint their identities, as shown by the considerable effort put into decorating and personalising most homes.

Table 2.1 Costs and benefits of 'awareness' systems [41]

Costs	Benefits
Obligations: Social obligations felt or created as a result of using the awareness system	*Thinking about:* Thinking about another and knowing one is thought about
Expectations: Expectations for communication raised or unmet as a result of using the awareness system	*Staying Aware:* The extent to which people feel aware of the daily life activities of others
Privacy: The extent to which awareness threatens privacy	*Connected:* The feeling of being connected or 'in touch'
	Sharing experiences: How much one feels other people are involved in his/her life, sharing experiences
	Recognition: The extent to which each other's feelings are understood
	Group attraction: The feeling of being part of a group

- *Households are sanctuaries.* The home is a private place where people take refuge from the pressures of work and where they can rest or play without scrutiny.
- *Family life is the household priority.* In homes, people are concerned first and foremost with other household members, followed by family members outside of the household and then, less importantly, friends and other relationships such as those of shared interest groups.

As a conclusion, it is important to remember the special characteristics of the home when designing services that will be used partly in the home environment.

2.3.1.3 Organising Time and Tasks

Research results highlight the need of the family members to keep in touch with each other specifically in order to organise schedules and tasks. There are three main categories of user behaviour identified: (1) *keeping track of each other's schedules*, (2) *making childcare arrangements*, and (3) *sharing domestic and other tasks*.

For example, Stephan Hoefnagels and his colleagues have described conceptual designs for systems that help working parents with children coordinate their schedules with each other [23]. The activities that need to be coordinated include transportation (e.g. car pool to work). The need to organise transportation is also discussed in [22]. The family members need to maintain their individual scheduling information as well as group plans with each other.

The FAMILIES project has addressed the impact of e-working (telecommuting, working from home) on the domestic life of families. This project found that even with one parent working at home, making childcare arrangements and setting up the equitable sharing of domestic/household tasks were still challenging [16].

2.3.1.4 Work–Life Balance

As mentioned previously, the focus in this book is to help people in their multiple roles as family members, people going to work or school, and so forth. The need to *juggle multiple roles* creates particular user requirements for family members. For example, as described in the previous sections, family members need to be able to keep in touch with each other while they are at work and school, both for scheduling and coordination purposes and for less tangible 'intimacy at a distance' reasons. In addition, family members need help establishing '*work–life balance*'.

In order to not exacerbate the perceived work–life imbalance, applications and services that allow people to work 'anytime, anywhere' need to take these issues into account. For example, family members will need support in protecting their family time from work encroachments, and vice versa.

The power to work from home or to work while mobile is an important aspect of life for many people, including creative knowledge workers whose work is not tied to one location. Some sources have pointed out that especially for these workers, the power to make decisions about where and when to work is an expected part of defining one's everyday life [18]. Services and applications that make it easier to work from home or while mobile may then be seen as empowering these workers, rather than simply exposing them to risks of overworking and so on.

Finally, another way of helping family members maintain work–life balance is to enable them to stay in touch with each other while the parents are at work (or travelling for work). Kim *et al.* describe the design of a picture frame system that lets parents at work stay in touch with their children [33].

2.3.1.5 Playing Together

In addition to the practical matters already discussed, much family interaction is about *having fun together*. For example, the interLiving project installed communication technology probes (e.g. a real-time two-way video system called the videoProbe, and an asynchronous messaging system called messageProbe) in families' homes in order to study emergent uses of the technology. After several such studies, the interLiving researchers observed that, while the technologies were used for practical purposes, they were also used in more playful ways [54].

These findings are here to point out that, while one is designing technologies that help people manage their multiple roles, one should not forget that one of these roles is in fact that of a *person who enjoys having fun with his or her family.*

2.3.1.6 Managing and Sharing Family-related Media

Finally, perhaps because of the suitability of information and communication technologies as vehicles for sharing digital information, several studies have focused on the management and use of *family media such as photos and videos*. For example, Kim *et al.* [33] described a conceptual design for a 'home media pond', in which users can deposit and collect family-related media.

Similarly, the Family Video Archive project [1] worked on the concept of a library of home movies that could be browsed and annotated by the members of the family. In general, projects like these are based on the ideas that families enjoy sharing their own media, and that these media serve an important role in the 'group memory' of the family.

2.3.1.7 Cultural Differences

Identifying the cultural differences in family life has been an active research topic in sociology and psychology for many years [5]. For example, Parsons argued in the 1940s that nuclear families in North America and Northern Europe tend to be more isolated from their kinship networks (extended families) than families in other regions [46]. While more recent research has not borne out that particular conclusion, there is general agreement that family life differs across cultures in several important ways [11]. For example, Georgas *et al.* conclude that:

> [It] would appear that there are significant differences in means among the cultures in terms of emotional distance, geographical distance, meetings, and telephone communications. This would support the argument that cultures vary in terms of significant differences in emotional distance to members of the nuclear and extended family, in terms of how close or far they live from these members, how often they meet these members, and how often they telephone them. [20]

These four factors – *emotional distance, geographical distance, meetings, and telephone communications* (as well as use of other communications media, such as SMS) – will affect the acceptance of mobile services and applications by families in the future. For example, families who have very close (emotional and/or geographical) ties with their extended family members will have different communication needs from families whose ties with their extended families are less close.

Another important set of cultural differences is in the acceptance and use of information and communication technologies (ICTs), including the mobile phone, the Internet, and so on. All of these differences must be taken into account as one plans and conducts user research, and especially as one designs new services and applications for the families.

2.3.2 From Scenarios to Requirements

For this book, a *scenario-based approach*, following the process described in Section 2.2.2, was used to portray the experiences that users will be able to have with future mobile applications and services. In this context, the word 'scenario' is used to mean a *narrative description of what the user does* and experiences when using a computing system [8].

Scenario-based techniques have been proposed in user-centred design research as effective ways to capture, analyse and communicate information about possible user needs and system functions [10,12,28]. The basic principle of this approach is that developing scenarios of use of the technology can help to bridge the gap between the description of the users' tasks and the design of new technology to accomplish these tasks. Several methods have been proposed to carry out systematic development and utilisation of each scenario [9].

In order to develop the set of high-level illustrative scenarios presented in the Appendix, several existing scenarios were gathered, including the WWI Consolidated Scenario: 'Around the World in a B3G Day' [3], the ITEA Software Roadmap [25], the Wireless Foresight [30], the MITF Flying Carpet 4G scenarios [44], ISTAG Scenarios for Ambient Intelligence [15], and multiple conference papers from leading conferences. These scenarios were analysed in order to understand:

- *the underlying user tasks or behaviours* shown in each scenario;
- the related *system capabilities* that enable the user task or behaviour;
- related *research questions* for that task or behaviour, relating to the technological feasibility, acceptability to users or other user research issues, and the economic environment.

The results of this scenario analysis were consolidated in order to identify the user tasks and behaviours that were in common across many of the scenarios. For example, if a user in one scenario was shown entering an event in her electronic calendar, and another user performed the same task in a different scenario, a single canonical task ('User edits calendar') was created, with the goal of creating a 'master list' of the user tasks/behaviours shown in all of the scenarios that were analysed. The objective of this activity was to survey all of the user tasks and behaviours that have been included in scenarios of future mobile applications and services in order to identify the tasks/behaviours that are most relevant for families' everyday lives.

Following this analysis of the existing scenarios, the master list of user tasks and behaviours was divided into the following 15 categories (listed here with selected considerations from the user studies):

1. Scheduling

- Users must be in control. Instead of automatically managing scheduler entries, the system should help the user in managing them. For example, the system should suggest viable alternative scheduling arrangements to the user who can then select one.
- Making calendar entries should be as easy as using a paper calendar.
- Managing shared events and changes should be as easy as making a mobile phone call.
- Within the family, scheduling is a communication method (e.g. like an extension of messaging) rather than being only for the purposes of scheduling (e.g. a separate feature).

2. Location Tracking

- Due to privacy issues [2], the user needs to be able to control: (i) when to turn her location function on and off (needs to be clear across multiple devices), (ii) the specificity/granularity of the location information ('she is in the downtown area', rather than 'she is in the restaurant'), and (iii) who is allowed to access (which level of) location data and when they are allowed to access it (consider different possible people: e.g. parents vs. authorities).
- Coordinates are meaningless; use more user-friendly concepts such as street addresses or known sites nearby.
- Rather than building an independent location tracking application, consider combining it with some other functionality. This is a building block, not a user feature.

3. Navigation and Way-Finding

- Support two types of navigation: *exploration – finding new things* (areas of interest in new places) and *point-to-point navigation* (exact instructions from A to B).
- Navigation and presence information should be combined for non-corporate multi-user applications; for example, a group of friends trying to meet (a moving target).
- The map should serve as an interface for further information about the things/locations shown on the map (e.g. like hypertext).
- The system should provide navigation information in a form that is good for all types of users (not just map-based navigation).
- Presentation of navigation information/instructions should take personal preferences and/or situational information into account.
- Allow print copies or saved electronic copies of itineraries/instructions (for backup).

4. Interaction with Commercial Services

- The user needs to be in control of whether they do or don't receive 'pushed' information (proactive suggestions made by the system, such as location and/or context-based alerts or profile-related offers).

- Users should be in control of who has access to information about their interaction with commercial services.
- Provide easy means to secure the device in case of theft or loss.
- Specific practical advantages allowed by the mobile/wireless usage situation.

5. Multimedia and Data Applications and Services – Shared Uses (Multi-Person)

- Users should be able to control who gets to see (and also annotate) their photos, media clips, documents, and other data. Multimedia sharing should be combined with group management features and/or with localisation services.
- Users expect to have mobile access to the kinds of functionality they have on the Internet (printing, broadcasting, storage, annotation).

6. Multimedia and Data Applications and Services – Personal Uses (Single Individual)

- Users expect to have mobile access to the kinds of functionality they have on the Internet.
- Users want the option to check the source of all data (again, taking control/privacy of all users in account – the needs/desires of the producer and consumer need to be balanced).

7. Interaction with Multiple Devices

- It should be easy to access the same data and services with different devices and in different locations, even within the same house.
- Some situations require interfaces other than keypad and screen interface.
- While the user is mobile, she should be able to take advantage of nearby resources (devices, accessories, etc.).

8. Personalised Environments

- Rather than automating the environment, provide suggestions.
- Personalisation should be playful/emotional.
- Design for the social aspects of personalised environments; for example, harmonising the preferences among a group of users (but privacy must be taken into account).

9. Personalised User Interfaces

- Users should be able to control the adjustments made to the interface of her device.
- The cost or damage done in case of misinterpretation (of the user's wishes/preferences by the system) should be low or nonexistent.

10. Personalised Content and Services

- The system should help the user discover new things that might be of interest, but the system should not be intrusive (users should be able to control when they get suggestions for new content and services, for example).
- If the system makes recommendations, it should provide some rationale so that the user can understand why this recommendation was made.

11. Group Formation and Membership

- Users form a group because it's needed for some purpose.
- Forming a group, adding/removing people from the group, leaving the group (permanently or temporarily) should be easy.

- Users must be able to see who is in the group.
- The logic of automation should be transparent: if groups are formed/expanded automatically, the user must be able to see the rationale, and the trust level among this group.
- Users must be able to control the personal information other group members can access.
- With a basic device (cell phone), one should get a minimal [group] service – basic features at least (inclusion).

12. Monitoring

- Access to monitoring information collected needs to be controlled (who and when) and the access rules understood by users.
- The monitoring system needs to be reliable so that users can trust the information it gives, especially when monitoring critical action.
- The overhead of wearing the sensors should be balanced by the advantages to the user.

13. Billing and Cost Control

- Billing and cost monitoring should be integrated into a single easy-to-use and trusted mechanism.
- The idea of paying with the device is appealing but should be easier than with current means (money, credit card, etc.).
- The specific details of billing are less important to the user than the absolute cost of the service.
- Users want to be able to switch between service providers.

14. Real-time People Interaction

- One should enable rich real-time interaction between people.

15. Messaging

- One should support multiple ways of messaging (examples of currently available solutions include email, conference calls, SMS, MMS and instant messaging).

Based on the high-level tasks and behaviours presented above, four scenarios (see the Appendix) were created: the ‘Monday’, ‘Friday’, ‘Sunday’ and ‘Olympics’ scenarios. Each story is a sequence of related episodes depicting the life of a hypothetical family. Each scenario focuses on a different aspect of life: planned activities during the work week, dealing with unexpected events during the work week, enjoying leisure activities on the weekend, and taking a special outing as a family. Each story has been divided into episodes, identified by a letter denoting the story and a number denoting the step of the scenario (e.g. M.1 for the first step of the ‘Monday’ scenario).

During the user evaluations and interviews, the high-level user requirements and guidelines for each of the task and behaviour categories were extracted (examples included in the list above). In addition, several general recommendations were made, including:

- Convenience-focused ideas were well-regarded in the user evaluation.
  - The system should help the user in solving some practical problem;, i.e., one should concentrate on real-world benefits for the user.
  - Absolute efficiency is not necessarily the goal; do not automate every process.

- Playful/fun aspects should also be considered.
- The users want the same level of service as on the Internet but want to be able to have it on the move and are worried that it will be unreliable or unusable (connectivity problems, small display size, mobile use will cost more than fixed use, etc.).
- Social aspects of the technology must always be considered. Human interaction and communication is crucial. Many seemingly rote everyday tasks serve other purposes, such as creating the relationships among the family members.

While combining the key learning from the scenario evaluations and expert knowledge, the following requirements were collected to guide the work with the mobile applications and services related development and technologies further.

General User Requirements

1. Users want to be in perceived control.
2. Users do not want to be tied to or limited by services, providers and devices.
3. Users want easy and intuitive devices, systems and services, with automation to achieve this (as long as perceived control is not violated).
4. Behaviour should be context-dependent.

User Experience

1. Users want systems to manage complexity for them.
2. Using future wireless systems should be consistent and intuitive and should not add complexity.
3. Services and content should be personlised to the user.
4. Content and services should not be tied to the device on which they are requested and consumed.
5. The user should be in (perceived) control.
6. The user will expect a certain level of automation and services that know about their context.

Provisioning New Services

1. Services should be intuitive and deliver the service in a form most appropriate to the user's context:
   - appropriate interaction mode (multimodal interfaces)
   - translation of content
   - discovery of services
   - services not necessarily tied to a network or device
2. Services should display appropriate level of privacy and trust.
3. The user is in control of service provision, automation, privacy and trust, communication and content choice.
4. Group services should be provided.

Seamless Access

1. Users should be in control of access and cost of access.
2. Telecommunication services, content and devices should not necessarily be tied together.
3. Access should be context (specifically location) aware.
4. Access methods should be automatically disclosed and discovered.

Quality of Service (QoS)

1. The level of QoS should be guaranteed.
2. QoS should be device-independent.
3. Fixed Internet levels of QoS are expected for wireless.
4. Devices should be able to multi-task.
5. Content should be matched to the device.

Security, Privacy and Trust

1. Users want to remain in control of their content.
2. Future systems which are context aware and adaptive have to build the trust of users.
3. Levels of trust with a group or group hierarchy have to be supported.

Deployment and Operation

1. Users should remain in control of services and delivery channels.
2. Users should remain in control of their profiles, which may be distributed and not under the control of a single provider.
3. Systems have to be reliable and trustworthy.

Migration

1. Users want to create and annotate content.
2. Users want consistent behaviour across devices, providers and systems.

2.3.3 From Requirements to Results

The core of user-centred design consists of planning and conducting user research activities. A wide set of tools can be used in the collaborative UCD process. Table 2.2 lists and briefly describes some tools used in this book.

The *UCD agenda* is a document jointly created by the user researchers and technical developers. It describes the general user-centred design approach and process to be followed in a development project.

For each application or service to be developed, one should nominate a *contact person* from the user research team. This person acts as the contact point between the user researchers (or user evaluations) and developers (or the actual application/sevice).

Again, for each application or service, one should compose a *user research plan*. A plan contains:

- a list of *user research questions* related to the application/service;
- a description of the *prototype* (or alike);
- a description of the *test setting*;
- a description of the *participants* (i.e. users), including rationale for their selection.

Typically the items are planned in parallel. For example, at the prototype design phase one already knows what sort of test environments one would like to have. While creating the reference applications of this book, during mock-ups and Probes 1 phases, one in practice started user research planning from (i) user research questions and (ii) description of the

Table 2.2 UCD tools

UDC tool	Explanation
UCD agenda	Aims for collaborative agreement on the UCD approach in the development project
Contact persons	Named person in the user research team for each application; and vice versa, named technical contact person for the user research team representatives to contact
Research plans	Based on collaboration between contact persons, the user research team prepares research plans related to each evaluation round
Generating research questions	Collaboratively identifying specific research questions for all technical/technology areas and end-user aspects to be included in the evaluation round
Wish lists	Lists sent by the user research team contacts to inform technical persons on the user point of view wishes on applications to be evaluated
Face-to-face meetings	Specific meetings taking place during collaboration
Deliverables (= evaluation results)	Evaluation deliverables authored by the user researchers

prototype. The former describes what one wants to learn from the study while the latter tells what is feasible to implement at that stage. The aim is to come up with a *balance between user research questions and the prototype that is used to address these questions*. Once the balance is found, a detailed plan, including description of the setting, test protocol etc. can be composed.

As soon as one roughly knows what the developers can build, it is possible to start to think of the test environment. Distinction between the prototype and the setting does not always make sense. For example, a prototype can consist of a handheld device and some ubiquitous technology that is in fact part of the environment. However, the description of the test setting goes beyond the immediate environmental factors.

Regarding the test settings, one can distinguish between four types of test settings and select the suitable settings for each of the evaluations:

- *Laboratory test*. For example, usability tests are conducted in a laboratory, in front of a partially functional prototype, steered by a facilitator and following a closed or open-ended protocol. The objective is to test how well the features of the system are offered to the user. One can also ask about the usefulness of the application/service.
- *Field laboratory test*. Laboratory-like semi-controlled trials can also be organised 'in-the-wild' (outside laboratories). This is likely to provide insight into environmental factors affecting the product usage (e.g. sunlight, rain, noise). However, to keep the setting together, the facilitator still needs to follow the test protocol and the user has only a limited set of (typically predefined) actions to choose from.

- *Field trial.* In field trials, the tested application offers enough functionality so that users can explore it without the detailed guidance of the facilitator. Usage time can be limited or can range to several weeks or months.
- *Snowball field test.* The distinction to the above is that users are not only recruited, but more users can participate at will. Knowledge about application/service spreads by word of mouth as people tell their peers about the possibility to join in. This, of course, requires that the hardware and software be available.

Selection between possible settings is, again, not a matter of selection at a predefined gate, but part of the interactive planning between user researchers and developers. It has to be aligned with research questions and to the prototype at hand. Regardless of the importance of field trials, it makes no sense organising them if the prototype is linear slideware.

Ambitious user research should strive for high-fidelity real-life settings. This means not only doing tests in the everyday environment of the user, but aiming at naturalistic and self-organising usage of the application. In order to study, for example, social usage of some mobile application, one cannot invite people into a usability laboratory and ask them to act 'as usual'. Similarly, for research economic and reactivity reasons, one cannot extensively interfere in their everyday life by, for example, spending days shadowing them with a camcorder.

Regarding user acceptance, given the technologies dealt with in this book, one can distinguish between four types of categories and select a reasonable balance of them to give a framework for each of the evaluations:

- *Traditional usability.* The functionality of the applications and services need to be accessible for the users. The products should be easy, effective and pleasurable to use. If the user is not able to use the product, it does not matter how beneficial its features are.
- *Mobile usability.* The user should be able to use the applications and services in her everyday physical environment and situations. If the user cannot use the product in real-life settings, it does not matter how well she performs in laboratory conditions.
- *Personal acceptance.* The applications and services should match the needs and desires of the individuals. If the user does not want to use the product(s), it does not matter how user friendly it is.
- *Social acceptance.* The applications and services need to be acceptable and attractive in a peer group of the user. They not only fit to personal preferences, but support communication and co-operative processes within and between groups of users. If the system is cumbersome for co-operative use or socially or culturally unaccepted, it does not matter how much a single user likes it.

Studying traditional usability related to products and services is usually not a problem, even though it requires time and resources. Usability testing is a mature discipline: there are well-tested methods and established industrial practices for this. Addressing mobile usability is slightly more complicated. It is, however, a minor obstacle. With some limitations, one can arrange laboratory-like tests in the field.

It is more complicated to go from personal acceptance to social acceptance. The technologies covered in this book enable people to be aware of, interact with, communicate with and coordinate activities with other people. Consequently, from the point of view of user research,

research and design challenges and potentials are in the social dimension of technology rather than in the personal dimension. However, the question of social acceptance is very problematic, as it equates to anticipating the outcome of time-consuming social processes – or *collaborative actions* as described by sociologist Howard Becker. The problem, as Becker [6] puts it, consists of the following:

- *People are active.* They are trying to do something; they are looking to the environment, in search of ways to accomplish whatever they are trying to do.
- They also recognise that *the other participants have their own agendas* that are not always in line with their own.
- *Reflexivity:* During human conduct, the actors *think of alternative ways for responding* to the actions of others.
- Knowing that the others are also doing similar type of reasoning, actors *adjust their doings to the imagined responses* of the others.
- The actors are *inaccurate* in anticipating the moves of the other participants. Therefore, they *need to continuously readjust* based on the information provided by the others.

As a result, although individuals have power over their personal action at all stages, *the overall process (of social action) is not in the control of any single individual.* Therefore, it makes no sense asking people (one by one) about the outcome. They simply are not equipped to answer before they have lived the reality we want to know about.

The *requirements for testing the social dimension* have been identified to be the following [36]. The application/service prototype:

- should be tested in the real environments of the user (instead of alienating laboratory conditions);
- should enable localisation:
  - translations (e.g. English/Italian/Finnish, according to test locations and type of users)
  - last-minute changes to content (e.g. because testing place or users may change)
- Should enable users to act (instead of just them to talking about it).
- Should enable users to do things together (instead of having one person using the system).
- Should enable users to use the system for a longer time (instead of collecting first impressions).

Applying the collaborative UCD process in a development project is not, however, just a one-time activity by creating a *UCD agenda* and application-specific *research plans* at the beginning, but it requires continuous and active co-operation between the user researchers and the developers to be successful. In order to enable this, the remaining UCD tools from Table 2.2 should be used:

- *Contact persons* need to create and maintain the link between the user research team and the developers from each technology/development area.
- *Face-to-face meetings* must be organised in order to fine-tune the jointly agreed UCD agenda and research plans, as well as to analyse the results of each evaluation round.

- Intermediate and final *deliverables* are needed to report the user research findings to the developers.
- Specific *wish lists* can be used to provide requirements from the user researchers so that the applications or services to be tested can actually be tested. A wish list can contain, for example, requirements for user interface improvements or application robustness. The contact persons should help in communicating these wish lists to the developers so that they understand the requirements and need for them.

Section 2.4 includes some final remarks and notions related to the UCD process. The actual user evaluation activities and results on the technologies as well as reference applications described in this book are explained in Chapter 8.

2.4 Conclusions

In this chapter, the user-centred design (UCD) approach has been described. The results [37–39] show that the UCD approach can have positive effects on several levels:

- It generates *clear improvements to applications*. This is often the most visible and expected result in a UCD process.
- It *helps the developers to prioritise and structure technical work* and manage their workflow and workload.
- It generates *new understanding related to the enabling technologies.*
- It creates a *significant amount of learning on personal and organisational levels, related to end-user orientation and UCD management*. As a result, both the technical developers and their organisations become more sensitive to the viewpoints of the customer.

UCD orientation is used to increase the likelihood that the customers will ultimately see the system as valuable; however, it cannot guarantee that this will happen. The equation has too many changing parts to guarantee anything. It would also be unsustainable to present only ideas that can be turned into product features – as many of the collected comments imply. For example, if the research findings show that users have serious issues with the application, it should be legitimate to present that finding even if the solution is not immediate visible. Ultimately, the developers, not the user researchers, are responsible for turning the findings, even unpleasant ones, into designs.

One critical success factor for the user-centric approach is the ability of user researchers to really participate as team members. This ensures discovery early on of the key items from the users' perspective and mutual understanding of their implications on the designs.

In addition, *unplanned last-minute development effort is often needed to make the concepts really testable with the end-users*. This often requires a lot of work (e.g. related to 'localisation'), which requires changes to the user-interface and content of the concepts for better match with the evaluation location and test users. To perform these kinds of activities in a synergetic way also requires fluent and frequent interaction, some of it face-to-face.

It is typical to underestimate the amount of work needed to get the concepts to function up to the level where they are really testable. This often results in delays in the evaluation schedule and respectively in reporting the results, endangering exploitation of the results in the next stage.

The purpose is not to come up only with suggestions that are feasible today. Some comments from the developers recognise that and see value also in these findings. User research findings often contain more questions than answers. This is the nature of design and planning activities in general.

Design activities are not only about finding solutions to predefined problems, but also about finding new problems. In practice what happens is that, based on a preliminary understanding something is designed, after which the designed artefact raises new questions [50].

One of the biggest challenges is the evaluation of the social dimension of applications. Many novel mobile applications are meant to be used within groups or as tools for group communications. Thus, testing them with individual persons does not always give meaningful results; on the other hand, arranging evaluations with large user groups in both time-consuming and expensive, and it sets high requirements for the stability and robustness of the application.

Finally, *there is no single method that guarantees generating good designs and great ideas.* One good design idea or driver may save the day even when the groundwork is not done properly. Similarly, even well grounded ideas can go seriously wrong due to 'insignificant details', such as the UI implementation level.

2.5 Acknowledgements

The following individuals contributed to this chapter: Andy Aftelak (Motorola LTD, UK), Luca Galli (Neos, Italy), Annakaisa Häyrynen (Elisa, Finland), Ulla Killström (Elisa, Finland), Esko Kurvinen (Helsinki University of Technology, Finland), Harri Lehmuskallio (Helsinki University of Technology, Finland), Mia Lähteenmäki (Nokia, Finland), Kevin Mercer (Motorola LTD, UK), and Antti Salovaara (Helsinki University of Technology, Finland).

References

[1] Abowd G., Gauger M., Lachenmann A.: 'Video Retrieval: The Family Video Archive: an annotation and browsing environment for home movies'. In Proceedings of the 5th ACM SIGMM International Workshop on Multimedia Information Retrieval, 2003.

[2] Acquisti A. and Grossklags J.: 'Privacy and Rationality in Individual Decision Making'. *IEEE Security and Privacy*, Vol. 3 (1), Jan/Feb 2005.

[3] Aftelak A.: 'Consolidation of the Scenarios from the Wireless World Initiative'. Wireless World Research Forum (WWRF) Meeting #11, Oslo, Norway, June 2004.

[4] Akyildiz I., Su W., Sankarasubramaniam Y. and Cayirci E.: 'A Survey on Sensor Networks'. *IEEE Communications Magazine*, Vol. 40, No. 8, 2002.

[5] Alasuutari P.: *Researching Culture: Qualitative method and cultural studies*. Sage, London, 1995.

[6] Becker H.: 'Interaction: some ideas'. Online: home.earthlink.net/~hsbecker/Interaction.htm.

[7] Brumitt B., Meyers B., Krumm J. *et al.*: *EasyLiving: Technologies for intelligent environments*. Handheld and Ubiquitous Computing, September 2000.

[8] Carroll J.M.: 'Introduction: the scenarios perspective on system development'. In J.M. Carroll (ed.): *Scenario-based Design: Envisioning works and technology in system development*. John Wiley & Sons, Inc., pp. 1–17, 1995.

[9] Carroll J.M.: *Making Use: Scenario-based design of human–computer interactions*'. Massachusetts Institute of Technology Press, 2000.

[10] Carroll J.M. and Rosson M.B.: 'Getting Around the Task–Artefact Cycle: How to make claims and design by scenario'. *ACM Transactions on Information Systems*, Vol. 10, No. 2, pp. 181–212, 1992.

[11] Cheal D.: *Sociology of Family Life*. Palgrave, 2002.

[12] Chin G. and Rosson M.B.: 'Progressive design: Staged evolution of scenarios in the design of a collaborative science learning environment'. In Proceedings of CHI'98, 1998.
[13] Conaty G. (ed.): 'Initial Scenarios, Requirements and Guidelines: User-centred approach for the design of future mobile services and applications'. IST-MobiLife Project Deliverable D6b (D1.1b), February 2006. Online: www.ist-mobilife.org.
[14] Digital Living Network Alliance: 'Use Case Scenarios White Paper'. 2004. Online: www.dlna.org/news/DLNA_Use_Cases.pdf.
[15] Ducatel K., Bogdanowicz M., Scapolo F., Leijten J. and Burgelman J-C.: *ISTAG Scenarios for Ambient Intelligence in 2010*. IPTS-Seville, 2001.
[16] FAMILIES project: 'Deliverable No 3: Results of family survey', 2002. Online: www.families-project.com/outmain.html.
[17] Finnie G.: 'Fixed–Mobile Convergence Reality Check', 2003. Online: www.heavyreading.com/.
[18] Florida R.: *The Rise of the Creative Class: and how it's transforming work, leisure, community, and everyday life*. Perseus Books Group, 2002.
[19] Gellersen H., Schmidt A. and Beigl M.: 'Multi-sensor Context Awareness in Mobile Devices and Smart Artefacts'. *Mobile Networks and Applications*, Vol. 7, No. 5, 2002.
[20] Georgas J., Mylonas K., Bafiti T., Poortinga Y., Christakopoulou *et al.*: 'Functional Relationships in the Nuclear and Extended Family: a 16-culture study'. *International Journal of Psychology*, Vol. 36, No. 5, 2001.
[21] Gould J.D. and Lewis C.: 'Designing for Usability: Key principles and what designers think'. *Communications of the ACM*, Vol. 28, No. 3, March 1985.
[22] Hindus D., Mainwaring S., Leduc N., Hagström A. and Bayley O.: 'Casablanca: Designing social communication devices for the home'. In Proceedings of the SIGCHI Conference on Human Factors in Computing Systems, 2001.
[23] Hoefnagels S., Geelhoed E., Stappers P., Hoeben A. and Van Der Lugt R.: 'Interactive Posters: Friction in scheduling and coordinating lives of families: designing from an interaction metaphor'. In Proceedings of the Conference on Designing Interactive Systems: Processes, Practices, Methods, and Techniques, 2004.
[24] ISO 13407 1999. Human-centred Design Processes for Interactive Systems.
[25] ITEA (Information Technology for European Advancement): 'ITEA Technology Roadmap for Software-Intensive Systems, edition 2'. *ITEA*, 2004. Online: www.itea-office.org/index.php.
[26] ITU (International Telecommunication Union): 'Trends in Telecommunications Reform 2003, Promoting Universal Access to ICTs, Practical Tools for Regulators, Executive summary'. ITU, 2003. Online: www.itu.int/publications/docs/Exec_Summary03.pdf.
[27] Kaasinen E.: 'User Needs for Location-aware Mobile Services'. *Personal and Ubiquitous Computing*, Vol 7, No. 1, 2003.
[28] Karasti H.: *'Bridging the analysis of work practice and system redesign in cooperative environments'*. In Proceedings of Designing Interactive Systems 1997, 1997.
[29] Karasti O. *et al.*: 'State-of-the-Art in Service Provisioning and Enabling Technologies'. IST-MobiLife Deliverable 5.2, 2004. Online: www.ist-mobilife.org.
[30] Karlson *et al.*: *Wireless Foresight. Scenarios of the mobile world in 2015*. Wiley, 2003.
[31] Kidd C.D., Orr R., Abowd G.D. *et al.*: 'The Aware Home: a living laboratory for ubiquitous computing research'. International Workshop on Cooperative Buildings, 1999.
[32] Killström U. (ed.): 'Marketplace Dynamics and Socio-economic Implications'. IST-MobiLife Project Deliverable D13 (D1.8), June 2006. Online: www.ist-mobilife.org.
[33] Kim S., Chung A., Ok J., Myung I., Kang H., Woo J., Kim M.: 'Communication Enhancer: Appliances for better communication in a family'. *Personal and Ubiquitous Computing*, Vol. 8, Issue 3/4, 2004.
[34] Kirk J. and Miller M.L.: *Reliability & Validation in Qualitative Research*. Beverly Hills, CA. Sage, 1986.
[35] Kruse E. and Carlsson A.: *A Journey to the Third Place: Market reality among early adopters*. Ericsson, 2003.
[36] Kurvinen E. (ed.): 'Results of Mock-ups Evaluation'. IST-MobiLife Project Deliverable D8 (D1.3), May 2005. Online: www.ist-mobilife.org.
[37] Kurvinen E. (ed.): 'Results of Service and Application Evaluation'. IST-MobiLife Deliverable D14 (D1.9), December 2006. Online: www.ist-mobilife.org.
[38] Kurvinen E., Häyrynen A. and Klemettinen M.: 'MobiLife UCD Process'. IST-MobiLife Deliverable D1.4c, January 2006. Online: www.ist-mobilife.org.
[39] Kurvinen E., Lehmuskallio H. and Häyrynen A.: 'Mashing Up Mobile: Lessons from the Field'. Wireless World Research Forum (WWRF) Meeting #17, Heidelberg, Germany, November 2006.

[40] Ling R.: *The Mobile Connection: the cell phone's impact on society*. Morgan Kaufmann, 2004.

[41] Markopoulos P., Romero N., van Baren J., IJsselsteijn, W., de Ruyter, B., Farshchian, B.: 'Keeping in Touch with the Family: Home and away with the ASTRA awareness system'. In Extended Abstracts of the 2004 Conference on Human Factors and Computing Systems, 2004.

[42] Medjahed B., Bouguettaya A. and Elmagarmid A.: 'Composing Web Services on the Semantic Web'. *International Journal on Very Large Data Bases*, Vol. 12, No. 4, 2003.

[43] Meyer S. and Rakotonirainy A.: 'A Survey of Research on Context-aware Homes'. Proceedings of the Australasian Information Security Workshop Conference on ACSW Frontiers 2003, Vol. 21, 2003.

[44] mITF (mobile IT Forum): 'Flying Carpet: Towards the 4th generation mobile communications systems, version 2.0'. mITF, 2004. Online: www.mitf.org/public_e/archives/Flying_Carpet_Ver200.pdf.

[45] Mynatt E., Rowan J., Craighill S. and Jacobs A.: 'Digital Family Portraits: Supporting peace of mind for extended family members'. In Proceedings of the SIGCHI Conference on Human Factors in Computing Systems, 2001.

[46] Parsons T.: 'The Kinship System of the Contemporary United States'. *American Anthropologist*, Vol. 45, 1943.

[47] Picard R.W.: 'Toward Computers that Recognize and Respond to User Emotion'. *IBM Systems Journal*, Vol. 39, No. 3/4, 2000.

[48] Project Oxygen: 'E21 Intelligent Spaces'. Online: http://oxygen.lcs.mit.edu/E21.html.

[49] Rheingold H.: *Smart Mobs: the next social revolution*'. Perseus Books Group, 2002.

[50] Rittel H. and Webber M.: 'Planning Problems are Wicked Problems'. In Cross N. (ed.): *Developments in Design Methodology*. John Wiley & Sons, 1984.

[51] Rosenmayr R. and Kockeis E.: 'Proposition for a Sociological Theory of Aging and the Family'. *International Social Science Journal*, Vol. 3, pp. 418–419, 1963.

[52] Rubin J.: *Handbook of Usability. How to plan, design and conduct effective tests*. New York, John Wiley & Sons, 1994.

[53] Säde S.: *Cardboard Mock-ups and Conversations: Studies on user-centered product design*. Doctoral thesis, A34, University of Arts and Design Helsinki (UIAH), 2001.

[54] Sarvas R., Viikari M., Pesonen J. and Nevanlinna H.: 'Mobshare: Controlled and immediate sharing of mobile images'. In Proceedings of ACM Multimedia, 2004.

[55] Sundblad Y., Beaudouin-Lafon M., Conversy S. *et al.* 'InterLiving project: Deliverable 1.3 & 2.3: Studies of Co-designed Prototypes in Family Context'. (2004). Online: http://interliving.kth.se/papers.html.

[56] Tollmar K. and Persson J.: 'Understanding Remote Presence'. In Proceedings of the 2nd Nordic Conference on Human–Computer Interaction, 2002.

[57] van de Kar E., Maitland C., de Montalvo U. and Bouwman H.: 'Design Guidelines for Mobile Information and Entertainment Services: Based on the Radio538 ringtunes i-mode service case study'. In Proceedings of the 5th International Conference on Electronic Commerce, 2003.

[58] WWRF Book of Visions Tafazolli R. (ed.): 'Technologies for the Wireless Future: Wireless World Research Forum (WWRF)'. Volume 2, ISBN 978-0-470-02905-3, Wiley, April 2006. 520 pp.

3

Mobile Services Architecture

Edited by Bernd Mrohs and Stephan Steglich
(Fraunhofer FOKUS, Germany)

Today's service architectures have critical hurdles and barriers, which limit their flexibility and the possibilities for providing services. Innovative services and applications require supplementary improvements of the underlying service infrastructure:

- The increasing complexity of applications has to be supported to simplify the service development and deployment phases.
- The actual requirements and preferences of users must be understood to ensure usability and acceptance.

This chapter provides a global picture for future service provisioning. It introduces an approach for the provisioning of next-generation mobile services, enabling services to be *context-aware* and *personalised*, and to support *individuals* and *groups*, while maintaining *privacy*. The basic service building blocks (the Mobile Services Reference Model) and a derived architecture specification with detailed functions (the Mobile Services Architecture) are described.

This chapter first discusses the main requirements and hurdles that have influenced the Mobile Services Reference Model and architecture definitions. Section 3.2 then introduces the model, which is a general reference model for service provisioning based on general application scenarios and detailed specifications of envisioned mobile services and applications. This model describes the main building blocks of the service environment and represents a template for the required service architecture and infrastructure from a technical perspective.

Enabling Technologies for Mobile Services: The MobiLife Book Edited by Mika Klemettinen

Sections 3.3 and 3.4 describe in detail the necessary service building blocks, functional elements, reference points and interfaces. The chapter also describes practical cases that show the service infrastructure and illustrate the interactions of functional components. In order to emphasise the mobility aspects, the specification of the mobile services architecture concentrates on solutions for the following areas in particular:

- service execution environments;
- innovative and flexible business models;
- vertical and horizontal roaming across heterogeneous network and systems;
- security and trust;
- user acceptance.

The architecture will thus provide for the selection of service platforms, operation and support systems, as well as for the development of innovative services and applications.

Section 3.5 presents an exemplary approach as to how the service building blocks and services can be realised. Section 3.6 covers aspects of service lifecycle management. Section 3.7 contains recommendations concerning scalability issues, and Section 3.8 is a short conclusion.

3.1 Requirements and Hurdles

When designing an architecture, multiple requirements from different perspectives need to be defined in order meet the expectations of all relevant stakeholders. When defining the Mobile Services Reference Model and the Mobile Services Architecture, the following high-level dimensions were considered:

- high-level user and stakeholder requirements;
- general system requirements;
- detailed technology requirements:
  - business model requirements
  - personalisation and multi-modal interaction requirements
  - trust and privacy related requirements
  - context awareness and context management requirements
- major hurdles in service provisioning.

The user requirements have been discussed in Chapter 2. This section first summarises the general system requirements and then the major hurdles in service provisioning. Detailed technology requirements are not discussed here, but their influence is visible in the resulting reference model and architecture.

3.1.1 General System Requirements

The architecture must be user-centric. The user-centred design (UCD) framework recognises that users are a central part of the software development process, and that their needs should be considered from the very early stages in the development process.

The architecture must support mobility. Support for users with mobile devices and mobility are important requirements for the software infrastructure.

The architecture must be context-aware. The system uses information about the state of the context of the user to adapt service behaviour to suit the user. Contextual information includes low-level data such as location, time, temperature and noise, as well as higher-level data such as the user's situation (such as 'in a meeting, with friends').

The architecture must support context management. Context data management is an essential part of context-aware systems. Managing context information efficiently is a challenge. Different types of context data originate from various distributed sources. These data must be gathered and made available to those components that need them. Context reasoning mechanisms must be created to elaborate and infer information from raw data.

The architecture must support adaptation. The basic principle of adaptation is simple: when the circumstances change, then the behaviour of an application changes according to the desires of a user, or more precisely, according to principles ascribed to her. Services and context information should be dynamically adapted in the future to the context by the use of an automated learning functionality. Learning the wishes and desires of a user is a crucial part of adaptation. Adaptation must also be proactive, which in turn requires predictability of the near future.

The architecture must support service personalisation. Personalisation includes the ability of the framework to acquire and manage personal information about the user, including preferences, and the ability to use this information to adapt an application's behaviour. End-users may affect device characteristics by configuring its parameters. Additionally, users may have global preferences affecting all services and specific service settings (e.g. their favourite language).

The architecture must consider privacy and trust issues. Privacy and trust are amongst the most important features to increase user acceptance of services. Privacy guarantees that personal data is revealed only according to the user preferences (expressed as policies). The trustworthiness of components, identities and information need to be guaranteed to establish a basic user/application trust level.

The architecture must consider group support. The architecture must support the creation of groups of users and functionalities to manage the users in the group (e.g. enter a group, leave a group, privacy and trust issues, etc.).

The architecture must support both managed services and non-managed ('ad hoc') *services.* The framework supports services that are controlled and maintained by a service provider (e.g. through a service portal), and services that are provided in an ad-hoc manner directly between users without the control of a third party (e.g. pure peer-to-peer services).

The architecture must be open to emerging value networks. The framework is not restricted to current provider–consumer value chains, but must be flexible regarding new ways of service provisioning – such as new ways of incorporating third-party service providers.

The architecture must support seamless service access via multiple access technologies. The framework must support service provisioning within the framework independent of the access technology. While the transfer between different access networks should in general be transparent for the application, in some cases an adaptation of the application is needed to ensure optimal service experiences.

The architecture must provide service and component lifecycle support. In addition to enhanced service features supporting user services, the framework must also include support for the service lifecycle – including development, deployment, operation and removal phases for services and their constituent components.

3.1.2 Major Hurdles in Service Provisioning

Analysis of today's major hurdles in service provisioning, including service deployment and delivery, has led to a number of challenges. The requirements have been summarised into the following categories:

- service roaming;
- authentication methods;
- business and operative related management information;
- network limitations;
- device limitations;
- billing and charging;
- interoperability, interworking and portability.

Wide acceptance of new mobile services and applications implies, for example, the capability of *service roaming*, which means the ability to access and use a service regardless of an end-user's geographical location (country) and current network access. From the end-users' point of view, service roaming has to be as simple as it is nowadays for the voice service of cellular networks, in terms of service subscription, usage, charging and invoicing.

Authentication of the user is usually needed. Authentication gives the user rights to access services. Nowadays, there are a number of authentication methods. Additionally, the common case is that each service provider has its own authentication method.

The *business and operative related management information* hurdle essentially concerns the flow of management information. The service architecture should support both managed services and non-managed (ad-hoc) services. (For managed services, mapping of the Mobile Services Architecture components to the IP Multimedia Subsystem (IMS), and especially of its functions related to *operational management*, are evaluated in Section 3.5.)

The *network limitations* hurdle involves problems and challenges arising from the inability of networks and systems to provide the required service level. There are three aspects: *connectivity limitations*, *network hardware infrastructure limitations* and *service infrastructure heterogeneity*. This category concerns the network level, so it is out of the scope of the *Mobile Services Architecture* specification: the connectivity is taken for granted.

Device limitations in the sense of resources such as battery and computing power have to be addressed. A mobile terminal acting as a *context provider* (CP) must be permanently collecting raw data and sending it through the network connection. This has an impact on the battery resources.

The main cause of the flexible *billing and charging* hurdle is that the service providers (SPs), public network operators (PNOs) and content providers can employ several charging models available in the industry. These challenges arise when a service provider provides mobile services to an end-user, and, at the same time, brings together different charging schemes and models for the end-user. This hurdle is partially out of the scope of the Mobile Services Architecture specification: 'billing and charging' is assumed to be provided by a core platform, such as IMS.

The *interoperability, interworking and portability* hurdle consists of three interconnected concepts. *Interworking* can be seen as an enabler for interoperability. *Interoperability* is the feature that allows the user to seamlessly access and use applications and services across different networks and terminals, independent of the specific implementation and of the specific terminal in use for the application. Interoperability and end-to-end interoperability can thus be seen as a requirement that stems from the user, although it has significant impact on the network, service and application development. *Portability* is a feature that allows developers to easily install their software components on different platforms, including different terminals. An application that is portable may not be capable of being integrated or interoperable with other applications. Interoperability between multiple platforms, services and applications is vital for commercial success.

Consideration of interoperability concentrates mostly on *application and service interoperation* (including the provisioning platform and underlying enablers). The key issues here are:

- the use of a common data model to exchange data and information among applications and application components;
- the use of standardised interfaces to access services;
- the interoperation between different platforms;
- the definition of the specific services, platforms and terminals for which interoperability is guaranteed.

3.2 Mobile Services Reference Model

Using the vocabulary of the Open Archival Information System (OAIS), a 'reference model' is:

> ... a framework for understanding significant relationships among the entities of some environment, and for the development of consistent standards for specifications supporting that environment. A reference model is based on a small number of unifying concepts and may be used as a basis for education and explaining standards to a non-specialist. [19]

A reference model is not directly tied to any standards, technologies or other concrete implementation details, but it does seek to provide a common semantics that can be used unambiguously across and between different implementations. Thus, the Mobile Services Reference Model is a template for composing architectures for mobile service environments through defining the basic service building blocks.

Section 3.2.1 describes specification principles that were followed while developing the reference model. Section 3.2.2 specifies the meanings and relationships of mobile applications to application services and the reference model including its functional components. Section 3.2.3 describes the service building blocks of the reference model.

3.2.1 Specification Principles

The specification of the Mobile Services Reference Model follows the loosely coupled principle for defining all reference model components. There are several definitions for the term 'component'. By definition of [20], a component is a unit of composition with contractually specified interfaces and explicit context dependencies only. It can be deployed independently and is subject to composition by third parties.

Furthermore, the reference model specifications follow the *Law of Demeter* (LoD) [11]. It was originally formulated as a style rule for designing object-oriented systems; 'Only talk to your immediate friends' is the motto. This style rule was discovered at Northeastern University in the fall of 1987 by Ian Holland. More generally formulated, it means: 'Each unit should have only limited knowledge about other units: only units closely related to the current unit.' In this general form, the LoD is a more specific case of the low-coupling principle well known in software engineering: Unnecessary coupling should be avoided and a method should have limited knowledge of an object model. This leads to aspect-oriented programming: one pulls out the method and transitively all its auxiliary methods into a separate aspect. This works best if the separate aspect has limited knowledge about the object model. The components may have a managed lifecycle associated with them, depending on the service management support associated with the platform they are run on.

In the Mobile Services Reference Model, components have clearly defined interfaces and no unnecessary couplings. The reference model is technology-independent.

3.2.2 Mobile Applications and Services

Important terms for understanding further descriptions are: *mobile application*, *mobile application services*, and *reference model functional components*. These terms can be described as follows:

- A *mobile application* is a computational entity that uses 1 ... *n* mobile application services.
- A *mobile application service* is a computational entity that uses at least one *reference model functional component* over specified interfaces. Application services may also use other application services.

When simply referring to 'service' in this book, this always means mobile application and application services. Reference model functional components – the service building blocks – are defined in the Mobile Services Reference Model.

Users interact with mobile applications. Mobile applications can be implemented by service developers for a special purpose or can be generic user agents pre-installed on (mobile) devices. Figure 3.1 illustrates how the user interacts with the mobile applications.

Users and groups can use multiple interaction modalities for input and output to interact with mobile applications. They do this through user agents, which could be general user agents delivered with the devices (e.g. a web browser) or specific applications. User agents are therefore located on devices, and are in charge of realising the user. To be able to support all kinds of mobile application services, these user agents have to support push and pop mechanisms for service interaction.

Devices can execute mobile application services or some functional components. Mobile applications, application services and components can also be used server-side, where they are accessed via networks such as GSM, UMTS or WiFi.

Mobile application services will not directly render a user interface. Instead:

- they could have a specialised per-device application that renders a user interface for them; or
- they use a general browser that invokes a server-side application that accesses the application service and transforms the output to a usable format.

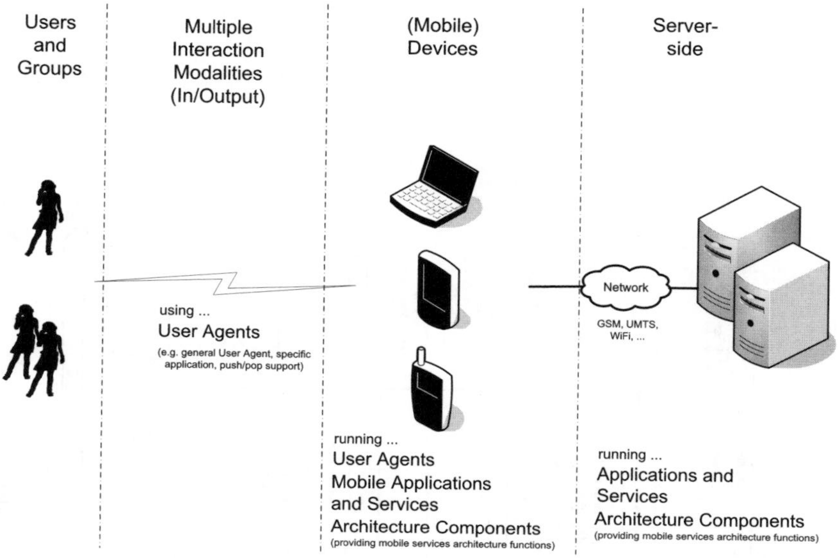

Figure 3.1 User–service interaction.

There are three different possibilities how mobile applications can be implemented:

- A general application is used on an end-user's device that is not directly using mobile application services, but interacts with a mobile application running server-side. An example for this general application is a user agent like a web browser delivered with mobile phones.
- A specific mobile application is used on an end-user's device that does the rendering of the user interface and accesses mobile application services running server-side.
- A specific mobile application is used on the end-user's device that is interacting with mobile application service on the end-user's device and server-side.

3.2.3 Service Building Blocks

Figure 3.2 depicts an architectural overview of the Mobile Services Reference Model [13].

The Mobile Services Reference Model is the architectural description of the functional components in mobile service environments. It is a template for composing architectures for mobile service environments, not directly tied to any standards, technologies or other concrete implementation details. The reference model also describes how the functional components interact and how they are structured in subcomponents.

On the left side of Figure 3.2, one can see the user that uses a (mobile) device, on which user agents are running, to access mobile applications and application services. The small box on the left side depicts *mobile applications and application services* (SER). Users interact with mobile applications, which use mobile application services to yield their functionality. Mobile application services themselves use the functionality provided by an implementation of the Mobile Services Reference Model.

The top of the figure depicts external *Data Sources* (DS) that deliver (sensor) data to user agents and to service infrastructure functional components.

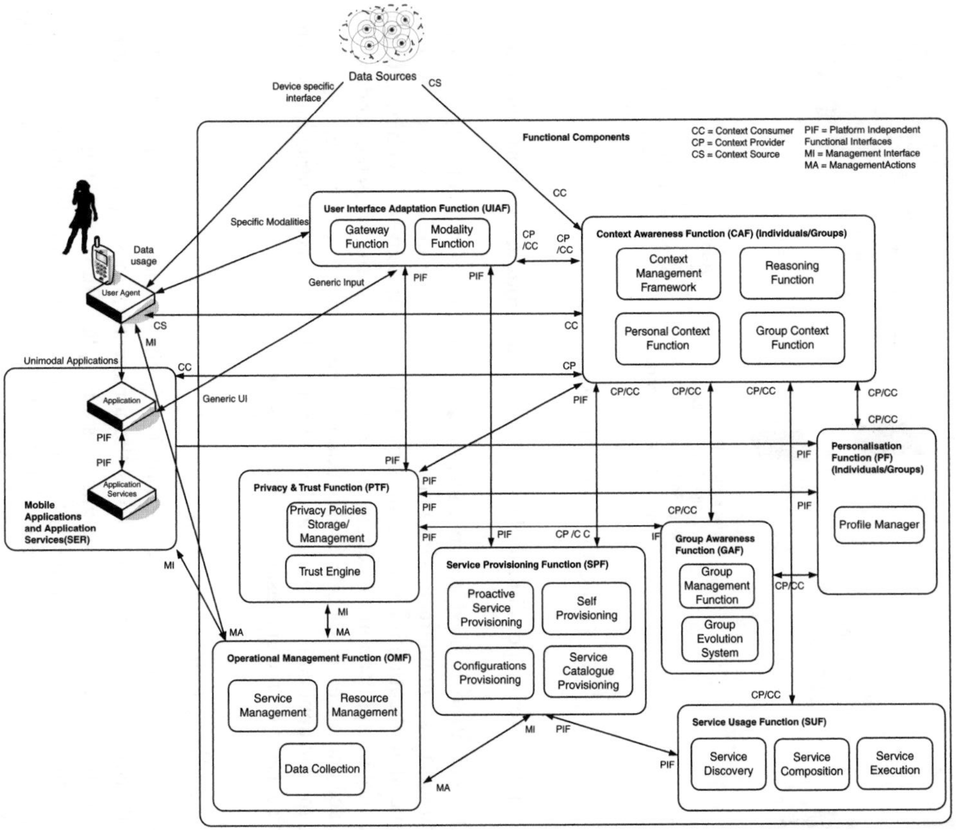

Figure 3.2 Mobile Services Reference Model.

The big box *functional components* depicts components of the reference model. They provide the basics for the development of a mobile application, such as functions that provide context awareness, personalisation, group management, and privacy and trust features.

All functions in Figure 3.2 have dependency relationships with others, which are illustrated using arrows. At the ends of the arrows, the following roles of the components are given:

- *Context Providers* (CP) are defined in the *Context Awareness Function* (CAF). CPs encapsulate context data for consumption by CCs.
- *Context Consumers* (CC) are also defined in the CAF. CCs consume context data provided by CPs.
- *Context Sources* (CS) are defined in the CAF as well. CSs are devices that deliver raw data, e.g. sensors.
- *Platform-Independent Functional Interfaces* (PIF) represent interfaces that do not rely on a specific technology.
- *Management Interfaces* (MI) represent platform-independent interfaces with the role of providing interfaces that allow management of the respective component.
- *Management Actions* (MA) represent management actions that the *Operational Management Function* (OMF) can execute.

Beside the core functionalities of a mobile service environment such as service deployment, storage, discovery and management, the Mobile Service Reference Model defines a set of features that is needed to tackle the needs of future mobile service development:

- A unified API for creating *personalised* applications and services is provided. Personal profiles can be distributed and activated depending on context, which is a novel approach in the personalisation area.
- For *context* data handling, a *context management framework* enables the development of context-aware applications and services. This is the first time in context data-related research when an open, generic and expendable context data structure and API are available.
- Concerning the *user interface*, the environment allows the development of device and modality independent services through specifying a service user interface description and a service adaptation framework. This way, users can employ services using multiple modalities at the same time.
- *Privacy* of users and *trust* in the service infrastructure are maintained through a novel approach that is also taking context information into account for privacy and trust management. Group features in the service infrastructure consider static as well as dynamic groups, and their trust relations. Using the group features, mobile application and service developers are now able to tailor their implementations to reach certain groups (e.g. to adapt the service behaviour for groups or address certain groups when sending content).

The Mobile Services Reference Model provides the following functions that cover the features described above for future mobile service development:

- *Context Awareness Function* (CAF). This function takes care of *raw, interpreted and aggregated context data*. It handles context data related to individual users and to groups of users. It supports the service developer by providing users' and groups' current context information through well-defined interfaces. It specifies the components *Context Management*

Framework (CMF) [7], *Personal Context Function* (PCF), *Group Context Function* (GCF) and *Reasoning Function* (RF).
- *Privacy and Trust Function* (PTF). Future mobile services and applications deal with data related to the user, which raises the fundamental issue of trust and privacy of the personal user data. Therefore, the PTF has been introduced to the Mobile Services Reference Model, ensuring privacy and trust through specifying a *Trust Engine* and defining privacy policies in *privacy policies storage/management*.
- *Personalisation Function* (PF). This function enables adaptation of mobile services and applications according to personal and group needs and interests [18]. It offers a *Profile Manager* to manage user and group-related profiles and preferences.
- *Group Awareness Function* (GAF). The GAF is the ingress point for all mobile application services and framework services related to group state information provisioning and to group lifecycle aspects. The GAF provides means for the management of group lifecycle aspects through the *Group Management Function* (GMF) as well as the automatic creation and deletion of groups through the *Group Evolution System* (GES).
- *User Interface Adaptation Function* (UIAF). Users may use different devices to access mobile applications, even concurrently, each device supporting different input and output modality services, different connectivity, and with different capabilities. To face this problem without creating a specific user interface representation for each device, mobile service developers are supported with user interface adaptation through UIAF. The *Gateway Function* deals with device discovery and capability management, while the *Modality Function* specifies the actual service adaptation functionalities.
- *Service Usage Function* (SUF). This function covers all aspects related to the service usage; in particular it covers every step in the 'timeframe' between service discovery and service offering. The SUF covers the components *Service Discovery*, *Service Composition* and *Service Execution*.
- *Service Provisioning Function* (SPF). This function deals with how services can be proactively offered to a user. It provides information about available mobile applications and application services in the *Service Catalogue Provisioning*. It enables the advertisement of applications automatically or proactively to users/groups, without any user/group request or interaction, defined in the *Proactive Service Provisioning* (PSP) components. Furthermore, it offers functionality to keep track of a user's service usage behaviour and to automatically recommend services to users based on this history data through *Self-Provisioning* functionality. Finally, service configurations can be pushed through *Configurations Provisioning*.
- *Operational Management Function* (OMF). This function performs operational management of mobile applications, application services, and related configuration of resources. *Service Management* focuses on the knowledge of services (access, connectivity, content, etc.). *Resource Management* maintains knowledge of resources (application, computing and network infrastructures) and is responsible for managing all the resources (e.g. networks, IT systems, servers, routers, etc.) utilised to deliver and support services required by or proposed to customers. *Data Collection* processes interact with the resources to collect administrative, network and information technology events and performance information for distribution to other processes within the enterprise.

A concrete architecture approach to the Mobile Services Reference Model giving detailed functions specifications is described in Section 3.3.

3.3 Mobile Services Architecture: Functions and Interfaces

This section presents function specifications of the Mobile Services Architecture. This architecture provides details of the service building blocks of the Mobile Services Reference Model described in Section 3.2.

3.3.1 Context Awareness Function (CAF)

3.3.1.1 Function Description

CAF takes care of raw, interpreted and aggregated context data. This function handles context data related to individual users and to groups of users. It supports the service developer by providing users' and groups' current context information through well-defined interfaces.

To determine the situation of a user and/or a group entity, and to enable context-aware services that provide relevant information depending on the user's/group's task, CAF defines representation, exchange, interpretation and reasoning of raw context data from various sources: terminal devices, sensors, the collaborative objects surrounding the entity and the entities' preferences. The use of a standard representation and exchange framework for context data abstracts from the need for each context source to define independent representation models for allowing raw context data to become context information.

CAF also provides reasoning functionalities for context inference and interpretation, including a recommender system [18]. A service developer will use high-level context information, containing low-level context data, to make the service application behave in a context-aware manner. For example, a service application might deliver personalised news to users and groups in particular situations.

3.3.1.2 Components and Interfaces

CAF specifies the components *Context Management Framework* (CMF) [7,25], *Personal Context Function* (PCF), *Group Context Function* (GCF) and *Reasoning Function* (RF). The overall functional architecture of CAF is shown in Figure 3.3.

In the following these components and their interfaces are explained in more detail.

Context Management Framework (CMF) components and interfaces

CMF defines components and their interfaces supporting discovery and exchange of context information, as well as the means to describe this information and to infer higher-level context information through reasoning. All CAF components described below rely on the CMF definitions. Among the numerous defined components and interfaces, only the two most relevant ones for the functional architecture are introduced here, namely the *Context Provider* and the *Context Broker*.

- *Context Provider interface*. A context provider (CP) is defined as 'a software entity that produces new context information from internal or external (context) information'. It implements interfaces defined by the *Context Representation Framework* (CRF) [25] to register itself to the *Context Broker* and to allow *Context Consumers* (CCs) to request context information.

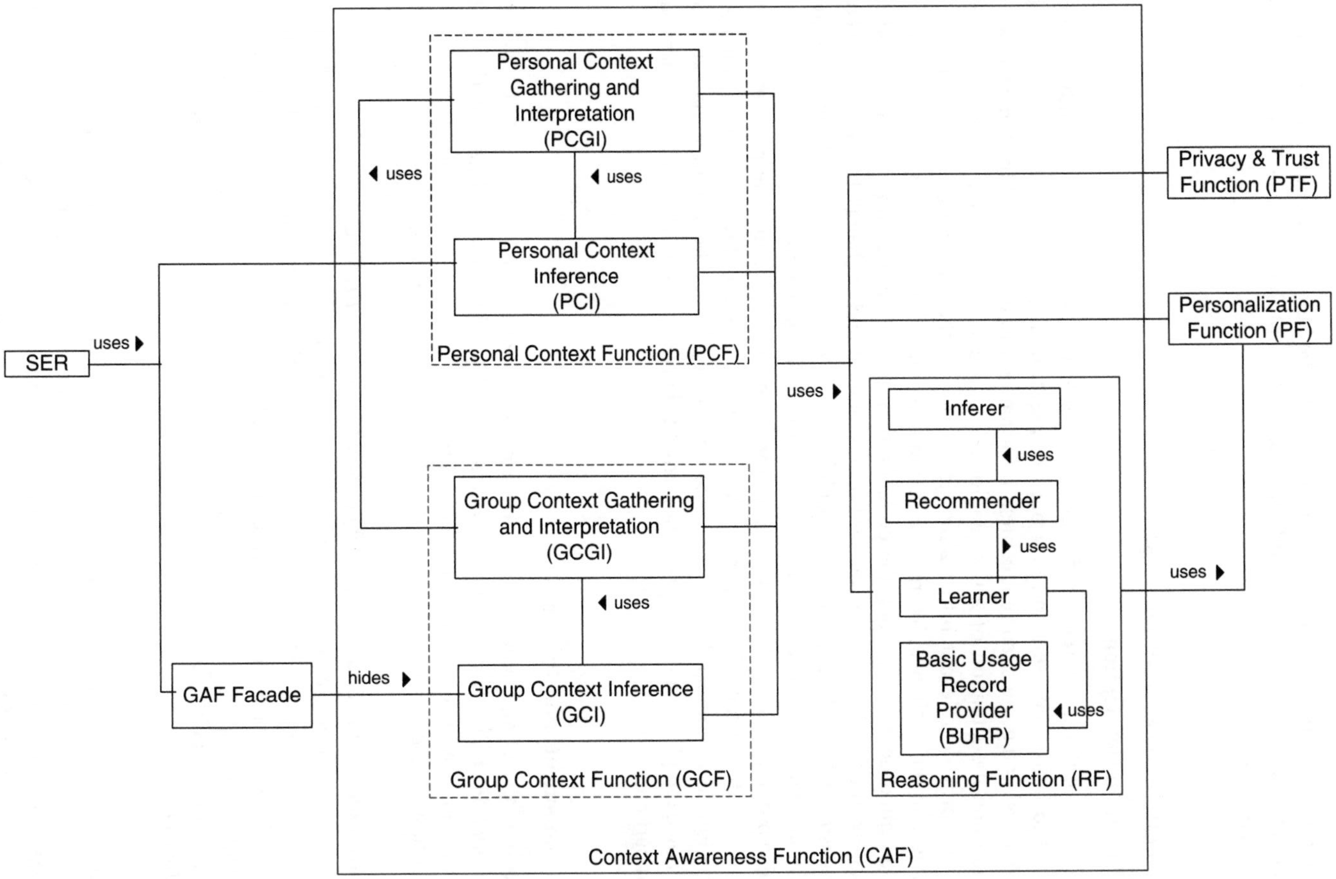

Figure 3.3 CAF functional architecture.

- *Context Broker interface*. A Context Broker (CB) is a registration and lookup service to enable the discovery of various context providers, their interfaces, the context information elements they can supply and the entities that play a role in (or: are related to) these context information elements. When a service or application needs context information, it asks the Context Broker for information from the appropriate Context Provider. In order to enable the CB to answer context information requests, all Context Providers have to register themselves to the CB. The CB is passively waiting for requests from CCs or CPs.

Personal Context Function components and interfaces
The *Personal Context Function* (PCF) deals with context information for a single user through two functional components: *Personal Context Gathering and Interpretation* (PCGI) and *Personal Context Inference* (PCI). It delivers as its output the *Operational Personal Context* (OPC) to interested parties such as services. OPC describes personal context data tailored for a user's current context. A personal context description is learned and produced through the co-operation of the PCF with RF. PCF provides, or allows the RF to learn each user context description by using a profile and model of the user. Once this context information is learned, it is stored in the *Basic Usage Record Provider* (BURP) of RF.

- *Personal Context Gathering and Interpretation*. The Personal Context Gathering and Interpretation (PCGI) component produces interpretations of the environmental context that surrounds the user; i.e. the user's personal context. The user's personal context consists of data related to her environment, such as location, environment relationships, present activities and preferences. The PCGI component manages the history of the incoming raw data and robust and grounded layered interpretations that are produced by categorisation schemes and are linked to an open upper ontology. Such an ontology provides suitable means to describe, in human understandable ways, user–application situational contexts. Besides those interpretations, PCGI records also provide the descriptions of the applied categorisation schemes for obtaining the interpretations with certain confidence measures that are associated with the PCGI's performance or accuracy. This information is also linked to the open upper ontology.
- *Personal Context Inference*. The personal context description is used by the *Personal Context Inference* (PCI) to produce an *Operational Personal Context* (OPC) and by the RF to learn casual models for the user. The PCI component produces an OPC containing highly relevant operational preference information for the individual user within the current user's context. PCI performs assumptions and conclusions on the basis of a user's context description provided by the PCGI, parts of the user's profile and the environment interaction model stored in PF. The PCI is needed to both contextualise and personalise mobile applications and services. Next, the PCI performs reasoning on the interpreted dynamic user situational context information and profile. In order to perform the inference, PCGI and PCI obey partially user (interaction) models (e.g. through following transaction policies that ensure the user's privacy concerns). Thus, these models are used to derive the proper user's situational contexts and to select related application-specific control actions. The OPC is provided to other mobile applications or services (SER) through the PCI.
- *Operational Personal Context*. The PCF is a Context Consumer of raw environmental data, user behaviour (history) data (i.e. what actions the user has selected in a particular situational context), service discovery data, and application request data from the UIAF.

Subsequently, the PCF processes this data in several steps yielding the so-called *Operational Personal Context* (OPC), which consists of semantic descriptions of the user's environmental and situational context along with confidence measures. In addition, the OPC includes profile information, such as preferences, and control actions. The control actions tell which operations the applications, services and/or the UIAF should take in the given situational context. The OPC is provided as an output to mobile applications and services, and to the UIAF. The OPC consists of the basic context information itself, as well as the control actions and the models including rules, which result from inference in PCI. Parts of the OPC are models, which have been inferred from the CC interface and from the PCGI module. At the same time, the OPC provides access to the context information itself, which has been used in the inference process. This information can be used by other components to, for instance, verify the inference process.

Group Context Function components and interfaces

The *Group Context Function* (GCF) deals with context information for users' groups. GCF consists of two functional components: the *Group Context Gathering and Interpretation* (GCGI) and the *Group Context inference* (GCI). Both GCGI and GCI rely on the PTF to verify whether access to context data (group context, OGC) is allowed and to validate input from different sources. The GCF produces the *Operational Group Context* (OGC). The group context description is produced in co-operation with the RF on the basis of context description for each member of the group provided by the PCF, and the group profiling data provided by the PF.

- *Group Context Gathering and Interpretation* (GCGI). The GCGI component produces interpretations of the environmental context of the group, i.e. the group context. The group context consists of various data related to the group environment, such as the group location, environment of the group, or group present activities. An example for a group context is the composition of the following facts: the group is in a car, travelling on a country lane at a certain speed, at a certain time, listening to country music. GCGI produces the group context description. To do so, GCGI utilises the member context description produced by the PCF, and the profile group information from the PF. GCGI interface inherits the CP interface. In addition to CP methods, GCGI component implements specific methods implemented in the GCGI interface.
- *Group Context Inference* (GCI). The group context description is used by the GCI to produce the OGC and by RF to learn casual models for groups. The OGC is used to further select suitable service categories, such as interface adaptation functions. To do so, the component uses group context produced by the CGCI along with schemas learned by RF. These schemas link the group context to particular actions or to sequences of actions. GCI performs assumptions and conclusions on the basis of group context descriptions provided by the GCGI, parts of the group profile and the group interaction models (casual models) stored in PF. The GCI is needed to both contextualise and personalise mobile applications and services. The GCI interface inherits the CP interface. In fact, it can be considered that GCI provides a particular context (e.g. the OGC). GCI interacts with PF to store or retrieve models learned for groups; for simplicity the same interface has been adopted.
- *Operational Group Context*. The OGC consists of the group context plus relevant preferences for the group that have high values in the actual context. An example for a group context is the composition of the following facts: group context (the group is in a car, travelling

on a country lane at a certain speed, at a certain time, listening to country music) + group preferences (group likes jazz music, vegetarian food). OGC is provided to other mobile applications or services (SER) through the *GAF Facade*.

Reasoning Function components and interfaces

The *Reasoning Function* (RF) provides a consistent mechanism for learning, recommendation and inference, and usage recording within the CAF. The RF consists of four functional components: *Recommender*, *Inferer, Learner* and *Basic Usage Record Provider (BURP)*.

- *Recommender*. The Recommender component of the RF can be regarded as the central interface component used by requesters in the process of requesting contextual reasoning. The Recommender provides standardised interfaces and allows links to different contextual reasoning components. It will return results (operational contexts or inferred facts) to the requester.
- *Inferer*. The Inferer component of the RF allows a given model as input and provides inferred results as output. The output results can be in form of actions supplied during normal PCI or GCI operation, or new facts required by CPs. The *Inferer* design allows different kinds of models to be used by various concrete implementations.
- *Learner*. The Learner component of the RF applies machine learning methods in order to construct or update the model based on new usage records. The results will be used to allow adaptation of new behaviour and hence modification of facts and rules. The updates will affect the next phase of inference.
- *Basic Usage Record Provider*. The BURP component of the RF is responsible for the creation and retrieval of usage records for users or groups. These records contain the usage behaviours history of an individual user or group. They can be used by the Learner to learn facts and rules related to those usage behaviours. BURP acts as a Context Consumer towards context interpreters. BURP will subscribe to context interpreters for specific types of context. Whenever that context changes, the BURP will get the data pushed to it. This in turn may trigger the collection of related context data from other providers. All the data will then be combined to a single usage record and stored in a database for later retrieval by interested components such as the Learner.

3.3.1.3 Data Structures

In this overview, the context element of the CAF is described (Figure 3.4).

A context element is the most elementary piece of context information. Each context element has metadata describing the Context Provider that delivered the data, a name, the entity that is the subject of this context element, and a time stamp denoting on what point in time this context information was captured. The value of the context element is a parameter tree, including the values of the parameters. A value can be accompanied by a measure for accuracy or confidence.

3.3.1.4 Cases of Use

CAF can be used by mobile service developers to:

- query the *Context Broker for Context Providers*;
- send and receive context data (context elements) to/from *Context Providers*;

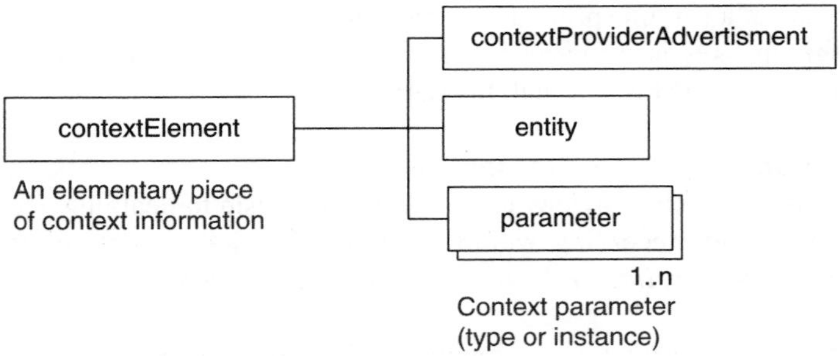

Figure 3.4 CAF context element.

- receive a bundle of personal context from the *Personal Context Function*;
- receive a bundle of group context from the *Group Context Function*;
- infer higher-level context information, learn service usage behaviour and recommend services.

3.3.2 Privacy and Trust Function (PTF)

This section specifies the *Privacy and Trust Function* (PTF). More detailed information is given in Chapter 6.

3.3.2.1 Function Description

Trust is considered here as the firm belief in the competence of an entity to act dependably, securely and reliably within a specified context. Actually, trust is considered twofold: first, trust can exist *between the users*; and second, trust can exist between *the user and the system*. Future mobile services and applications deal with data related to the user which raises the fundamental issue of trust and privacy of personal data. Therefore, PTF has been introduced, ensuring privacy and trust through specifying a *Trust Engine* and defining privacy policies.

3.3.2.2 Components and Interfaces

PTF consists of the *Trust Engine* (TE) subcomponent and *Privacy Policies Storage/ Management*.

The TE has a layered structure and it acts in a pervasive way in the enforcement of the privacy rules defined by the user and the existing system policies. It inspects every interaction in the system involving user data and allows or denies the access to the data based on the specified policies (system and user policies). There are two kinds of trust engine: the *User Trust Engine* (User TE) and the *Context Provider Trust Engine* (Context Provider TE). In addition, the *GAF Trust Engine* is a special kind of context provider TE for handling group context and profile data.

The User TE

The User TE acts as the *Policy Decision Point* (PDP). A user can create and set policies regarding the access to her data. The policy relates to one user, but may contain rules for different pieces of data, retrieved from different CPs. If a user has multiple roles in the system, she may have multiple User TEs associated with her, one for each role (e.g. the roles of a mother and an employee).

The User TE takes care of uploading the relevant parts of the policies to a policy repository or directly to *Policy Enforcement Points* (PEPs); that is, Context Provider TEs having custody of data items that are handled by the respective CPs. In case a policy is altered later on, the Trust Engine is responsible for propagating these changes to relevant targets holding policies that are affected by the changes.

The Context Provider TE

The Context Provider TE acts as the *Policy Enforcement Point* (PEP) for individual context and profile data. When a CP receives a data request, it forwards it to the Context Provider TE. This Trust Engine retrieves the policy related to this request from the policy database, and determines if the data request can be allowed.

Since the Context Provider TE does not know which members are in which groups, processing the request requires interaction with the GAF, if the policy states that access to a data item can be allowed only if the requesting user is a member of a group that the data are shared with.

If a change in a policy occurs, which can happen only after involvement of the user, the Context Provider TE is notified of this fact by the User TE.

The GAF TE

A special type of Context Provider TE is the *GAF Trust Engine*. It is also a Policy Enforcement Point (PEP), but in this case for group context and group profile data. The behaviour of the Trust Engine is very similar to the behaviour of the Context Provider TE, but this Trust Engine, while being a dedicated Trust Engine for the GAF, can decide whether the group-related request is to be allowed. In general, only the GAF Trust Engine has access to the group membership lists.

At the start of the device, the user must have full control over the running applications. Each running application will be launched under the full control of the TE software stack, through a dedicated API layer. The user interface is also a special application under control of the TE. All applications running on behalf of the user have an identity which can be traced back to the user's *Trust Seed*, which acts as the foundation of all of the user's trust associations in the system. The Mobile Services Architecture is not aiming to limit the ways in which the Trust Seed can be implemented. Several initiatives are willing to address this issue.

All the applications, installed on the mobile device or running outside the mobile device, will interact with components of the Mobile Services Architecture through the use of Application Programming Interface (API) layers depending of the granted access rights and policies defined by the user.

3.3.2.3 Data Structures

When a data access request is checked, the CP interface provides only two outcome options: success, which results in data delivery; or failure, which results in non-delivery. The rules language can be, for example, based on a *Resource Description Framework* (RDF) [24], designed as triples describing *the entity requesting the operation, the data to be delivered, and the entity the data is about*. For each triple, a triple describing the desired result is added: *continue, fail, or deliver*. The user chooses how the system should handle the data that belongs to her. Those choices that the user makes are built into a policy. When a request comes in, trust engines evaluate the request, and deliver one of the three results. The user can choose what data to share with guests – other users with whom she has not formed a group. Trust engines will only allow one guest model in the policy.

For groups that the user is a member of, there is no limit on how many models the trust engine creates, because of the fact that a user can be part of numerous groups.

3.3.2.4 Cases of Use

This section presents two important use case descriptions for the Privacy and Trust Function (PTF).

Requesting data

Applications request data in a controlled manner. Policy Enforcement Points (PEPs) prevent unauthorised access to personal data for privacy reasons and only trusted applications are allowed under the policies specified.

Whenever a user wants to get access to data that resides inside the mobile service infrastructure, the PEP decides whether the user is allowed to receive this data or not. The following steps are carried out for every request to the CP:

1. The user requests data from the CP by sending a(n authenticated) data request. The target of the request can be either the user's own or another user's data item.
2. The CP then forwards the request to the PEP to check if the user is authorised to receive the requested data.
3. The PEP checks the policy repository for the end-user's privacy policies.
4. If the policies authorise the user to have access to the requested data, a 'true' is sent to the CP, otherwise a 'false'.

The CP checks the response given by the PEP and reacts accordingly. If the response is 'true', the CP gets the data from the data repository and sends it to the requester, else not.

Setting policies

Privacy and trust policies can be set only if authorisation is granted. Policies at start-up are set by the user on her local device, and the most interesting phase occurs when the policy set is deposited in encrypted form elsewhere, to be easily reused without re-entering. The remotely stored set should always be synchronised with the local one, and user intervention is called for when a discrepancy is detected – a checksum procedure can be easily implemented for detecting such discrepancies. Since trust relationships change, the start-up policy might allow some application-specific policies to be changed also after start-up.

When the user acting in a role has chosen her preferences, the user interface sends the data to the PDP (User TE) that stores them in the *Privacy Policies Storage* for this user ID. Therefore, every user ID has an own table of policies that is uploaded appropriately to the relevant servers which the PEP (CP TEs) check. Relevant servers refer to a set covering the user's environment – places where user data are stored as well as groups the user has chosen to join. Every time someone in a user group wants data that is related to that user ID, the group policies and the user policies are merged before the CP can make a decision and release the data that has not been specifically given over to group storage for that specific group. There are special requirements on the trusted path used for configuration of privacy and policies – an ordinary application cannot impersonate this user interface.

3.3.3 Personalisation Function (PF)

This section specifies the Personalisation Function. More detailed information is given in Chapter 5.

3.3.3.1 Function Description

The PF enables adaptation of mobile services and applications according to personal and group needs and interests by offering a *Profile Manager* to manage user and group-related profiles and preferences. Service adaptation means transformation of service content based on individual, service-specific user/group models within a given context. Service-specific user/group models (i.e. user/group profiles) can be created either manually or via profile learning (part of the *CAF Reasoning Function*). These profiles are used as a basis for building sophisticated, service-specific recommender systems. The service adaptation itself belongs to the service implementation. However, the *Personalisation Function* provides well-defined interfaces to support the management of user/group models (i.e. *creating, modifying, deleting* and *querying*).

3.3.3.2 Components and Interfaces

The PF consists of one major functional element, the *Profile Manager*. Figure 3.5 shows the corresponding overall class diagram.

Profile Manager

The Profile Manager component, embodied in the *Individual Profile Manager* or the *Group Profile Manager*, applies the identified data model to all personalisation data that should be managed by the PF. Therefore, it needs to include all classes of the data model and to provide an interface to access these objects – the so-called *Profile Manager* interface. It offers functions to create, retrieve, modify and remove entities, profiles, profile entries and qualifiers. In order to retrieve any personalisation information, an *InteractionContext* object is needed to issue a request to the Profile Manager. The *InteractionContext* consists of an entity with its profile and schema and context qualifiers.

Relation to the Privacy and Trust Function

A set of privacy enforcement rules can be applied by a Policy Enforcement Point (PEP), so that a service solely receives preferences that belong to the service which issued the query, or that are explicitly allowed by the user or group. Figure 3.6 shows the corresponding class diagram.

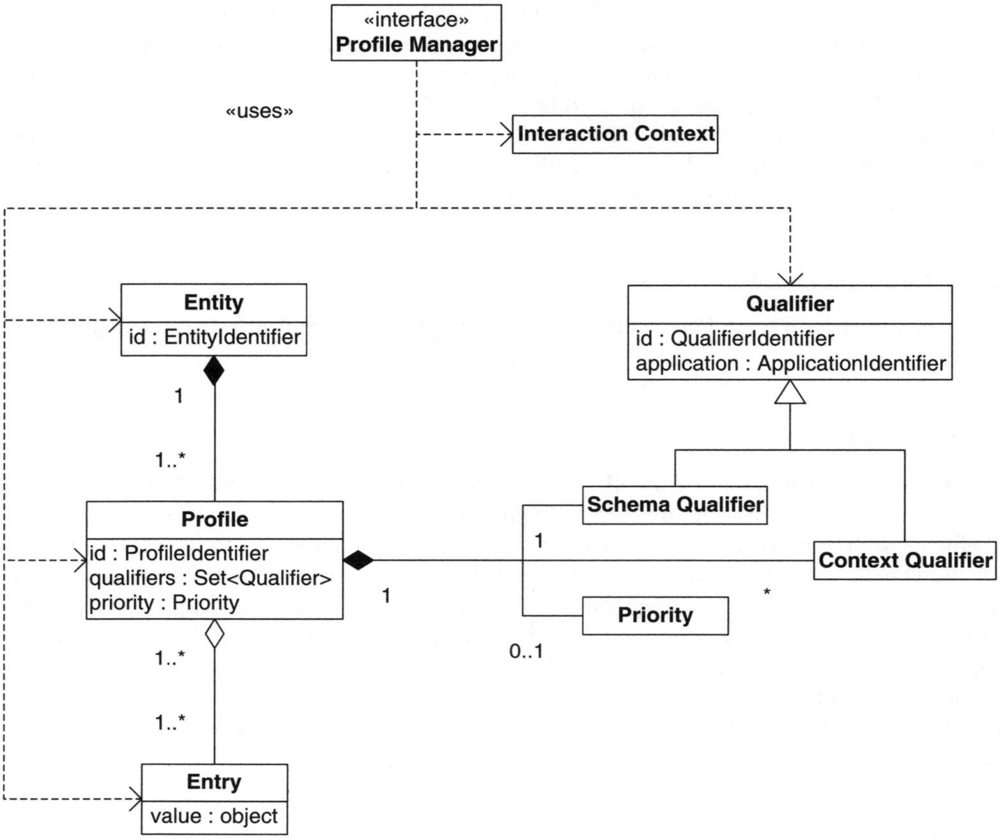

Figure 3.5 PF class diagram.

3.3.3.3 Data Structures

The PF data structure consists of three major objects: *Entry*, *Entity* and *Profile*. It describes how any kind of personalisation data can be described. Figure 3.7 shows the corresponding overall class diagram.

An Entry object is a single unit of personalisation data (such as entity facts and rules or entity preferences and interests). An Entity object manages all basic information about a user or a group. This information consists of static data such as a short description or the full name of the user or group. To represent this information, the previously introduced Entry objects will be used. An Entity object is also a *ProfileFactory*, which means that all Profile objects that belong to this specific Entity object can be created and added by the Entity object itself. A Profile object groups specific units of personalisation data (aka *Entry* objects) to describe an entity. Besides units of personalisation data, a Profile contains a unique ID, qualifiers and an optional priority. Qualifiers can be separated into two different classes, so-called *schema qualifiers* and *context qualifiers*. A schema qualifier is mandatory and context qualifiers are

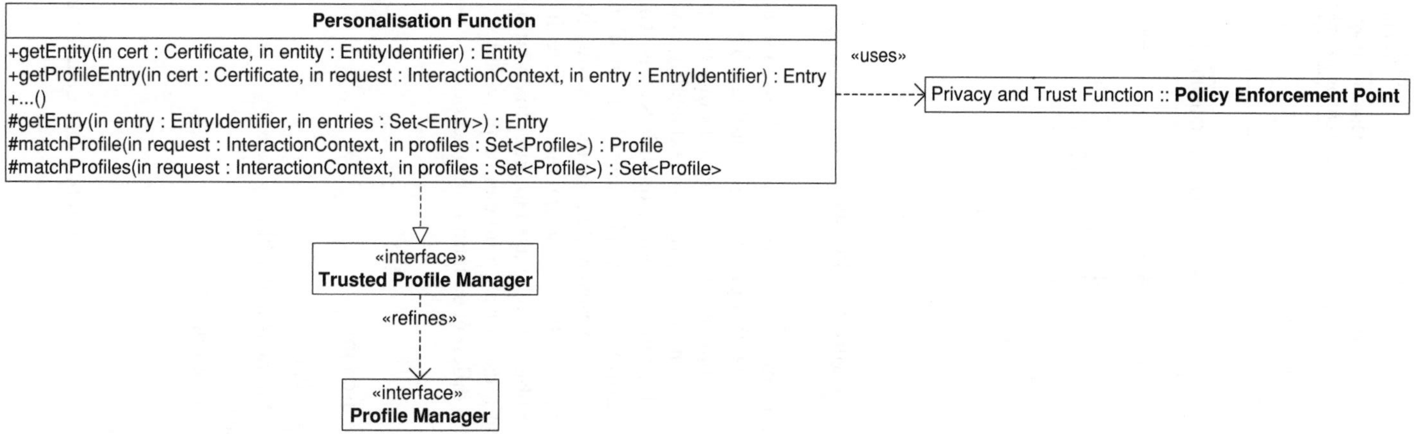

Figure 3.6 PF class diagram showing the PTF integration.

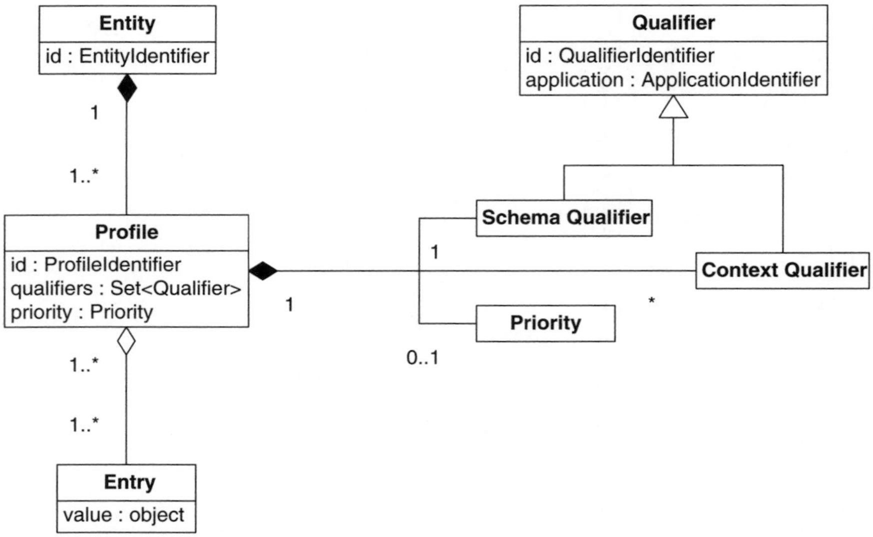

Figure 3.7 PF data model.

optional. A Profile object contains one schema qualifier. It is used to describe all information belonging to a Profile object by semantics given by an ontology. This ensures the reusability of stored Entry objects by various applications in using ontology mapping techniques.

A Profile object can contain one or more different context qualifiers which describe different situations associated with this Profile object. Context qualifiers consist of different information; for example, they might feature information about location, time, available devices, or available modalities. A given situation description must match one of these qualifiers in order to consider a Profile valid for this specific situation. In the case that there is more than one matching Profile object, the priority of the Profile object can be used in order to determine which Profile object has to be used. This means that a user or a group can be provided with customised services that take different preferences in different situations into account.

3.3.3.4 Cases of Use

The PF Profile Manager can be used to:

- create individual and group profiles;
- store individual and group preferences and arbitrary data;
- read individual and group data;
- update and delete individual and group profiles.

3.3.4 Group Awareness Function (GAF)

This section specifies the Group Awareness Function. More detailed information is given in Chapter 6.

3.3.4.1 Function Description

The GAF is the ingress point for all application services related to group state information provisioning and to group lifecycle aspects. The GAF provides means for the management of group lifecycle aspects as well as the automatic creation and deletion of groups. Unlike personal data about individual users, the configuration and definition of a group may evolve over a short timeframe. Groups may update their member lists – they are gaining new members or losing existing members. In addition, groups may be active or passive. *Active* means that some of the group members co-operate or undertake activities within the group at a time, while *passive* means that group members do not co-operate.

3.3.4.2 Components and Interfaces

The GAF consists of the following components: *Group Management* (GM), *Group Evolution* (GE) and *GAF Façade*.

Figure 3.8 shows the GAF functional architecture, details GM and GE sub-components and shows the main relations/dependencies with other service components.

Group Management (GM)
Group Management deals with the automatic management of all group lifecycle aspects. It governs all the processes that occur when a group is created, deleted, or when a group changes states: active, inactive, and vice versa. GM also interfaces services, applications and components to the group profile of the PF. Incoming requests for retrieving or updating group information are handled by the GM that checks permissions to access private data. A group is characterised by its group profile and the corresponding group template. The group profile contains all information related to a group, such as group membership list, policies, and preferences of the group. Group templates contain predefined elements and policies to be included in the group profile when a new group is created. A group is created by instantiation of a group profile according to a distinct group template which serves as a guideline for information allowed in the group profile.

Group Evolution (GE)
The Group Evolution component proposes the grouping of a set of users according to the type of group (defined by templates), personal preferences and users' social relations. Furthermore, GE monitors the group's behaviour to learn new or modify existing group templates. The GE component uses GM functionality for the storage of the group templates.

GAF Façade
The GAF Façade exposes the various methods related to groups to mobile applications and services (SER) in a unique interface. Therefore, the GAF Façade hides not only interfaces and methods provided by GM and GE components, but also the group-related methods provided by the *Group Context Function* (GCF) and *Personalisation Function* (PF).

GAF components' layout
GM consists of four components. The *Group Manager* component is responsible for maintaining the lifecycle of a group (e.g. group creation, management and deletion). The *Template Manager* component is responsible for storing and managing the group templates. The

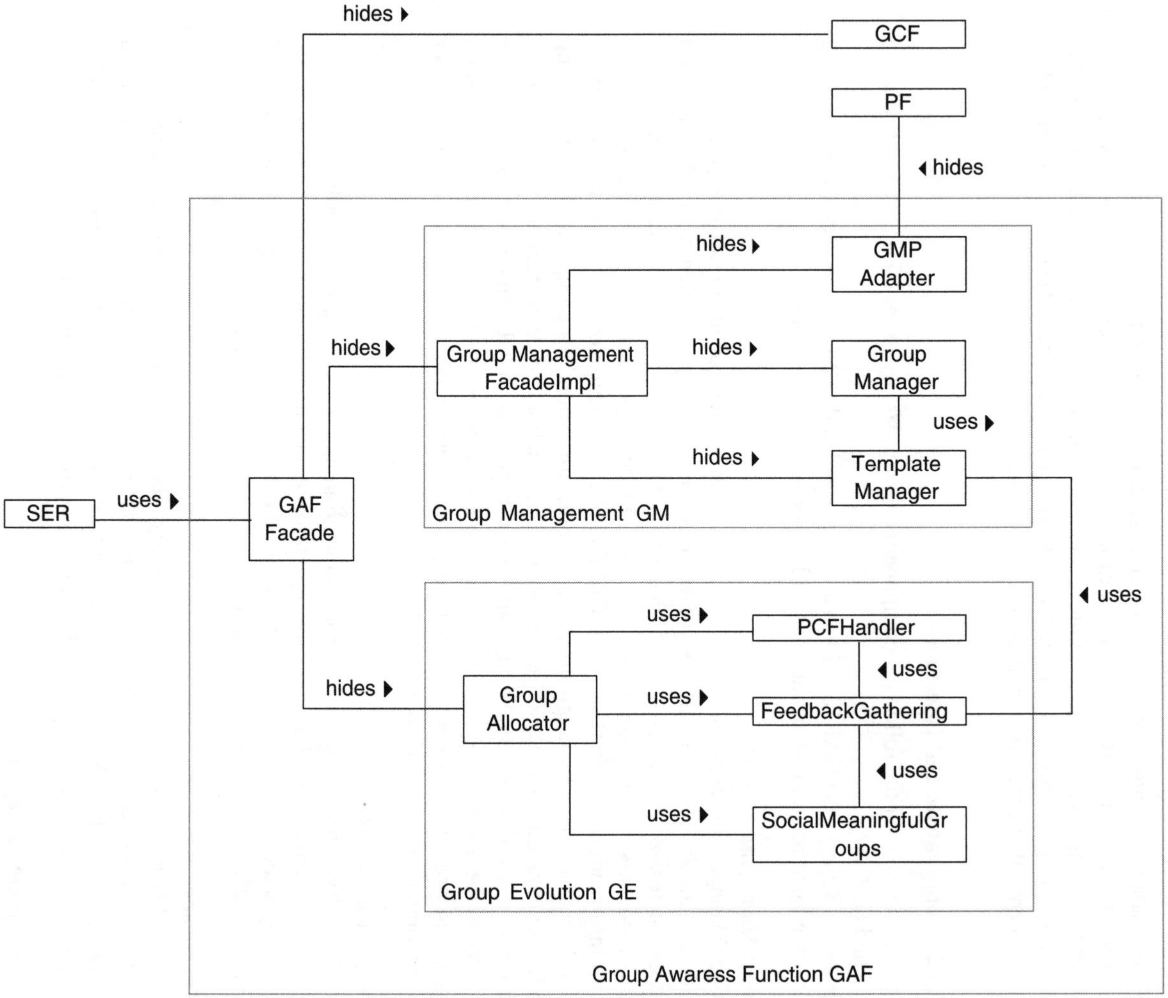

Figure 3.8 GAF GM and GE functional architecture.

GPM Adapter component is responsible for facilitating the access to group profile entries. The *GroupManagementFacadeImpl* component hides GM internal components and the PF group-related components, and implements the visible GM interface towards the other mobile applications and services (SER) through the GAF Façade.

GE consists also of four components. The *Group Allocator* component receives a list of users and the type of group application defined by the group template, and suggests possible groups that could be created and the allocation of users to groups. The decision process for the identification of possible groups and the allocation of people to them takes into account user preferences with respect to a certain group type (*PCFHandler*), the user's relations with other persons (*SocialMeaningfulGroups*) and user satisfaction (*FeedbackGathering*). The *SociallyMeaningfulGroups* component learns users' relations by observing user interactions and behaviour in certain circumstances; socially meaningful groups are learned by observing users' repetitive behaviour. The *PCFHandler* component handles the exchange with the Personal Context Function (PCF). *PCFHandler* requires from PCF the group parameter priorities for each user that the Group Allocator is requested to assign to a certain group. Group parameter priorities are specific for each user. Each user could have a different value for each different group application. The *FeedbackGathering* component gathers implicit or explicit user satisfaction. User satisfaction is used in order to fine-tune behaviour of the Group Allocator component (i.e. to provide more satisfying groups to the users). Additionally, it decides whether the creation or updating of group templates and commands template modifications to the GM occur.

GAF Interfaces

The GAF operational interface is compliant to the CP in order to access the *Operational Group Context*. Additionally, it provides methods for group lifecycle management, template management, user grouping recommendations, feedback gathering and to access group profiles.

3.3.4.3 Data Structures

Group Management data structures

The data structure of GM consists of the following main classes: *Group Templates*, *Group Profiles*, *Groups* and *Members* for groups. Each Group is associated to one Group Profile, and each Group Profile is associated to a Group Template. The same Group Template can be associated to more than one Group Profile. The *Group* is a logical relationship between users. Users are bound together by a criteria of membership specified in the *Group Policy*. Each group is associated with one and only one Group Profile. The *Group Manager* is responsible for group creation, management and deletion. A group consists of a set of identifiable users. The users are represented by the *Member* class. A new member can be added to a group if her policies and constraints match policies and constraints of the group. The Group Profile consists of the information data defining a particular group instance (e.g. members, group attributes and group preferences), and contains all information relevant to the group. The Group Template governs the Group Profile information structure, mandatory information, and is associated to the Group Profile during profile creation. The Group Template serves as an initial starting point for the creation of new groups. When requiring a new group, the requester chooses a template, closest to the rationale of the group it needs to create (e.g. buddy list, distributed collaboration work group, video conferencing group). The Group Template is used to instantiate a Group

Profile which contains the mandatory group information as defined in the template. It may contain additional information specific to the group's purpose. The same Group Template can be assigned to various groups (group profiles). The Template Manager is responsible for the storage of Group Templates. The creation and updating of Group Templates is decided by the GE.

The *Group Principal* is the entity responsible for the group. There is one and only one Group Principal for each group. The Group Principal is responsible for managing group metadata (e.g. policies, access rights), membership, group creation, disposal, suspension and activation.

Group Evolution data structures

SocialMeaningfulGroups are groups of people that are learned by observing the user behaviour in relation to other persons. SocialMeaningfulGroups are continuously updated. The *SocialMeaningfulGroup Manager* learns user's social relations and buddies and creates the social meaningful groups.

RecommendedGroups are created by the *Group Allocator*. The Group Allocator receives a list of users and the type of group application (group template identifier) and decides possible groups and the allocation of users to the various groups. User records with user preferences are retrieved from the PF. This data is used by the Group Allocator to decide the allocation of users to groups. This data can be considered as a subset of user profile data.

3.3.4.4 Cases of Use

GAF offers functions for:

- creation and group member addition;
- group updating and deleting;
- automatic group creation and management based on group templates.

3.3.5 User Interface Adaptation Function (UIAF)

This section specifies the User Interface Adaptation Function. More detailed information is given in Chapter 5.

3.3.5.1 Function Description

Users may use different devices to access applications, even concurrently, each device supporting different input and output modality services, different connectivity, and with different capabilities. To face this scenario without creating a specific user interface representation for each device, the Mobile Services Architecture must support user interface adaptation. The UIAF provides functionalities to allow service developers to make services available through multiple devices using multiple modalities. The determination of the devices and their supported modalities to be used as user interfaces is also a necessary feature of the UIAF. It relies on the detection of surrounding devices and their capabilities, and also on user context and preferences, in order to decide which combination of devices and modalities will provide the best user experience.

3.3.5.2 Components and Interfaces

The UIAF consists of the two functional subcomponents, the *Gateway Function* and the *Modality Function*. Determination of the devices and their supported modalities to be used as user interfaces is done by the Gateway Function. The Modality Function realises the actual fusion of user input and fission application output, implementing content adaptation features.

Gateway function

The Gateway function is composed of *Device and Service Discovery* and *Device Agents*. Device and Service Discovery is in charge of discovering the devices and their modalities surrounding the user. It relies on a set of discovery adapters that are in charge of performing the actual discovery operation on a specific network adapter. The Device Agent represents a modality service and is linked to a modality description. It provides the methods to receive input from and emit output to the represented modality.

Modality function

The Modality function is composed of *Input Functionality*, *Control Functionality*, and *Output Functionality* including *Transformers*. The Input Functionality orchestrates the different subtasks of the input processing. It basically receives input from a Device Agent, uses a specific *Recogniser* on it, and possibly requests fusion on different input streams to provide processed input to the application and input model to the Control Functionality for further delivery to the CAF. It maintains a list of Recognisers. The Control Functionality is in charge of pushing context information coming from the Input Functionality or the Device and Service Discovery to CAF. It also maintains the device list upon notification by both the Device and Service Discovery and the CAF/SPF of new modality services outside BAN/PAN scope. Finally, it is responsible for processing application requests for input, output or control.

The Output Functionality orchestrates the different subtasks of the output processing. It receives output from the application and context information from the CAF, determines the best modality choice for actual output to the user, and uses Transformer instances to process modality output flows before delivering them to the desired Device Agent. Transformers receive a media element and certain conversion parameters as input. If the conversion is possible, it returns a media element that fits all conversion parameters. The conversion is done in the form of an on-the-fly converter because this enables the UIAF to cache the media elements, and no temporary files have to be deleted. The input parameters of a Transformer are all parameters that are needed to deliver the media element. This means that the needed capabilities that didn't match the output device and modality have to be passed to the Transformer.

Example: If an image is too tall, the height that the device can handle has to be passed to the Transformer. The Transformer tries to do all transformation steps to satisfy all requested target parameter values.

Interfaces towards other components

The interfaces of the UIAF towards other components are represented in Figure 3.9. They are realised through:

- the Device Agent interface for the user devic;
- the Control Functionality for the interface to the application, based on the *ApplicationRequest* data model (see Section 3.3.5.3);
- the Output Functionality for the interface to the CAF. The interface between UIAF and CAF follows the CC/CP model.

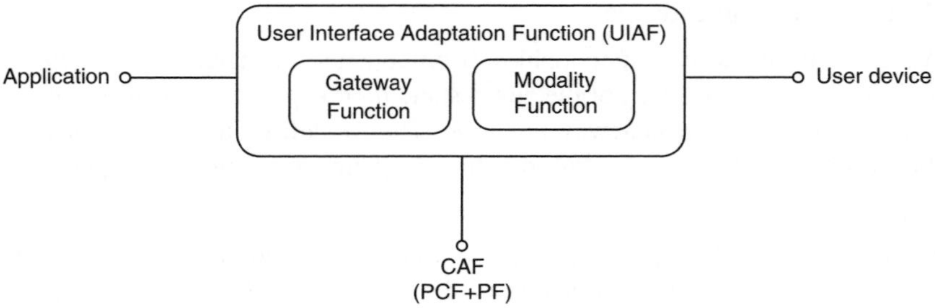

Figure 3.9 UIAF interfaces towards other components.

3.3.5.3 Data Structures

Device and modality

To take decisions on which devices and modality to best use in rendering a user interface, the UIAF relies on an accurate description of devices and modality services available to the user.

The device and modality description data model is represented in Figure 3.10.

This model is based on the different modality devices (*Physical Modality*) that constitute a user interface device. These physical modalities are regrouped under one *Host Device* that is the actual device the user sees; e.g. a PDA host device is composed of a screen, stylus, speaker, microphone, which are here considered physical modalities. Finally, Device Agents are the software entities that realise a particular modality service, and can make use of multiple physical modalities. The same physical modality can also be used by several Device Agents. Therefore, a Host Device description includes the unique identifier of the device along with capabilities information, the list of physical modalities that this device contains, and the related modality services. Physical modalities in turn can be described using capabilities parameters, the latter being simple attribute (name, value) pairs. The Device Agent description includes a unique ID, the modality class to which this particular modality service implementation belongs (e.g. textual input for a keyboard or video output for a TV screen), and a set of capabilities related to the modality service. It also contains a pointer to a Device Agent that can be used to access the service.

Application request

The interface between applications and the UIAF is realised through the exchange of XML messages. *Application requests* are thus of four kinds:

- *Modality output requests* are used to request an output to the user. They consist of a modality type element to give a hint about the desired output modality (from the application's point of view), so that the UIAF can decide which device it should use for output, or if some

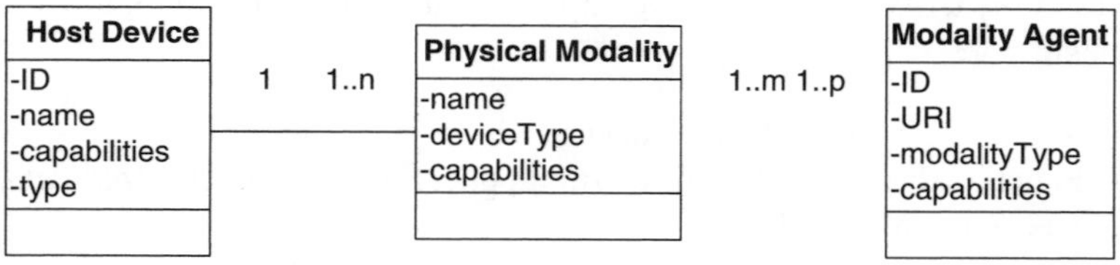

Figure 3.10 UIAF device and modality descriptions.

modality transformation should take place. The other element describes the content to be output to the user, either inline in the request or in the form of a URL.

- *Control requests* are used when an application wants to send a control message to a given modality identified by the request ID. The command element of this request depends on the modality agent implementation.
- *Modality subscription requests* are used by applications to register notifications from a modality; i.e. user input or signalisation information. The modality is designed either by a request ID, or through a modality type.
- *Disposal requests* allow the application to unregister as a consumer of modality events for the modality identified by the request ID.

3.3.5.4 Cases of Use

The use cases for the UIAF to gain device independent mobile application input and output are:

- device and modality handling;
- service output request processing;
- service input request processing.

3.3.6 Service Provisioning Function (SPF)

This section specifies the Service Provisioning Function. More detailed information is given in Chapter 4.

3.3.6.1 Function Description

The SPF deals with how services can be proactively offered to a user. It provides information about available mobile applications and application services. It stores application and application service information in terms of description, properties and semantic information. The SPF supports configurations made available by the service provider concerning user terminals for the correct usage of a specific application. Next, the SPF has a functionality to advertise applications automatically or proactively to users/groups, without any user/group request or interaction. Finally, the SPF provides functionalities to manage the user-service profile and service subscriptions. This allows the user's service usage behaviour to be tracked and automatically recommends services to users based on this history data.

3.3.6.2 Components and Interfaces

Service Catalogue Provisioning

Service Catalogue Provisioning provides information about available applications and application services. It stores application and application service information in terms of description, properties, semantic information, etc. It also stores information used by the OMF.

Configurations Provisioning

Configurations Provisioning supports configurations made available by the service provider concerning user terminals for the correct usage of a specific application. New advertised applications may need ad-hoc configuration files in order to be played on users terminals. In this case, it is necessary that users' terminals automatically receive the necessary information in order to upgrade their own configuration.

Proactive Service Provisioning

Proactive Service Provisioning advertises applications automatically or proactively to users/groups, without any user/group request or interaction. Proactive Service Provisioning could be triggered in two different cases:

- A new application, never released before, becomes available in the system. In this case Proactive Service Provisioning is named 'Advertising' and, according to static associations (even based on historical context information), it is possible to select users who could be interested in certain service categories. Associations used for the selection might also be dynamically deduced.
- Either the context or the preferences of a specific user, who previously asked for the list of available applications, has changed. As soon as the Proactive Service Provisioning is notified about such changes, the current situation is derived and analysed and the list of available applications suitable for that user in that situation is automatically pushed to the user herself. Based on such information, for example, a certain location (e.g. a bus stop) can trigger an application to buy tickets.

Self-provisioning

Self-provisioning provides functionalities to manage the user-service profile and service subscriptions.

3.3.6.3 Data Structures

Service Descriptor

A service is described by the Service Descriptor and its related meta-information. *Service Catalogue Provisioning* is a repository of services. The Service Descriptor is essential for enabling Service Discovery, Service Composition and Service Catalogue Provisioning.

A Service Descriptor is represented as an XML document containing the following information:

- *Name*. This element represents the name assigned to the service at creation time. A service *must* have a *name* element.
- *Version*. This element represents the version of the service. This versioning is usually managed by the service provider and is necessary for managing service upgrades on the network. A service *must* have a *version* element.
- *Definition*. This element contains the URL of a formal description of the service, such as a WSDL document. This element has also an optional attribute named *Type* which can be used to describe and categorise the content type of the element. A service description *must* have one or more *definition* elements.
- *ID*. This element represents a unique identifier of the service. The identifier can be automatically generated by the publishing system, or provided by the publisher itself. A service *may* have an *id* element.

- *Vendor.* This element represents the name of the service provider. A service *may* have a *vendor* element.
- *Description.* This element represents a language-dependent message to be displayed on the user's terminal to describe it. The attribute *lang* specifies the language in which the description is expressed. A service *may* have one or more descriptions, to support different languages.
- *Information.* This element enriches the language-dependent description contained in the *description* element. It provides a reference to additional information such as web pages that describe the service, or links to subscription or activation pages. Each information element may have a type attribute to describe and categorise its content (e.g. *help*, *price*, *tech*). A service *may* have one or more information elements.
- *Semantic.* This element is an XML document or the URL of the document that describes the semantic of the service. A service *may* have a *semantic* element.

3.3.6.4 Cases of Use

The most important use case for the SPF is the service push, whereby the Mobile Services Architecture pushes a service to users upon a context change.

3.3.7 Service Usage Function (SUF)

This section specifies the Service Usage Function. More detailed information is given in Chapter 4.

3.3.7.1 Function Description

The SUF covers all aspects related to service usage; in particular it covers every step in the timeframe between service discovery and service offering. The SUF covers the components *Service Discovery*, *Service Composition* and *Service Execution*.

Service Discovery allows discovering services related to the user's current context and user preferences. It depends on the SPF to get information on available mobile applications and application services for discovery and composition.

If the requested application service is not available by itself but can be composed from mobile application services, a service composition may take place – at runtime – in order to provide the requested application service:

- *Application services* are services that could represent both a complete, executable service, and could also be combined together with other available elementary application service parts in order to provide a composed service. Its descriptor is stored in the Service Catalogue Provisioning.
- *A composed application service* is obtained by composition of different application services. Its descriptor is stored in the Service Catalogue Provisioning.

When services are deployed in the Service Execution component, they can finally be started.

3.3.7.2 Components and Interfaces

Main sub-functions of the SUF are *Service Discovery*, *Service Composition* and *Service Execution*.

Service Discovery
Service Discovery is the ability to discover all available applications and application services already registered into the Service Catalogue Provisioning sub-function of the SPF, according, if necessary, to the user's current context and user preferences. This function supports manual or static service discovery.

Service Composition
If the requested application service is not available by itself but can be composed from application services, a service composition may take place in order to provide the requested application service. Service composition can be performed at runtime.

Service Execution
This function offers methods for service deployment and execution.

3.3.7.3 Data Structures

As relevant data structures are shared with the Service Provisioning Function, refer to Section 3.3.6 for more information.

3.3.7.4 Cases of Use

The two main use cases for the SUF are:

- Available service request, in which the SUF processes a discovery request from another component.
- Automatic service composition, in which the SUF performs automatic service composition to fulfil a request.

3.3.8 Operational Management Function (OMF)

This section specifies the Operational Management Function.

3.3.8.1 Function Description

The component specifications within these sub-functions and their interworking are based on *eTOM of the Telemanagement Forum* [20]; please refer to *eTOM* for details. Furthermore, the data model used is based on the *TMF SID-model* [21]. Thus, the relationship of *eTOM* to the Mobile Services Architecture is described here. The OMF performs operational management of mobile applications, application services and related configuration of resources.

The OMF manages the knowledge of services (access, connectivity, content, etc.) and includes all functionalities necessary for the management and operations of communications and information services required by or proposed to customers. The knowledge of resources is kept (application, computing and network infrastructures) and all resources (e.g. networks, IT systems, servers, routers, etc.) are managed that are used to deliver and support services required by or proposed to customers. Administrative, network and information technology events and performance information are collected for distribution to other processes within the enterprise. The OMF supports the lifecycle management of services during operation. To perform this, it needs access to information about services stored in the SUF. This fact has to be reflected when modelling the physical distribution of the components, in the system model.

3.3.8.2 Components and Interfaces

The OMF includes three main sub-functions: *Service Management*, *Resource Management* and *Data Collection*.

Service Management
This process grouping focuses on the knowledge of services (access, connectivity, content, etc.) and includes all functionalities necessary for the management and operation of communications and information services required by or proposed to customers. The focus is on service delivery and management as opposed to the management of the underlying network and information technology. Some of these functions involve short-term service capacity planning for a service instance, or the application of service design to specific customers or managing service improvement initiatives.

Resource Management
This process grouping maintains knowledge of resources (application, computing and network infrastructures) and is responsible for managing all these resources (e.g. networks, IT systems, servers, routers, etc.) used to deliver and support services required by or proposed to customers. It also includes all functionalities responsible for direct management of all such resources (network elements, computers, servers, etc.) used within the enterprise. These processes are responsible for ensuring end-to-end delivery of the required services. The purpose of these processes is to ensure that the infrastructure runs smoothly, is accessible to services and employees, is maintained, and is responsive to the needs, whether directly of indirectly, of services, customers and employees. This process also has the basic function to assemble information about the resources (e.g. from network elements and/or element management systems) and then integrate, correlate and, in many cases, summarise that data to pass on the relevant information to service management systems, or to take action in the appropriate resource.

Data Collection
This process interacts with the resources to collect administrative, network and information technology events and performance information for distribution to other processes within the enterprise. The responsibilities also include processing the data through activities such as filtering, aggregation, formatting and correlation for the collected information before presentation to other processes. Client processes for this information perform fault and performance analysis of resources and services. These include resource performance management and service quality management.

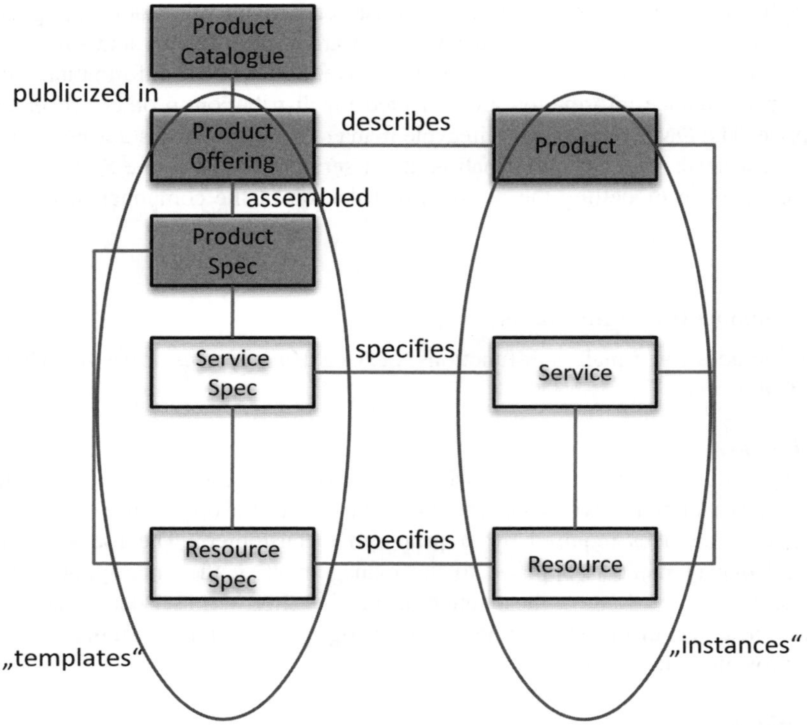

Figure 3.11 OMF specification relationships to specification instances.

3.3.8.3 Data Structures

The operational management high-level data model is based on the *TMF SID-model* [21], with a few additions.

Figure 3.11 describes the data model at a high level. This data model includes two aspects: *service and resource specifications* and *service and resource data*. Specifications are basic data templates, values and references. Service and resource data include instance data (current values, history, etc.). The data model is also distributed between service and resource blocks.

3.3.8.4 Cases of Use

The *OMF Customer Care System Interface (Reports & Actions)* provides trouble reports in the management system to find solutions and reports back to the user whatever service quality problem issues are ongoing. In some cases, OMF should report problems directly to end-users through *OMF End User (Report)*, but this is not the normal behaviour – the normal method is through a customer care system or service. Through the *OMF Service Provisioning Function Interface (Control & Changes)*, the OMF sends different types of requests (or rather triggers) for service configuration and activation. These triggers typically include different types of control and configuration data which change the service configuration. This interface also synchronises product information (data) between management and service provisioning. Through the *OMF Service (Measurement Data, Management Status) Interface*, operational management collects different types of management and performance data, problem tickets, logs and status information.

3.4 Mobile Services Architecture: Functional Components Interworking

This section describes the Mobile Services Architecture functional components interworking by describing the most important use cases of these functions.

The typical use of the Mobile Services Architecture contains the following use cases:

- service deployment;
- service monitoring and maintenance;
- service discovery;
- proactive service provisioning;
- multimodal service usage;
- trusted personalised, context-aware services;
- joining a group.

Using these seven use cases, interworking of the functional components is described. For each use case, a brief description is given followed by a message sequence diagram or diagrams to illustrate component interworking.

3.4.1 Service Deployment

First of all, the Mobile Services Architecture offers the functionality to deploy new services to the system. The deployment can happen:

- *manually*, using SUF service deployment functions; or
- *automatically*, through the *Gateway Function* of the UIAF that can discover services in the vicinity.

For automatic service deployment, the UIAF uses the functions provided by the SUF.

The service deployment functionality is offered by the SUF through the *Service Execution* component. The deployment is done by calling *deployService()*. After calling this method, the service is ready for execution. The SUF also communicates with the SPF to insert the service to the service catalogue using *insertService()*, where a list of all services is kept. The service catalogue is a repository of all services. Having a service in this catalogue does not mean that it can be executed – for this the service needs to be deployed.

In a more complex case, the CAF might request a certain service using the SUF's *request-Service()* method, which deploys and executes a new composed service based on the services existing in the service catalogue. Services are selected from the catalogue taking context information into account, and composed using context information. Figure 3.12 illustrates the message exchange for this complex service deployment case.

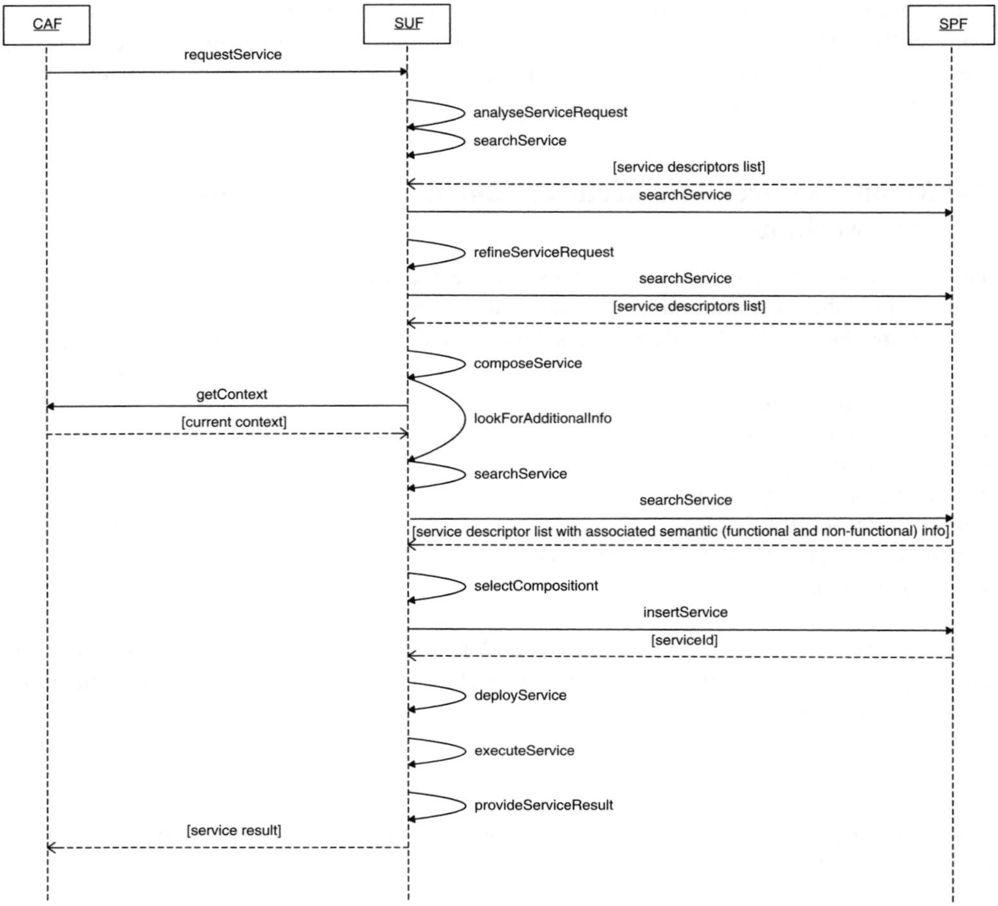

Figure 3.12 Component interworking – complex service deployment.

3.4.2 Service Monitoring and Maintenance

Services that are deployed and running can be monitored and maintained. This functionality is provided by the OMF. Figure 3.13 illustrates that the operator can start SLA monitoring using functionalities provided by the *Service Management* component of the OMF that uses functions of *Resource Management* and *Data Collection*. Here, the Mobile Services Architecture does not provide its own specification, but relies on the processes defined in the enhanced *Telecom Operations Map (eTOM)* standards of the *Telemanagement Forum* [20] and on the data model *Shared Information and Data Model (SID)* defined there [21]. Figure 3.13 shows that functions of the Mobile Services Architecture have to offer functionalities to collect performance data and service usage data.

For maintenance of terminal device configuration, the SPF provides the *ServiceConfiguration* interface.

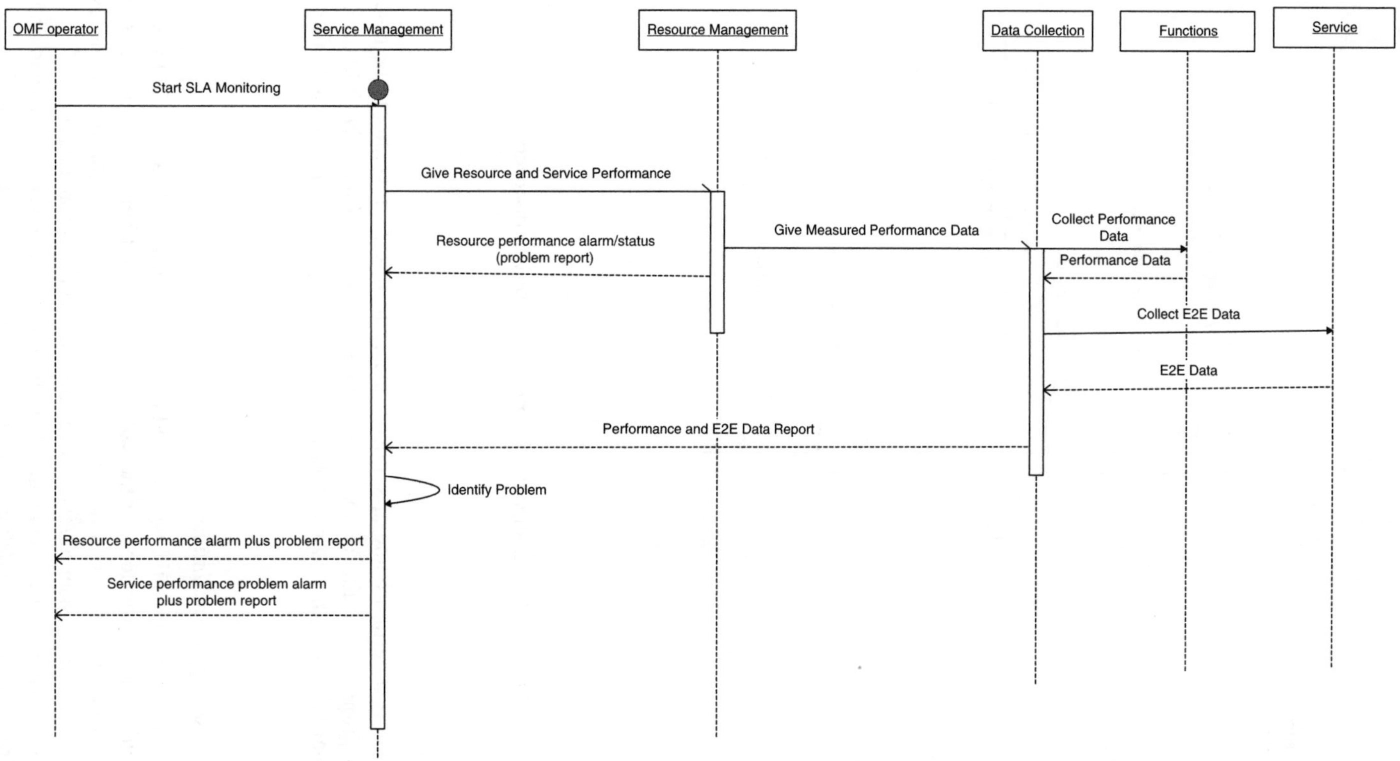

Figure 3.13 Component interworking – service monitoring and maintenance.

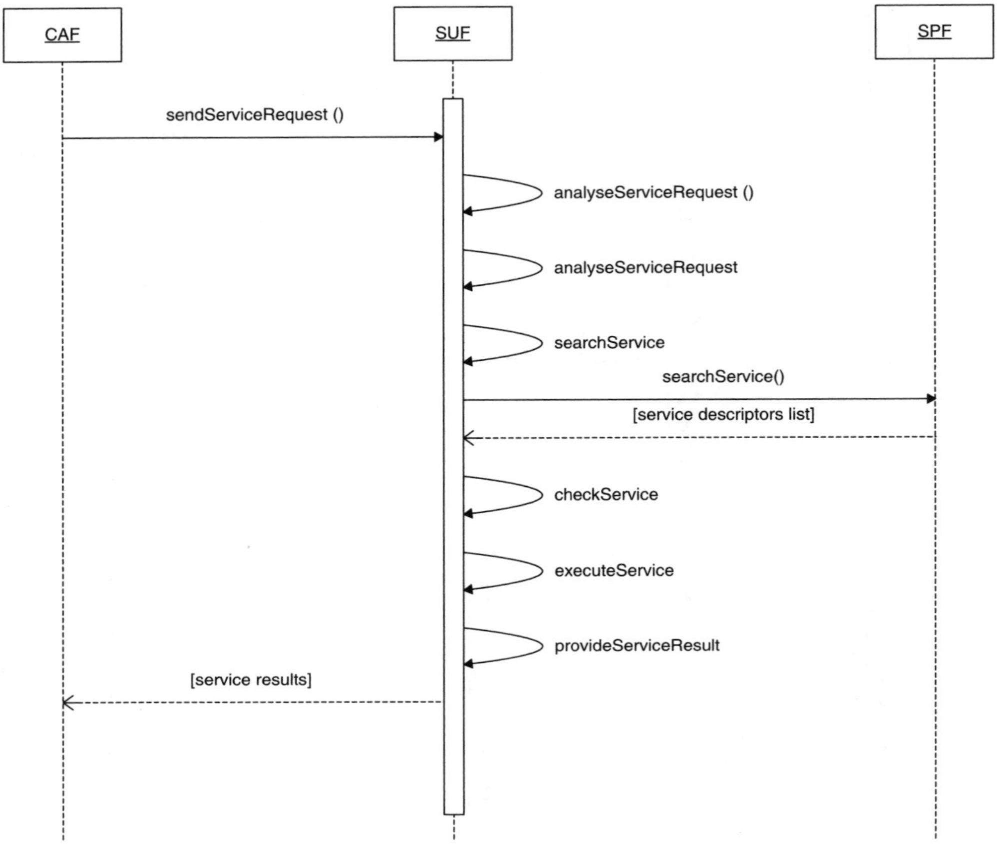

Figure 3.14 Component interworking – service discovery.

3.4.3 Service Discovery

When a user enters a certain context, services related to this context should be discovered. When a subscription to a certain context at the CAF is set using *subscribeToContext(),* service discovery is started when the user enters this context. Figure 3.14 illustrates this service discovery.

3.4.4 Proactive Service Provisioning

After discovery, the service could be executed directly, or just made available for execution, such as to provide a list of services for the user to select from. Even if the service is executed automatically, its execution is also dependent on the service-usage profile of the user (see *SelfProvisioning*); i.e. if the user agreed to execute the service automatically.

Figure 3.15 shows the message sequence diagram for proactive service provisioning, where the SPF subscribes to context changes. When users enter a certain context, a list of services

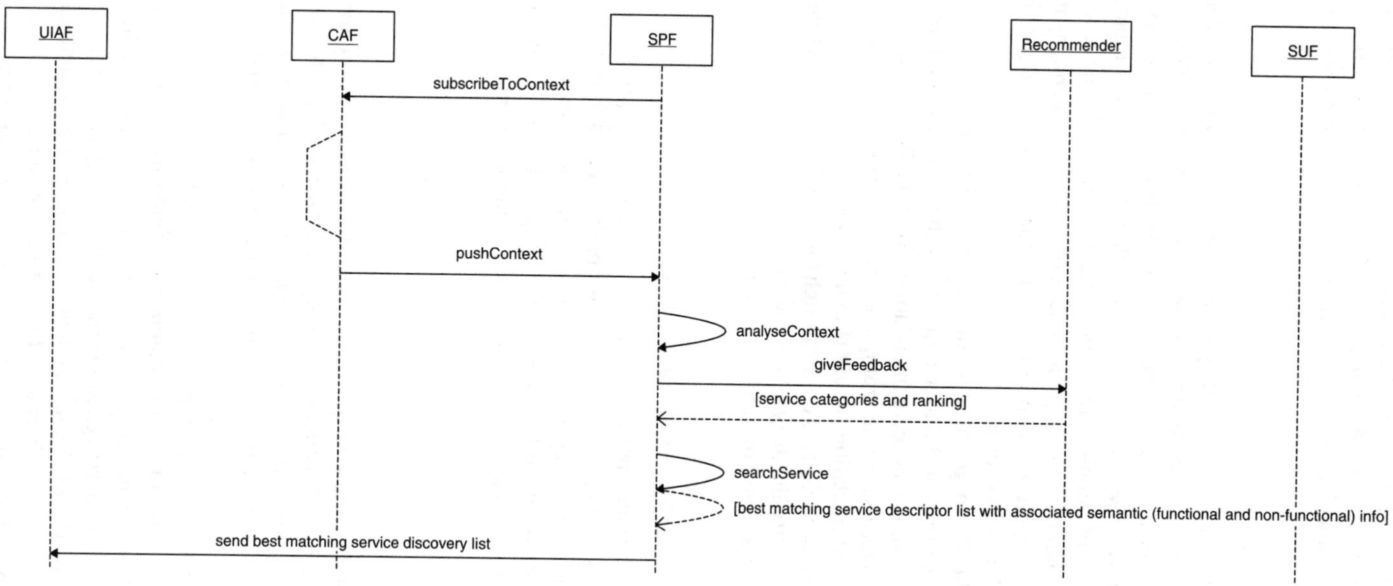

Figure 3.15 Component interworking – proactive service provisioning.

is automatically recommended to them. The selected service in a certain context is sent to the *Recommender* of the *Reasoning Function* of the CAF to store the user's behaviour and influence the service recommendation for the next proactive service provisioning.

3.4.5 Multimodal Service Usage

When users finally start using a service, they send some input to the User Agent running in her device. Additional User Agents running on devices in the environment can also receive user input. All the User Agents pass the input to the UIAF that first fuses these input streams to one definitive user input towards a mobile application. Sending this input, it requests an application request from the mobile application that describes the application output in a device-independent user interface description. The application is triggered by this call and can execute some operations (application-specific). It has the possibility to get the *Operational Personal Context (OPC)* from the PCF.

The PCF accesses the PF to get user profile entries. When the application has finished its application logic execution for the user request, it returns the *Application Request* to the UIAF. The UIAF requests modality recommendations from the PCF to finally make the application output available to the user using multiple modality and devices. The recommendations describe which devices and modalities to use (Figure 3.16).

Discovering devices in the vicinity is done through the UIAF, as this function uses *pushLastContext()* of the CAF to notify the discovery of new devices. These devices are then available for mobile services to use as input or output channels.

3.4.6 Trusted Personalised, Context-aware Services

When the PF is used by mobile applications and services to read a profile entry (see *getProfileEntry()*), the PTF is used to check the policies for profile access. Figure 3.17 illustrates the reading of a profile entry with policy checking for an exemplary user, Bob.

The same policy check needs to be done when an application requests context data from a *Context Provider* (CP). Figure 3.18 illustrates the component interworking for this use case.

3.4.7 Joining a Group

The Mobile Services Architecture supports service provisioning not only for individuals but also for groups. Figure 3.19 illustrates the component interworking for an exemplary car sharing scenario *MobiCar* (see the reference application description in Chapter 7), where cars are recommended based on the user's preferences and the available groups and their preferences.

The precondition for the user to join a *MobiCar* group is that cars are available to be shared. We assume here that, for each available car offered for sharing purposes, a relevant group is created. The *Operational Personal Context (OPC)* is read from the PCF of the CAF containing the current user preferences in a certain context. If the user adjusts the preferences, they are stored in the PF. The *MobiCar* application evaluates the possible cars (each car is a group) available for the user and does a ranking of it. The ranking is done with the help of the *Group Evolution* on the basis of the social relations of the user (social meaningful groups,

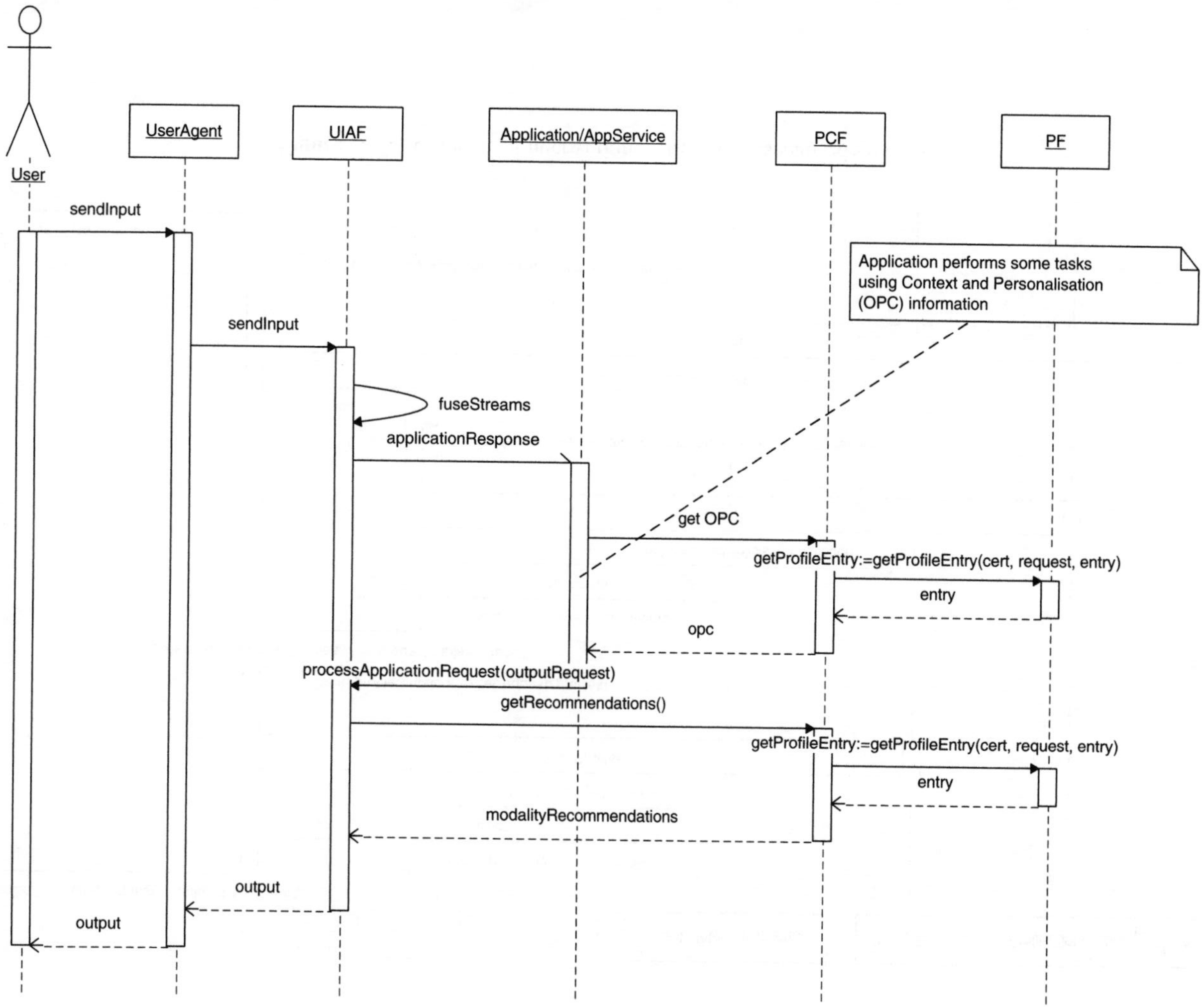

Figure 3.16 Component interworking – multimodal service usage.

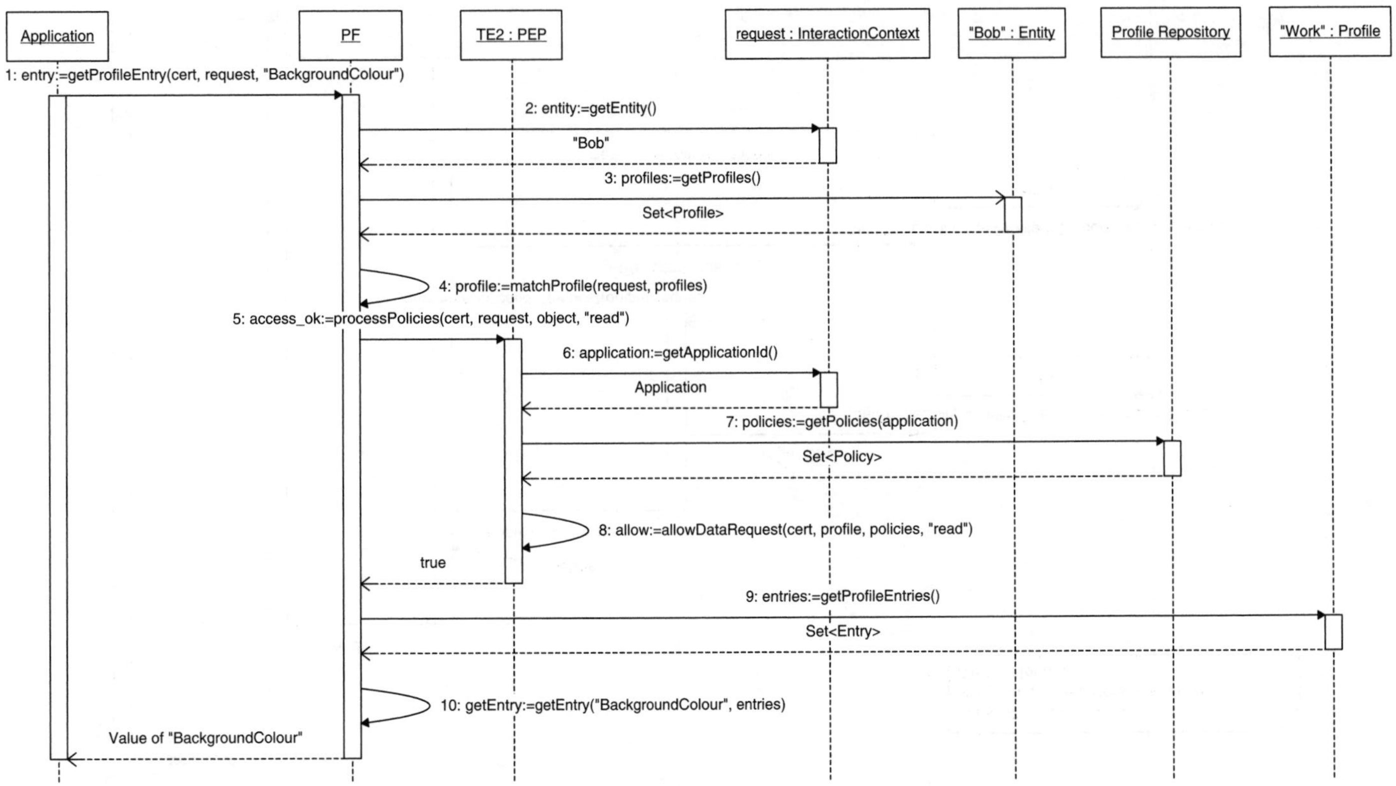

Figure 3.17 Component interworking – trusted personalised services.

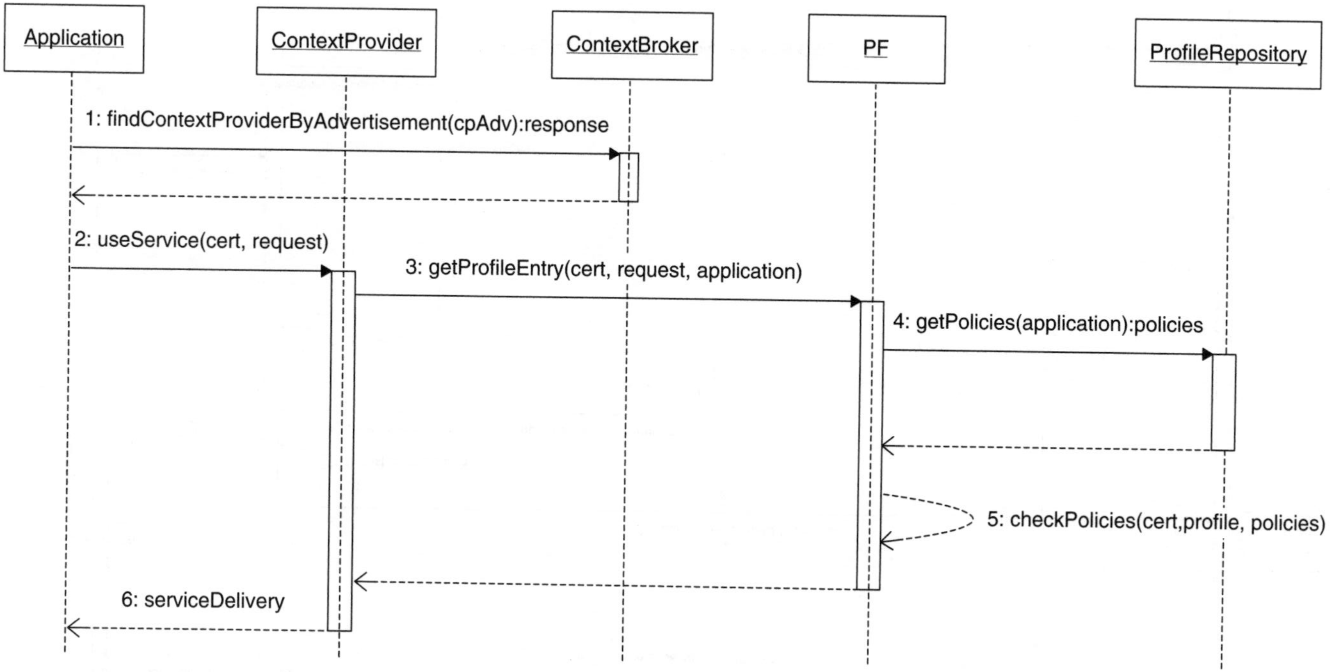

Figure 3.18 Component interworking – trusted context-aware services.

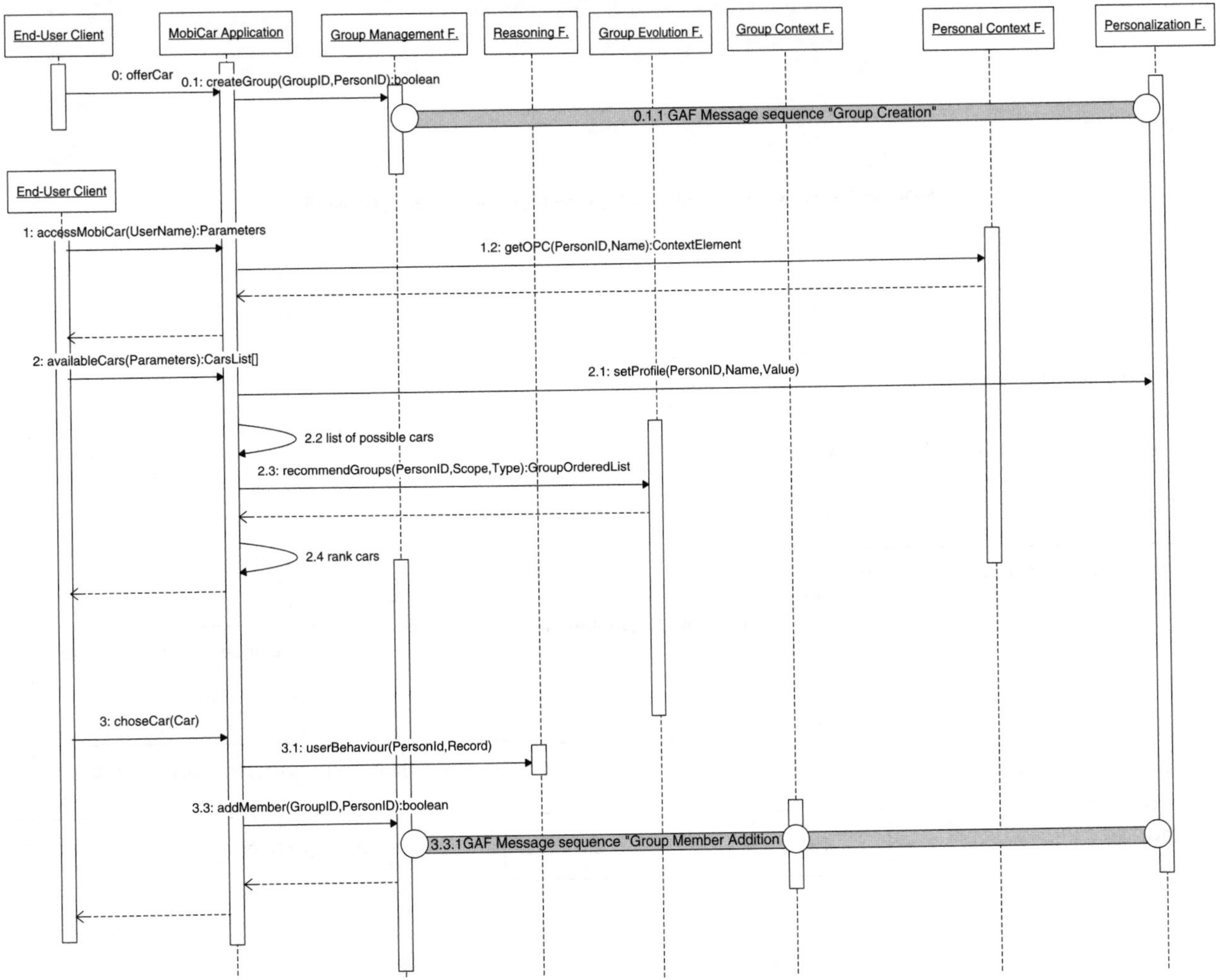

Figure 3.19 Component interworking – joining a group.

groups of people the user is already involved in), group and personal preferences, and user feedback.

The user receives the prioritised list of cars and performs the selection. Note that the final selection is done by the user, while *MobiCar* only proposes the ranking. The user is added to the group using *addMember()*. The result of the user selection is sent to the *Reasoning Function* (RF) for further learning, and the user is added to the group.

3.5 Mapping to the IP Multimedia Subsystem (IMS)

One of the most important objectives of the 3G (Third Generation) networks is the merging of cellular networks and the Internet, two of the most successful paradigms in communications. The IP Multimedia Subsystem (IMS) [2] is the standardised solution of the 3rd Generation Partnership Project (3GPP) [5], which supplies *Core* and *Enabling Services* in order to permit the creation of sophisticated advanced multimedia services and to improve the service creation capability with shorter development times. In this sense, IMS is one of the key elements in the 3G architecture that enables ubiquitous cellular access to all the services that the Internet provides. Some examples, depicted within the IMS paradigm, are: accessing favourite web pages, reading email, watching a movie, using instant messaging and presence, or taking part in a video conference. Moreover, the IMS defines the standard interfaces to be used by service developers. Operators can then take advantage of a powerful multi-vendor service creation industry, avoiding potential dependency on a single vendor to obtain new services.

IMS mainly leverages the Session Initiation Protocol (SIP) [8] as the session control protocol. Since the SIP is based on HTTP, SIP service developers can use all the service frameworks developed for HTTP, such as the Common Gateway Interface (CGI) and Java servlets. In addition to the session control protocol, there are a number of other protocols that play important roles in the IMS. One example of them is *Diameter* [9], chosen to be the *AAA* (*Authentication, Authorisation* and *Accounting*) protocol in the IMS.

IMS core services provide, for example, session control relying on the SIP, single sign-on (i.e. single user authentication for all SIP services via the *Home Subscriber Server* (HSS), which contains subscription related data and user's security information), QoS control, media gateway elements for interworking between the SIP and Public Switched Telephony Network (PSTN) services.

On the top of IMS core services, there are the so-called *Enabling Services* that enlarge the core set functions with advanced services such as presence, flexible charging, service-based policy enforcement, location service, instant messaging, etc. IMS services can be accessed through standardised interfaces (e.g. OSA/Parlay) by applications and services that can be provided by third parties or residing in the IMS application servers.

Dynamic handling of multimedia sessions in IMS relies actually on the intelligent behaviour of the *Application Servers* (ASs), which enhance service usage through adaptation giving users the feeling of being individually addressed (customising their services according to personal needs or current situation). For this reason, IMS enables applications on the application servers to get access to the IMS core through several reference points.

Both IMS and the Mobile Services Architecture aim at the provisioning of a service framework in order to easily create advanced applications and services for (mobile) users. However,

IMS is a commercial system founded on 3GPP/GSM service logic and targets the short term for service provisioning. In contrast, the Mobile Services Architecture has a mid- to long-term operative schedule addressing specifically the support of different mobile access networks, access communications technologies and modes, multimodality, a wide space for service provision, and the specification of service components and interfaces.

From the technical point of view, IMS and the Mobile Services Architecture have a few common functionalities, while some of the functionalities are specific either of IMS or the Mobile Services Architecture only. Some of the IMS functions are related to the Mobile Services Architecture components, including presence information and location (related to CAF), security mechanisms distributed in IMS elements (related to PTF), profiling data distributed in various databases of the IMS (related to the PF), as well as service usage data collection and service management (related to the OMF). Therefore potential candidates for further common functionalities are the *Context Awareness Function* (CAF), *Personalisation Function* (PF), *Privacy and Trust Function* (PTF) and *Operational Management Function* (OMF).

Figure 3.20 illustrates the relevant reference points between the Mobile Services Reference Model and IMS.

It is clear that the next generation of telecommunication services will be characterised by several features, all of which have to be well supported due to the numerous imposed requirements and increased competition in the information networking arena. The main features of such services – among other aspects – are *context awareness*, *mobility awareness* on heterogonous networks, *adaptation* to network and to devices changes, *personalisation*, *AAA-capability*, as well as *privacy and trust* considerations.

IMS does not yet specifically support features covered by the *User Interface Adaptation Function* (UIAF), and there are no standardised services for service discovery and service composition as introduced by the *Service Usage Function* (SUF). On the other hand, IMS provides charging, session control, multimedia gateway (media gateway user and control) for interworking with other networks (e.g. ISDN, PSTN, and QoS), which do not directly have any correspondence in the presented Mobile Services Architecture.

With respect to the definition of the Mobile Services Architecture, as described in the following, two main approaches for the mapping of its functional components to IMS can be envisaged: separated realisation of the interworking (Section 3.5.1) and tight integration of the Mobile Services Architecture into the IMS system (Section 3.5.2). However, a more realistic approach is probably a flexible combination on these approaches, which is described in Section 3.5.3.

3.5.1 Approach 1: Separated Realisation of Interworking

In the first scenario, IMS is seen as a standalone system that could provide some services to mobile services and applications, but following its own path for the evolution and specification. In this case, mobile services and applications as well as the functional components of the Mobile Services Architecture can access IMS services (e.g. profiling, presence and location services). The functional components would be basically realised in the IMS *Application Servers* accessing and using the other IMS components (e.g. MGW, HSS, etc.), and services through the standard IMS interfaces.

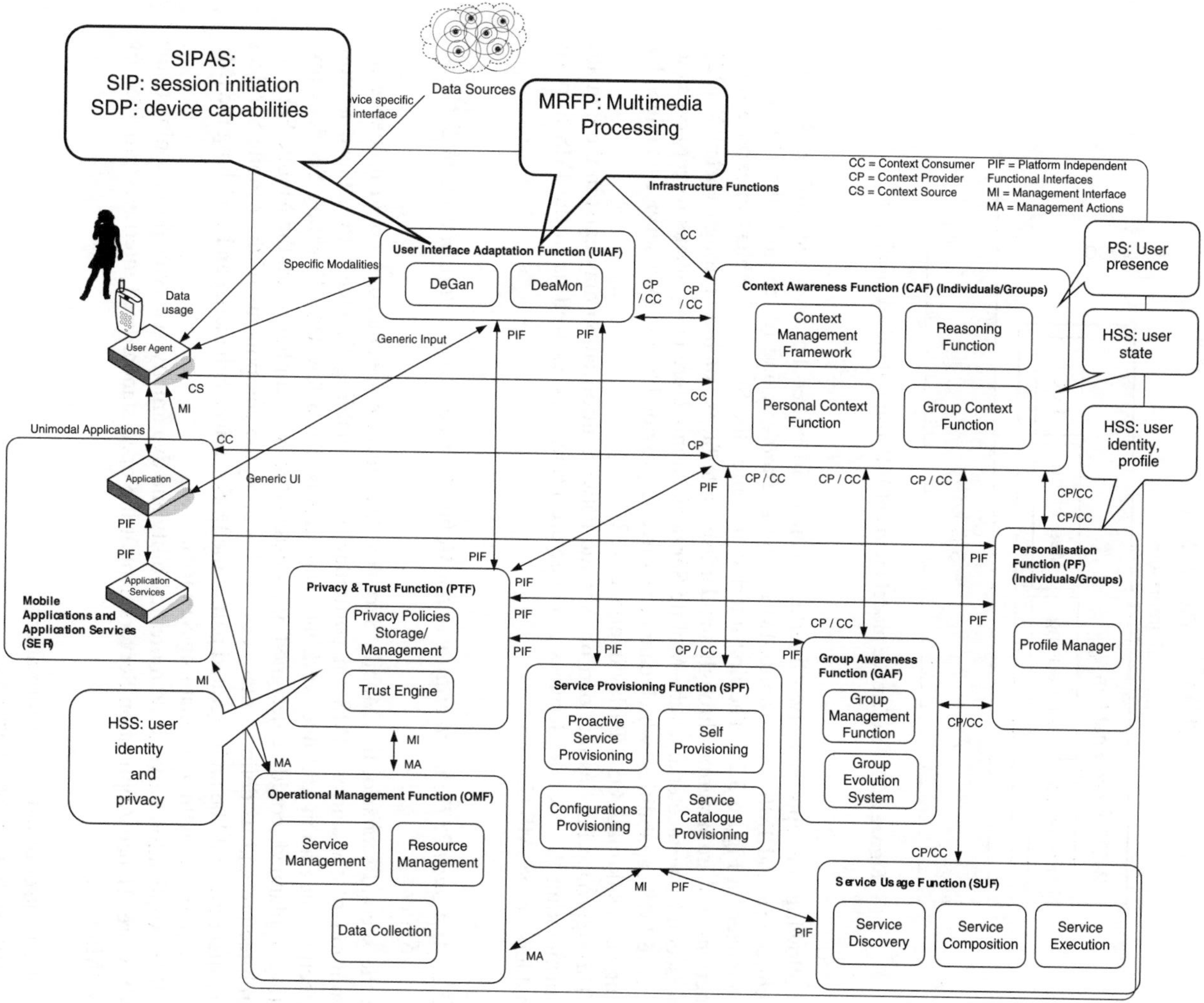

Figure 3.20 IMS and Mobile Services Reference Model exemplary points-of-mapping.

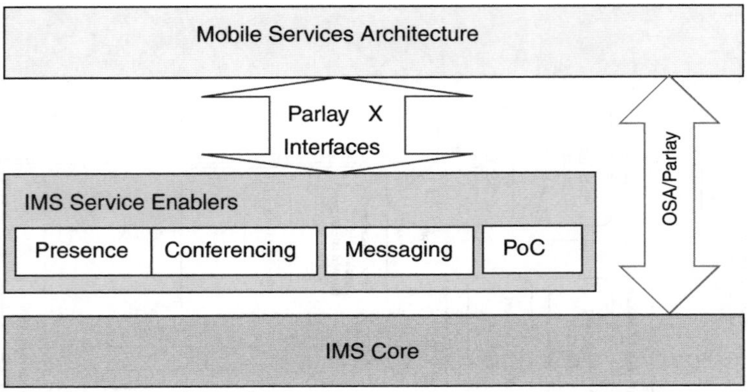

Figure 3.21 Reference model on top of IMS using standard interfaces.

Following this approach, two interfaces can be identified between the Mobile Services Architecture and IMS: (1) towards IMS *Service Enablers*, and (2) towards the IMS *Core*, as depicted in Figure 3.21. The first interface requires the Service Enablers to expose some functions and information through standardised interfaces, such as through Parlay X [22]. The second interface enables the Mobile Services Architecture components to get access to the core IMS capabilities, which cannot be directly provided by the IMS service enablers, by using access through OSA/Parlay interfaces [22].

This approach implies that part of user and network information may exist in the domain implementing the Mobile Services Architecture framework as well as in the IMS Service Enablers or the IMS Core.

3.5.2 Approach 2: Tight Integration of the Mobile Services Architecture into the IMS System

In the second scenario, the Mobile Services Architecture specification could influence the evolution of the current IMS (in particular of its Enabling Services) to achieve a more effective integration and interworking. This means that a few IMS Enabling Services (existing or planned) could be designed or redesigned to be compliant with the Mobile Services Architecture and its defined interfaces. In this case, the components of the Mobile Services Architecture could be an integrated part of the IMS services enhancing and enriching the IMS service portfolio with new Enabling Services.

IMS Application Servers add more flexibility and dynamicity in service provisioning to the IMS Core. Hence Application Servers exploit the open standardised interfaces provided by the IMS Core to:

- retrieve user or network information;
- make use of network capabilities;
- influence session handling;
- retrieve subscribed network events;
- behave as the termination point for session signalling.

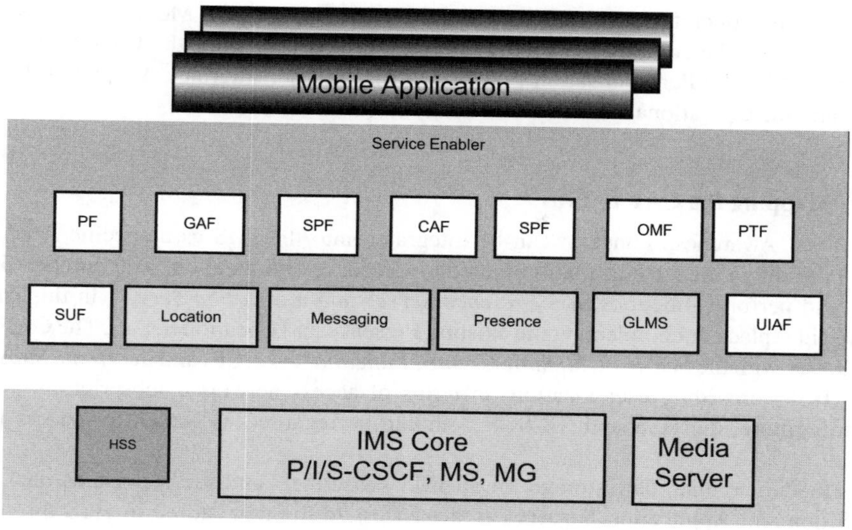

Figure 3.22 Mobile Services Reference Model components versus IMS/OMA enablers.

Whereas IMS defines a basic framework for delivering services, the *Open Mobile Alliance* (OMA) [15] is developing the main elements of this framework, namely the *IMS Service Enablers*.

The Mobile Services Architecture can be conceived as a comprehensive template for developing and deploying mobile services. Therefore it could complement the Service Enablers which already exist in IMS (e.g. for charging, presence, conference, messaging), with new functionalities, as depicted in Figure 3.22. The Mobile Services Architecture could be then realised as a set of new Service Enablers for the development of a wide range of new applications.

The components of the Mobile Services Architecture should support interfacing with the existing IMS Services and OMA Service Enablers. Such integration enables an efficient interworking between the IMS and the Mobile Services Architecture since it allows an important reuse of existing IMS features, both in terms of Service Enablers but also in charging and handling the IMS-based user identities (for AAA-related functions).

However, this tight coupling does require some specific adaptation on the architecture's components to be integrated into the IMS architecture, such as implementation of reference points.

3.5.3 Approach 3: A Flexible Combination

In addition to the two mapping approaches proposed above, the most promising solution after careful evaluation is a flexible combination of both scenarios. This solution embraces the advantages of the two approaches and minimises the disadvantages: some components fit better with the first approach but others with the second approach.

Accordingly, concrete recommendations for the mapping of the Mobile Services Architecture following the combined mapping approach will be given for the Context Awareness Function (CAF), the Personalisation Function (PF), the User Interface Adaptation Function (UIAF) and the Operational Management Function (OMF).

3.5.3.1 Mapping the CAF to IMS

The Context Awareness Function can be integrated into the IMS as a specific *Application Server* providing context information including presence and location. This implies that the CAF could perform functions of a *Presence Server* and a *Location Server*. In this case the CAF would replace or complement the existing Presence and Location Server. The CAF has to work closely with the *Serving Call State Control Function* (S-CSCF) and the *Home Subscriber Server* (HSS) providing user location (in terms of access network), subscription, policies, etc. Furthermore, the HSS and S-CSCF both can be regarded as context providers for the CAF.

The HSS is the main data storage for all IMS subscribers and service-related data within an IMS domain. Application Servers can get access to the data stored in HSS through the *Sh* reference point using a data handling procedure or a subscription/notification procedure. Dynamic data are the more relevant for the CAF than static data. According to [1], the following data can be requested from the HSS:

- Repository which contains the transparent data (understood syntactically but not semantically by the HSS, this allows the AS to store service specific data).
- Public user identities, the S-CSCF serving the user, filter criteria.
- IMS user state which exposes the user's current state in the IMS domain. There are four registration statuses:
  - REGISTERED
  - NOT_REGISTERED
  - AUTHENTICATION-PENDING: pending while being authenticated
  - REGISTERED_UNREG_SERVICES: unregistered but an S-CSCF is allocated to trigger services for unregistered users.
- User state contains the state of the user in the circuit-switched or packet-switched domains.
- Location information contains the location of the user in the circuit-switched or packet-switched domains (e.g. cell global identification).

The S-CSCF supports the registration of IMS users to services, and hence it can notify the CAF about a successful registration of a particular IMS subscriber. To inform the CAF about this, filter criteria should have to be set within the subscriber profiles stored in the HSS. This notification procedure is called a third-party REGISTER request.

There are some further aspects to be considered for this mapping, related to:

- the CAF integration, which should probably be coupled with the mapping of the Personalisation Function (PF);
- mapping of the user identifier between IMS and CAF components;
- publishing of different user presence data to various watchers.

3.5.3.2 Mapping the CAF Presence Enabler Specification to the IMS Presence Architecture

The IMS presence architecture [4] defines three related main functionalities: *Watcher Subscription*, *Subscription to Watcher Information*, and *Presence Publication*.

For the Watcher Subscription function, a watcher in the CAF specification is naturally mapped in the CAF specification to a Context Consumer (CC) of a particular Context Provider (CP); e.g. a Presence CP would be the natural mapping of the IMS Presence Server (the IMS presentity can be mapped to an entity within the Presence CP taking the entity as the subject of a presence context element). In the CAF specification, a Presence CC (i.e. a service or an application) needs to ask the *CAF Context Broker* to get connection information to the appropriate Presence CP (using the *findContextProvider* method). Using the returned connection information it is able to subscribe to specific presence context information at the selected Presence CP using the proper interface method provided by it. Finally, the Presence CC is able to get the required presence data from the Presence CP, in both pull and push modalities, as preferred. The *Privacy and Trust Function* (PTF) is normally involved in this process to guarantee its overall correctness with respect to the presence information content access rights.

For the Subscription to Watcher function, this mapping with respect to the CAF specification implies that an entity, which was registered (added) to the CAF Presence CP, should be able to get the information about all the Presence CCs that have been subscribed to get the presence data related to such an entity. However, this use case is not covered by the current CAF architecture specification and should be realised separately.

For the Presence Publication function, a *Presence Context Entity* can be registered (added) to the Presence CP according to the CAF specification. Such an entity is then the subject of a presence context element; i.e. the *Presentity* of the *IMS Presence Architecture* specification. The Presence CP offers a specific interface method (*addEntity*) to perform such operation.

3.5.3.3 Mapping the Personalisation Function (PF)

In this example, the Personalisation Function (PF) is mapped within the IMS to the Home Subscriber Server (HSS) – according to approach 1. Due to its nature, the HSS is more appropriate for containing static data than dynamic data. However, data related to the PF in the Mobile Services Architecture are composed of both static and dynamic data. A possible solution is to use different places to manage the static and dynamic personalisation data:

- Static data will be stored within the HSS. This data could be modified by the Application Servers only through the *Sh* interface as defined in the IMS specification [3].
- Dynamic data (for example represented by user profile data and user context data) can be stored within the OMA XML Document Management Service Enabler (OMA XDM), or can furthermore be handled by other existing enablers. The OMA XDM enabler is able to:
  - Provide both XML Configuration Access Protocol (XCAP) and SIP interfaces, for remote manipulation of data as well as notification of changes, and it is integrated at AAA level with all IMS identities. Besides providing access to Application Servers, the OMA XDM enabler also allows a secure manipulation of data for users.

– Manage user-related service profile data (and even user context data).
– Manage authorisation policies and rules on a per-user and a per-service basis towards individuals or groups.

Additionally, some dynamic data can be managed by other enablers (e.g. the OMA Presence Enabler [16]), which do not have any concept of static repository of data besides the *Presence Server Database* for the presence information. However, it could easily manage some context information as extensions to the presence data. This enabler facilitates, for example, both users and other network entities (e.g. Application Servers), to manage information on behalf of the users.

3.5.3.4 Mapping the User Interface Adaptation Function to IMS

IMS users may have more than one device at their disposal and thus IMS allows users to register with their home domain via one or more devices. The users can use their services through all these registered devices. IMS terminals will vary widely in their capabilities and in the types of devices they represent. It is important for the applications and services to learn the capabilities and characteristics of the used terminals. Therefore IMS provides mechanisms to determine the capabilities and the modalities of user devices.

During registration, the User Agent (IMS client) indicates its capabilities that are carried in the SIP REGISTER message initiated by the IMS terminal. This information is conveyed as parameters of the *Contact header* field within the REGISTER message. The User Interface Adaptation Function (UIAF) of the Mobile Services Architecture can in this way obtain knowledge about these capabilities by being notified of a third-party REGISTER request from the IMS Core. This standard IMS mechanism allows any IMS Application Server to be notified of registration procedures of users, based on an appropriate trigger to be defined.

On the other hand, many current and future services in the context of IMS require the processing or generation of multimedia data. For this, IMS defines the *Multimedia Resource Function* (MRF). The media processing capabilities of the MRF can be used by the *Presentation Adaptation* component of the UIAF specification. Additionally, the *Modality Fission* component can make use of mixing functionalities of the MRF, to address multiple channels when presenting media content to the user. Examples of functionalities covered by the MRF and supported by the UIAF are:

- multimedia conferencing (e.g. mixing of audio and video);
- text to speech conversation (TTS) and speech recognition;
- real-time transcoding of multimedia data (i.e. conversion between different formats).

3.5.3.5 Mapping Operational Management Function to IMS

Service Lifecycle Management topics are currently considered as work in progress in the OMA Service Provider Environment (OSPE). The IMS specification instead does not yet cover this topic. Assuming a service-based charging and accounting, the infrastructure currently used in the IMS can be used also for the realisation of the Mobile Services Architecture.

3.6 Service Lifecycle Management

This section introduces key stakeholders and key requirements for service lifecycle management. However, this book does not go into details of service lifecycle management support in the context of the Mobile Services Architecture.

Six key stakeholders can be identified in the context of Mobile Services Architecture:

- The role of the *end-user* goes beyond the usage and consumption of services and applications. For example, the Internet as an interactive platform actually provides several opportunities to influence or directly to participate in service design and delivery. The end-user can provide valuable feedback to improve the performance of mobile services.
- The *mobile network operators* (MNOs) focus on developing and maintaining the physical network. They can also adopt the role of service provider and provide value-added services (e.g. voicemail, MMS, etc.) to end-users. MNOs may recognise that the Mobile Services Architecture assists them in following their business objectives supporting the realisation and management of innovative context-aware services. Therefore, especially the service lifecycle aspects of Mobile Services Architecture are relevant to them.
- The *service providers* (SPs) acquire and serve customers by reselling the network capacity provided by a network operator, as well as providing their own additional services in co-operation with MNOs. An SP's main priority is to launch a new service package quickly and efficiently and withdraw it whenever needed. Thus SPs expect the Mobile Services Architecture to support the requirement of service lifecycle management.
- The *platform provider* develops the commercial off-the-shelf management components, which are bought and used by MNOs and SPs. Therefore it is an overriding requirement on platform providers to support service lifecycle management by implementing necessary interfaces of the Mobile Services Architecture in their platform.
- The *component developer* is a part of the service provider and platform provider domains and defines interfaces for functions such as user profile storage, authentication, privacy control, etc. These interfaces can be defined and invoked using internationally adopted standards. This reduces the need for service integration when a new service enabler is added to an existing service environment.
- The *standards developer* develops standards for service management functions. These standards are used mainly by platform providers and MNOs to help ensure that their products are interoperable and conform to industry best practice. In this sense, standards developers act as a mediator between the component developers and standardisation bodies and keep them abreast with industry best practice, while providing standardisation bodies valuable input.

OMA outlines seven steps to service lifecycle management. The case model illustrated in Figure 3.23 shows the relationship between the stakeholders and the lifecycle management steps and then further decomposes those general steps into more concrete use cases.

Step 1 – Idea stimulation

This is the process of stimulating the creation of the concepts for an application. It also involves analysis of market needs and identification of opportunities for new services. Specific

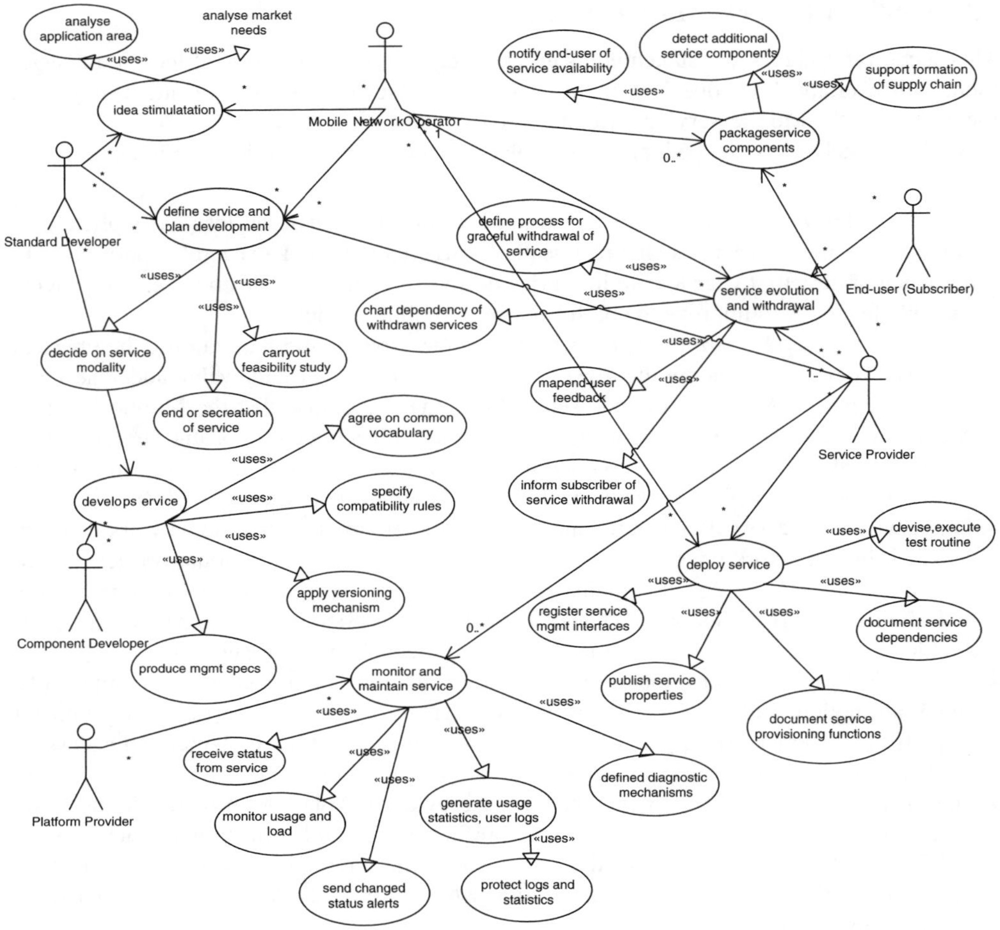

Figure 3.23 Service Lifecycle Management use case model.

requirements for this step are as follows:

- Analyse the application area.
- Analysis market needs.

Step 2 – Service planning and definition
Having analysed market, needs, etc., opportunities for new services are identified in this step. Specific requirements for this step are as follows:

- Carry out a commercial feasibility study.
- Decide on service modalities.
- Give formal endorsement for the creation of new services.

Step 3 – Service development

This is the process of implementing and testing the service components. Key requirements are:

- Produce requirements and design specifications for service components and the management infrastructure.
- Configure the environment for running and testing the services.
- Define the vocabulary that is agreed by all partners developing infrastructure components.
- Apply a versioning mechanism to component supporting services management.
- Specify rules for forward and backward compatibility.

Step 4 – Service deployment

This refers to the process of deploying the service in the SP's environment. Specific requirements for this step are:

- Register all interfaces used in management of operations such as context gathering, personalisation, performance monitoring, trouble reporting.
- Declare and document dependencies of services in the execution environment.
- Document all the functions (e.g. messaging, context source, location capability) and components needed for a complete and successful provisioning of the services.
- Devise and execute test routines before making the service offers public.
- Publish service properties needed for subscribing to and using the service.

Step 5 – Service packaging

This is the process of offering the service to the customer. Commercial packages are defined, aiming at concrete user segments, with concrete service features and billing conditions. Specific requirements for service packaging are:

- Notify end-users (e.g. utilising user preferences) when a service becomes available.
- Automatically detect additional services and components and provide support for the ability for automatic subscription to these services.
- Support the forming of a service provider's supply-chain and provide a service package to the end-users.

Step 6 – Service monitoring and maintenance

This involves understanding how well a service is performing technically and commercially. Main specific requirements for this step are:

- Generate anonymous user statistics and logs and protect them by privacy rules agreed between the end-users and SPs.
- Support the receiving of notification from the service of its status (available, withdrawn, etc.) automatically and in real time.
- Monitor usage and load, as well as detect problematic behaviour and predict problems.
- Define diagnostic mechanisms to detect out-of-order or malfunctioning components.

Step 7 – Service evolution and withdrawal
The end of a service requires its termination. If a service offer must be stopped, subscriptions have to be properly dissolved. Requirements for this step are:

- Define a process for graceful withdrawal of the service and chart the dependency between services being withdrawn.
- Inform end-users about service withdrawal in accordance with the Service Level Agreement (SLA).
- Map end-users' feedback on to service features and provide it to the service developer.

3.7 Architecture Scalability

Scalability of the Mobile Services Architecture is one of the critical factors ensuring the future continuous provision of robust, contextual, personalised and multimodal mobile applications and services to individual users or groups of users despite their growing numbers, the differentiation and the increase of their application and service requests and demands, and the growing number and differentiation of the service offerings as advertised by service providers. In the following, one scalability case study is presented, giving ingredients and ideas for planning and performing studies through simulation also in general.

3.7.1 Scalability Performance Metrics

The Mobile Services Architecture is designed as a Service-Oriented Architecture (SOA) and the scalability design will assume that the physical network it runs on is scalable. Distance, speed–distance scalability and Quality of Service (QoS) are out of the scope of this analysis.

One important question when thinking about scalability in the Mobile Services Architecture is: How many users can a Context Provider (CP) serve? A CP is basically a web service which runs on a commodity PC or server hardware. Further, the web service might execute tasks locally or remotely upon request. These tasks might bind certain hardware resources (CPU power, HD storage, etc.) for each request; but not every request is sent by a different user, and maybe the task needs also some historical data of the user.

The Quality of Context (QoC) concept used here states that through testing or experience the system administrator gains enough knowledge on her system behaviour to report a certain *Number of Users* (NoU). The NoU in this case means not only the current requests the web service must handle. Instead, the NoU is a definition which includes more than just the request per IP address. It can be assumed that the usage behaviour of one user who follows a certain user type can be defined and forecasted.

The scalability of the Mobile Services Architecture is influenced foremost by the NoU. The scalability aspects can indirectly be determined by measuring utilisation, response time and throughput of components as a function of NoU. Note that the NoU associates to the number of service usage models that are available and the Mobile Services Architecture can distinguish. Furthermore, NoU associates to various numbers of workloads and potential resource configurations involved in realising mobile applications and services usage scenarios.

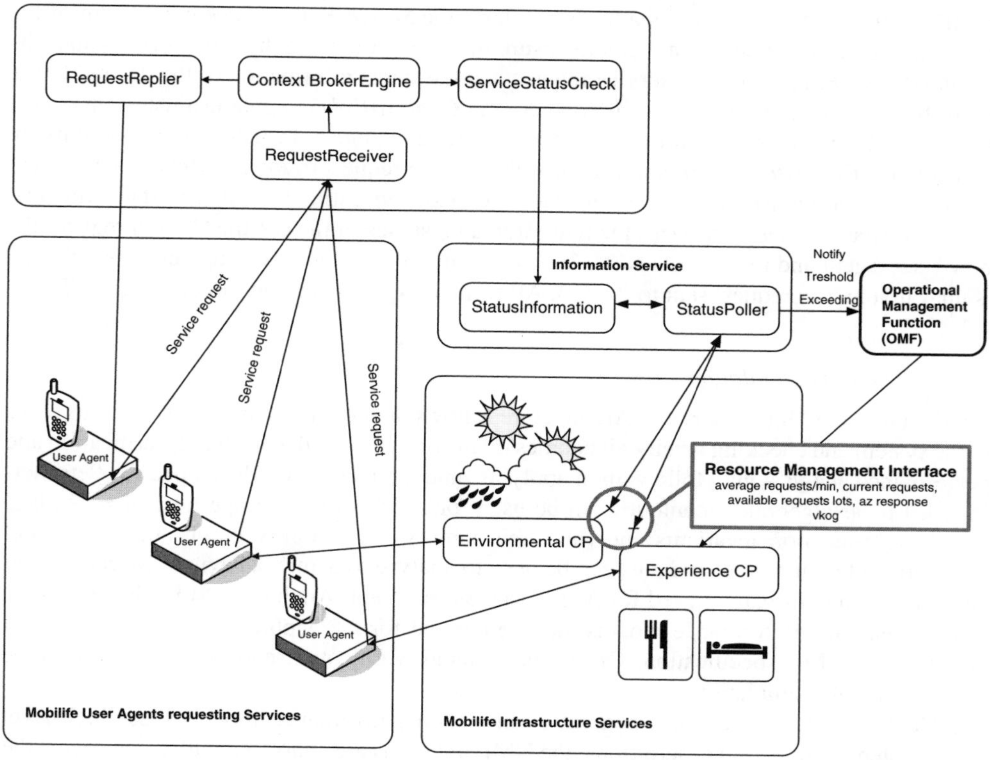

Figure 3.24 Mobile Services Architecture Scalability and Resource Management Model.

3.7.2 Scalability Components

The Scalability and Resource Management Model illustrates additions to the existing Mobile Services Architecture to enable scalability of the system.

As Figure 3.24 illustrates, the *Context Broker* (CB) instance is first queried by the clients instead of directly querying the service. The User Agents send their service requests to the CB where they are received by the *RequestReceiver*. Further the request is passed to the *ContextBrokerEngine*. Here the CB decides what to do with the current service request. The decision is based on a recent representation of all available CPs in the *ServiceStatusCheck* component.

StatusInformation is the result of the concept of *ServiceStatusCheck*. The *InformationService* regularly polls all active CPs and receives their status information which contains the QoC performance metrics. To calculate the NoU, system parameters (CPU, RAM, HD, etc.) need to be divided through the demand of a single user request. This assumption takes into account also that a user, depending on its type, does not necessarily send permanently a service request. Besides the QoC performance metrics, *InformationService* watches over all parameters of a CP. If one or more parameters exceeds a certain threshold, it will notify the

OMF. Further, *InformationService* feeds regularly the *ServiceStatusCheck* component of the CB with the current status of all CPs to ensure that the CB acts on the most recent data. The *ContextBrokerEngine* gets the service request and the status information of the desired CP as input for the decision making process, where the *ContextBrokerEngine* evaluates whether it can forward the request to *ContextBrokerEngine* or to a redundant Instance. If both options are unavailable, *ContextBrokerEngine* forwards the service request to *RequestReplier* which will send a reply stating that the service is unavailable to the requesting User Agent. If *ContextBrokerEngine* has decided to forward the request to an instance of the CP, the CP will register the new User Agent and increase its NoU. Only at the initial request at start-up, and when the CP is unavailable, is a request sent to the CB. All further requests are sent directly to the CP.

3.7.3 Scalability Simulation

Validation of the Mobile Services Architecture follows on feeding a number of fictitious cases to the system and checking whether it handles them correctly. On the one hand, verification and performance analysis may follow upon applying analytic techniques developed for Petri nets [6,23]. Linear algebraic techniques can be used for verification of properties, such as place invariants, transition invariants and (non-)reachability. Coverability graph analysis, model checking and reduction techniques can be used to analyse dynamic behaviour. Markov-chain analysis techniques can be used for performance evaluation. Analogous to Pooley and King in [17], analytic performance models may be derived for the Mobile Services Architecture based on its UML specification. On the other hand, verification and performance may be demonstrated by simulation.

UML-Psi [12] automatically generates simulation performance models from high-level UML system architecture descriptions; the UML diagrams are annotated with (a subset of) the UML Performance Profile [14]. The Mobile Services Architecture was designed using UML. Annotating UML diagrams with stereotypes and tagged values, according to a subset of the annotations defined in the UML Performance Profile, a mapping can be made from the UML model to a *UML-Psi* performance simulation model.

The activity diagram in Figure 3.25 is derived from the sequence diagram of Multimodal Service Usage, described in Section 3.4.5. This specific system interaction sequence was selected because it covers most functional components of the Mobile Services Architecture. Thus, the typical actions in the system are reflected.

Beside this model, *UML-Psi* needs a resources model as input, describing physical hardware components (i.e. servers, networks, clients) where the Mobile Services Architecture components are running. They are annotated with values like CPU speed, memory size, network speed, etc. The actions of the activity diagram are assigned to these hardware components, so that they are actually simulated as running on those hardware resources. Additionally, the particular actions are annotated with parameters that describe their computations. Parameters for the simulated environment are based on measurements of an existing application that implements parts of the Mobile Services Architecture, and on assumptions of hardware equipment for servers and network. The application that was used for measurements was the *ContextWatcher* [10] (see Chapter 7). It is a reasonable approach to use this application for determining simulation parameters, as its usage scenarios employ a lot of CPs that are the scaling components.

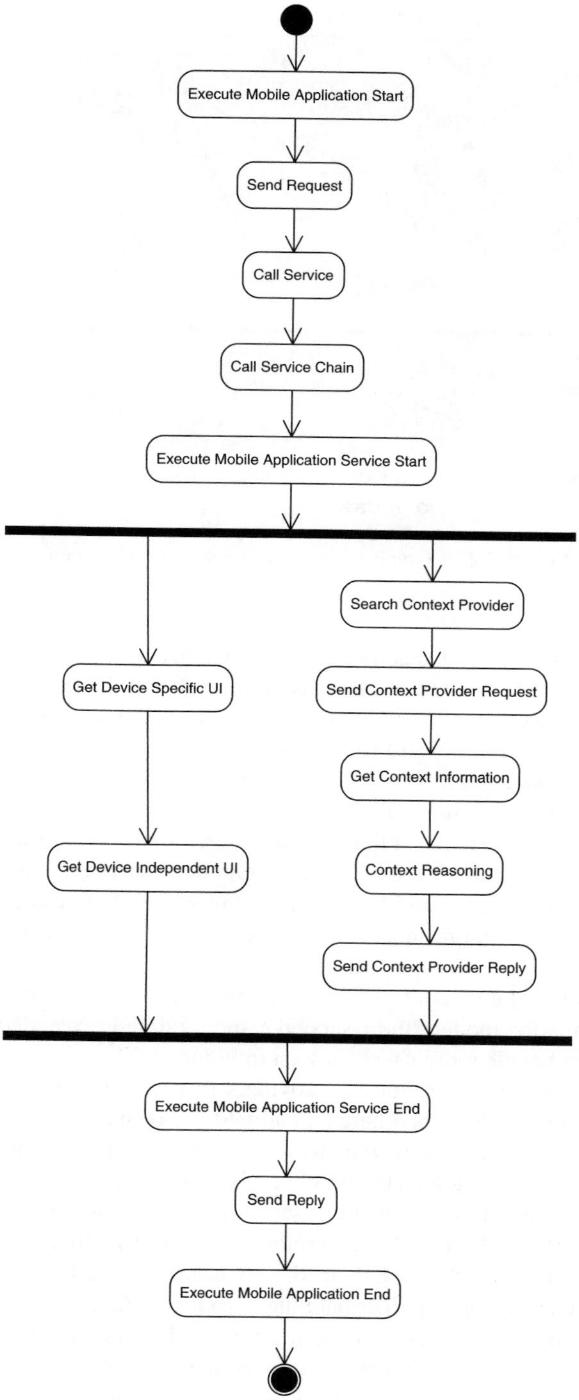

Figure 3.25 Activity diagram for scalability simulation.

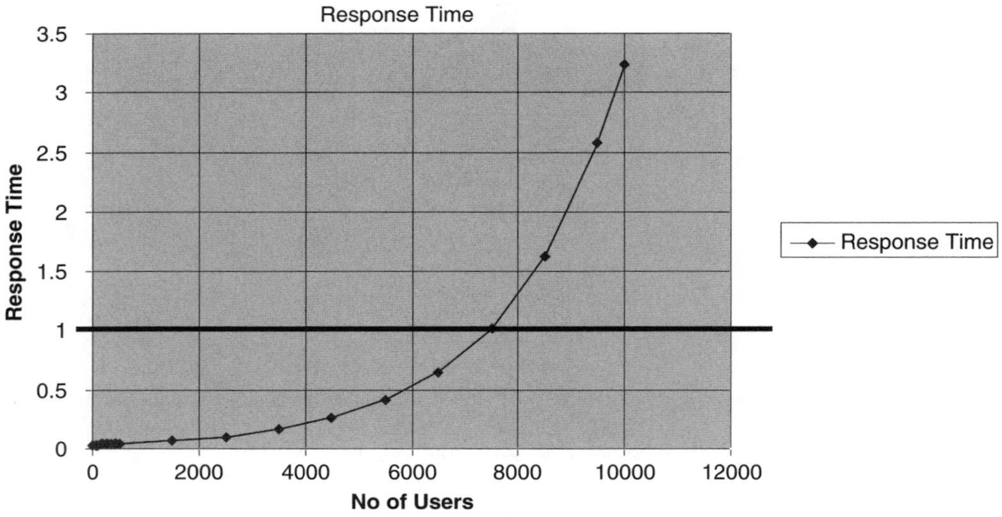

Figure 3.26 Graph showing response time with one CP.

After the hardware resources and activities are modelled and annotated with parameters, the simulations can be started. The NoU is increased for each simulation step. For each step, the response times are observed and stored in a data table. The following parameters that *UML-Psi* reports in a simulation run are useful to measure in this case [12]: *Resource(responsetime)*, *CompositeAction(responsetime)*, *Resource(throughput)* and *Resource(utilisation)*. All these values measure the response time of the architecture and, as an example, *CompositeAction(responsetime)* is used here to set the NoU in relation to the response time.

In the following figures, the measured and extrapolated response times for CP access of simulation runs with between 1 and 90,000 users are given. The results show that already, for 7500 users, a delay of more then 1 second response time is exceeded. This is seen in Figure 3.26.

If there are ten times the amount of CPs available (*PArate* [12] of the resources is raised ten times in the simulation model), the 1-second response time is exceeded with 60,000 users. Thus, the system can handle more users, as seen in Figure 3.27.

This result is only an example. Service providers that want to test the scalability of their system can take the model described here and adopt their own assumptions for the hardware setup. Then, they need to decide to how many concurrent users they want to provide services, and find out through the simulations the number of CPs they need for this.

The described approach to implementing scalability in the Mobile Services Architecture and the simulated scalability model measurement have shown that, to cope with a higher NoU, an increase in the number of CPs in the environment is sufficient. The CB with the described scalability extension can distribute the workload efficiently over the available CPs. How many CPs are needed to serve a certain number of users has to be measured with an adapted setup of the environment, either using tools like *UML-Psi*, or doing test runs with the physical resources. The simulation approach has to be performed by service providers if they

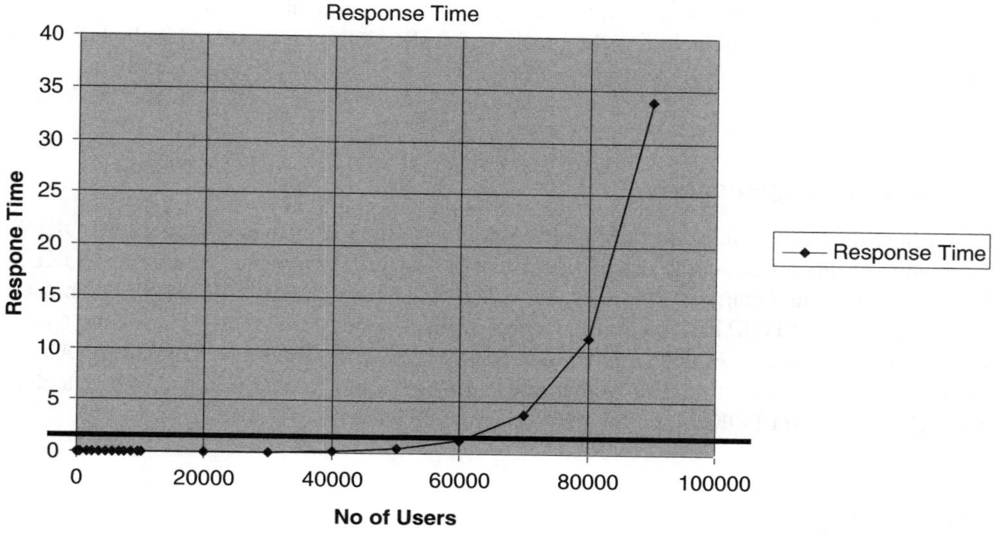

Figure 3.27 Graph showing response time with ten CPs.

want to know how many users can be handled. Running the test with physical components is not possible before the relevant parts of the environment are implemented. Even then, tests with some exemplary components still do not reflect the real-world environment, but are an assumption gathered from just this one case.

After determining the dimension of the system environment needed, it can be realised and made productive. It is possible to monitor the system behaviour through OMF data logging and raise the capacity of the system when it is finally reached, instead of doing a well-dimensioned setup from the very beginning.

3.8 Conclusions

The Mobile Services Architecture introduced in this chapter identifies a number of functional components (*Functions*) and describes their interworking through the components' interfaces. The main functional aspects covered in the specification cover several areas relevant to the provisioning of the envisioned services, including context awareness, privacy and trust, personalisation, group awareness, and multimodal user interactions.

Besides these functional aspects, some relevant non-functional aspects that are related to operational management and scalability issues are covered in the architecture specification. For realisation of the Mobile Services Architecture into a real working system, discussion of mapping and integration into the IP Multimedia Subsystem (IMS) provides initial approaches and recommendations. However, as IMS is only one potentially relevant candidate; the mapping and realisation approaches can also focus on other technological approaches such as selection of pure Internet-related concepts including Web 2.0.

Further elaboration of the concepts the functional components of the Mobile Services Architecture are based on, including their technical realisations, are presented in the following chapters.

3.9 Acknowledgements

The following individuals contributed to this chapter: Bharat Bhushan (Fraunhofer FOKUS, Germany), Mathieu Boussard (Alcatel-CIT, France), Alexander Domene (Fraunhofer FOKUS, Germany), Renata Guarneri (Siemens SpA, Italy), Denis Leclerc (Alcatel-CIT, France), Alessandro Mamelli (HP Italiana, Italy), Bernd Mrohs (Fraunhofer FOKUS, Germany), Christian del Rosso (Nokia, Finland), Christian Räck (Fraunhofer FOKUS, Germany), Alfons Salden (Telematica Instituut, The Netherlands), Jukka T Salo (Nokia, Finland), and Stephan Steglich (Fraunhofer FOKUS, Germany).

References

[1] 3GPP: *IP Multimedia (IM) Subsystem Sh interface, Release 6, Technical Specification Group Core Network*. 3GPP TS 29.328 V6.5.0 (2005-03).

[2] 3GPP: *IP Multimedia Subsystem (IMS)*. 3GPP, TS 23.228.

[3] 3GPP: *Technical Specification Group Core Network and Terminals; IP Multimedia (IM) Subsystem Sh interface; Signalling flows and message contents (Release 7)*. 3GPP TS 29.328, September 2006.

[4] 3GPP: *Technical Specification Group Services and System Aspects; Presence Service; Stage 1 (Release 7)*. 3GPP TS 22.141, December 2005.

[5] 3rd Generation Partnership Project. Online: www.3gpp.org.

[6] Ajmone Marsan M., Balbo G., Conte G. *et al.*: *Modelling with Generalized Stochastic Petri Nets*. Wiley series in Parallel Computing, Wiley, New York, 1995.

[7] Floreen P., Przybilski M., Nurmi P., Koolwaaij J., Tarlano A., Wagner M., Luther M., Bataille F., Boussard M., Mrohs B. and Lau S.: 'Towards a Context Management Framework for MobiLife'. 14th IST Mobile Summit, Dresden, Germany, June 19–23, 2005.

[8] IETF: 'RFC 3261 - SIP: Session Initiation Protocol'. *The Internet Society*, June 2002.

[9] IETF: 'RFC 3588 - Diameter Base Protocol'. *The Internet Society*, September 2003.

[10] Koolwaaij J., Tarlano A., Luther M., Mrohs B., Battestini A. and Vaidya R.: 'ContextWatcher: Sharing context information in everyday life'. IASTED International Conference on Web Technologies, Applications, and Services (WTAS2006), Calgary, Canada, July 2006.

[11] Lieberherr K. and Holland I.: 'Assuring Good Style for Object-oriented Programs'. *IEEE Software*, pp. 38–48, September 1989.

[12] Marzolla M., Balsamo S.: 'UML-PSI: the UML Performance Simulator'. Technical Report CS-2004-2, April 2004 Dipartimento di Informatica Università Ca' Foscari di Venezia.

[13] Mrohs B., Steglich S., Räck C., Klemettinen M. and Del Rosso C.: 'The MobiLife Service Infrastructure'. IST Mobile Summit 2006, June 4–6, Myconos, Greece.

[14] Object Management Group (OMG): 'UML Profile for Schedulability, Performance and Time Specification'. Final Adopted Specification ptc/02-03-02, OMG, March 2002.

[15] Open Mobile Alliance. Online: www.openmobilealliance.org.

[16] Open Mobile Alliance: 'OMA Instant Messaging and Presence Service V1.3 Approved Enabler'. January 2007.

[17] Pooley R.J. and King P.J.B.: 'The Unified Modeling Language and Performance Engineering'. *IEEE Proceedings–Software*, Vol. 146, pp. 2–10, February 1999.

[18] Räck C., Steglich S., Arbanowski S.: 'AMAYA: A Recommender System for Ambient-Aware Recommendations'. ICOMP 2005, Las Vegas, Nevada, USA, June 27–30, 2005.

[19] Sawyer D. and Reich L.: 'Reference Model for an Open Archival Information System (OAIS)'. CCSDS 650.0-B-1 (January 2002), pp. 12.

[20] TeleManagement Forum: 'Enhanced Telecom Operations Map (eTOM), The Business Process Framework'. TMF approved version 4.0, March 2004.
[21] TeleManagement Forum: 'Shared Information/Data (SID) Model: Concepts, Principles, and Business Entities - GB922'. TMF approved version 3.0, July 2003.
[22] The Parlay Group Specifications. Online: www.parlay.org.
[23] van der Aalst W.M.P.: 'The Application of Petri Nets to Workflow Management'. *Journal of Circuits, Systems and Computers*, Vol. 8, No. 1, pp. 21–66, 1998.
[24] W3C: 'RDF Primer'. W3C Recommendation, February 2004.
[25] Wagner M. (ed.): 'Final Report about Ubiquitous Mobile Applications and Services: Including updated specification for context management technologies'. IST-MobiLife Project Deliverable D32 (D4.6), December 2006. Online: www.ist-mobilife.org.

4

Context Awareness and Management

Edited by Patrik Floréen (University of Helsinki, Finland)
and Matthias Wagner (DoCoMo Euro-Labs, Germany)

The work presented in this chapter concentrates on context awareness, context management technologies and machine learning for application development towards context-aware mobile services and applications that take into account the full scope of user context. The work is centred on the *Context Management Framework* (CMF), which provides a common framework and component view to building context-aware applications. This framework includes mark-up vocabularies for context that facilitates the annotation of context at different levels of abstractions, context ontologies on the basis of W3C's Web Ontology Language (OWL) to describe context and infer the user's situation at a qualitative level, as well as algorithms for reasoning on context information and a framework for making recommendations based on learned preferences. Building on the CMF, a *Proactive Service Provisioning* framework is presented to formally describe and classify mobile services for personal recommendation. This framework serves as the basis for proactive services such as the Proactive Service Portal (see Chapter 7).

The structure of this chapter is as follows. Section 4.1 describes the CMF. In particular, a *Context Representation Framework* (CRF) is described with *Context Providers* (CPs), *Context Consumers* (CCs) and *Context Brokers* (CBs). Section 4.2 describes data gathering and simulation, and Section 4.3 describes machine learning for context awareness. Section 4.4 provides insights into the Proactive Service Provisioning framework. It covers service categorisation, recommendation-based service discovery and self-promoting services. Reference applications

Enabling Technologies for Mobile Services: The MobiLife Book Edited by Mika Klemettinen

relating to these themes, most notably Proactive Service Portal, ContextWatcher and Personal Context Monitor, can be found in Chapter 7. Some open source software and other material related to the above mentioned applications are available online [34].

4.1 Context Management Framework

4.1.1 Introduction

Context, according to a definition by Dey and Abowd [31], is:

> . . . any information that can be used to characterise the situation of an entity. An entity is a person, place, or object that is considered relevant to the interaction between a user and an application, including the user and application themselves.

Context can be described at different granularities, or levels. For example, GPS coordinates can be transformed to a street address, which can be given a semantic interpretation by the user, such as 'home'. Context awareness brings the additional intelligence to mobile services and applications that make them more useful and user-friendly.

The *Context Management Framework* (CMF) [39] is designed for the discovery of, exchange of, and reasoning on context information. The CMF is a set of components, which are connected at run time, that together provide the relevant context information for the service or application, using sensing and interpretation mechanisms.

The interdependencies of the CMF components are depicted in Figure 4.1. *Context Sources* (CSs) deliver raw context information, such as GPS coordinates or calendar data available on a mobile device. *Context Providers* (CPs) are software entities that produce context information from internal or external information. The internal working of a CP might range from simply wrapping a body sensor for a single user to a full-fledged inference reasoner that combines

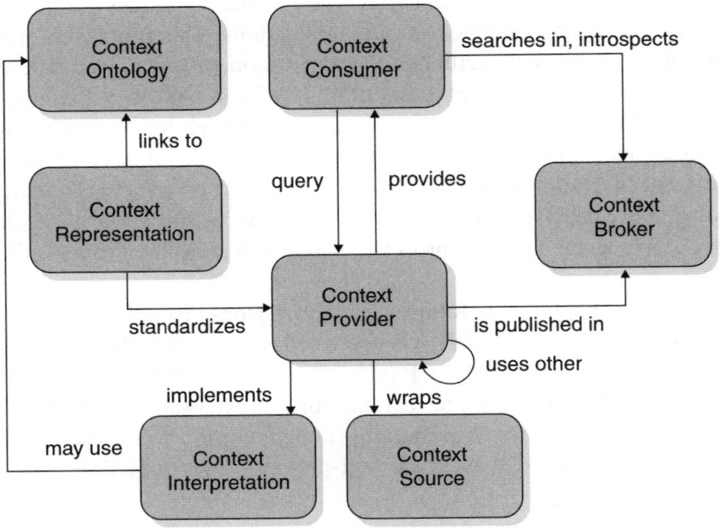

Figure 4.1 Components of the CMF.

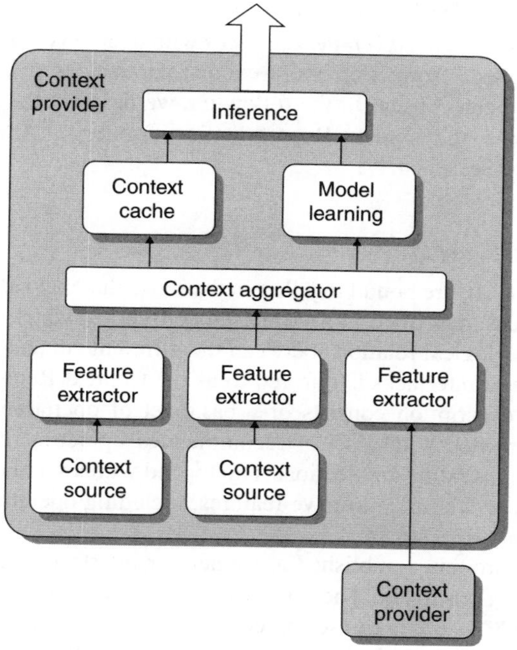

Figure 4.2 Context Provider.

information gathered from other CPs. An example CP undertaking *context interpretation* is described in Figure 4.2; a more detailed description of context interpretation and reasoning can be found in Section 4.3. Many different Context Providers can co-exist. A software entity that uses CPs as input is called a *Context Consumer* (CC). A CC can itself provide input to (multiple) other CCs, and thus be a CP for them. What binds the CPs together is that they all implement a minimal set of common interfaces, described in XML-based definitions, which make use of a *context ontology* describing the logical relations between the different context concepts in OWL-DL. This minimal set of common interfaces is called the *Context Representation Framework* (CRF). The CCs discover the correct CPs using a *Context Broker* (CB), which takes care of registration and lookup of CPs and provides a single point of entry for users of context.

A detailed description on design issues is provided in [39].

To summarise, the main tasks for the CMF are:

- enabling the discovery of Context Providers;
- standardising context exchange between Context Providers and Context Consumers;
- supporting easy context reasoning by allowing reasoning components to be added to an application in a plug-and-play manner;
- supporting the construction of different constellations of CPs to provide high-level situational information, such as a pipe-and-filter approach where context reasoners re-use context information from different CPs, and provide the newly derived information via a similar CP interface.

Context information can be represented in many ways. In this book, a straightforward approach that combines *XML-based context services* with deep links to an *upper-level context ontology* for more detailed information about context parameters and entities, as well as to support reasoning, has been chosen. In the following, we describe the XML-based interface definitions and messages, the Context Representation Framework (Section 4.1.2), and the upper context ontology (Section 4.1.3).

4.1.2 Context Representation Framework

All components in the CMF are bound together in the sense that they all implement a minimal set of common interfaces, described in XML-based definitions, which make use of a context ontology describing the logical relations between the different context concepts in OWL-DL [10]. This minimal set of interfaces is referred to as the Context Representation Framework (CRF). The interface description consists of a basic set of operations that are required to be implemented, augmented with a set of recommended operations to ease the life of the developers of context consuming applications, and topped with a set of optional operations for management purposes or with nice-to-have features, including operations to support context visualisation and context RSS streams.

The main task is to provide a published agreement or interface contract between Context Providers and Context Consumers. The contract is the only information a CC has prior to the binding with the CP in order to use the context service functions. The most important messages in the representation include the CP advertisement, the context query, the context element and the context subscription. With these messages, a Context Provider can implement an interface with a minimal set of operations to request an advertisement, to retrieve or push context information, or to subscribe to context changes. All services may support SOAP as well as HTTP GET and POST bindings, and are described in WSDL to ease the integration efforts [25].

A typical workflow example would be a Context Broker harvesting all Context Provider advertisements from the set of published CPs on a regular basis. When a context query comes in from a Context Consumer, the Context Broker would consult its cached knowledge base, and provide the CC with a link to a CP that matches the specified constraints and can potentially answer the query. The CC then queries this CP for context information of a specific user, and is provided with an array of context elements describing this user's context.

In a workflow like the above, many different messages play a role. In the following, the most important messages that are needed for interaction with a Context Provider are mentioned.

- *Context Provider advertisement.* Each CP should be described by an advertisement that uniquely describes the functionalities of the CP, the types of context information it can provide and the relevant entities playing a role in this information. Examples of entities are agents (person or groups), locations and objects. With the advertisement, one can describe CPs that can deliver the temperature of a collection of rooms, or the location of a specific group of users, together with a URI pointing to the service that can deliver this information. One CP may deliver different types of context parameters, which can be nested to create structures of context parameters, and which can be linked to the central context ontology to gain common understanding and enable ontology reasoning. Also, the entities may have their link to the context ontology, to further describe the characteristics of an entity (e.g. the ontology link for a room might describe the capacity of the room, and all objects normally contained in that room, including objects which can be sensed automatically via their radio interfaces).

- *Context query.* A context query describes what type of context information is requested from which CP, relating the information to specific entities. If not stated otherwise, the query refers to the last known context state, such as someone's actual location. However, a Context Consumer may also specify a filter, to select specific context information covering a certain time period from a CP with some kind of caching. These filters can be posed on all attributes of the context parameters, including the confidence or value of a parameter. A summary can be added to the query to perform operations on the result set returned from the filtering operation. The result of this operation is then provided as a context element to the CC. As an example, a CC can obtain the maximum speed of user Marko when his street information matches the $[AE][0\text{-}9]^{+}$ pattern (highways) with an accuracy of less than 1 m/s, in only one request.
- *Context element.* A context element is an elementary piece of context information. Each context element has metadata describing the Context Provider that delivered the data, a name, the entity that is the subject of this context element, and a timestamp denoting at what point in time this context information was captured. The value of the context element is again a parameter tree, as in the advertisement, this time containing the values of the parameters. A value can be accompanied by a measure for accuracy, confidence or other quality-of-context data.
- *Context subscription.* A context subscription is used by Context Consumers to subscribe to a Context Provider for notifications about context changes or for updates of context information at regular time intervals. The conditions and/or the length of the time interval can be specified by the CC, as well as the location of its *context push interface* (described in more detail later in this section) that it uses to receive the context elements, when the conditions of the subscription are met. The context subscription contains a filter and a summary that specifies *what* context data will be sent, and a condition that specifies *when* the context data will be sent to the specified interface. This means that the CP should dynamically bind with the CC's context push interface when the given conditions are evaluated positively.
- *Filter, conditions and summary.* The conditions specify a certain situation in terms of a set of logical rules, based on the attributes and values of the context parameters. A simple condition works on a specific parameter (e.g. speed) and a specific attribute of this parameter (e.g. the value or the timestamp). This attribute can be logically compared with a value, which results in a binary (true or false) decision. Complex conditions are of course compounds of simple conditions. A summary can be added to the query or to the context subscription. The effect is that a summary operator will be performed over the result set of the filtering (e.g. to compute the average of a context parameter in a certain time period).
- *Resource information.* To enable the scalability of the service infrastructure, CPs need to describe their current work load and utilisation information. This information is used by the Context Broker (specifically the *Information Service* that collects together all CP resource information) to distribute the workload to enable system scalability.

All these messages play their roles in the interface definition for Context Providers, Context Consumers and Context Brokers. The interface for a CC is relatively simple, because it is only needed to post the results of a context subscription, but the interfaces for the CB and CP are more extensive. The CB interface is targeted towards discovery of CPs, and the CP interface is targeted towards obtaining context information relevant for the CC and in different flavours.

The CP interface offers many different methods that can be categorised in the following functional groups.

- *Management.* Functionalities are needed for the administrator to maintain the Context Provider as well as user and relation management, and client software updates. Non-standard methods can be added here (e.g. to schedule a certain computational intensive task overnight).
- *Discovery.* Functionality is needed to discover the capabilities of a Context Provider, including entities and context types.
- *Context push.* Simple methods push newly obtained context information towards a Context Provider for storage.
- *Context pull.* A set of methods obtains relevant context information for a user and her authorised relations, including both context queries and context subscriptions in two different fashions: XML-based for powerful applications, and CSV-based for devices with limited bandwidth and/or limited processing capabilities.

The ContextWatcher and other applications described in Chapter 7 are built on top of a distributed network of (sometimes mutually dependent) Context Providers that are CRF compliant. The CRF specification has been proven to support such a distributed and heterogeneous environment [60], both for the interaction between the Context Providers and a Context Consumer (in this case a mobile client written in Python) and for interaction between Context Providers to support reasoning with and augmenting of context information.

4.1.3 Context Ontologies

Ontologies [105] are formalisms whose purpose is to support humans or machines to share some common knowledge in a structured way. They allow the concepts and terms relevant to a given domain to be identified and defined in an unambiguous way. In this book, ontologies are used to define basic contextual categories and the (logical) relations among them to ensure interoperability in communications with and between different Context Providers. The axiomatic descriptions of context elements, such as personal situations (i.e. Working, AtHome, etc.) can directly be used by logical inference engines to realise high-level context reasoning.

It is important to note that the ontologies described hereafter are not proposed as the main representation format for all aspects of context modelling, as ontologies are limited to the formulation of qualitative aspects and the available inference engines are generally weak in handing large amounts of data efficiently. Instead, elements of the XML-based context metamodel described above can be linked to elements of the ontologies to represent qualitative aspects of context information [39].

4.1.3.1 Mobile Context Ontologies

Most CPs link context elements representing qualitative data to corresponding elements of the Context Ontologies. For example, the *Location Provider* (see next section) allows users to associate location categories to important places (i.e. location clusters) using vocabulary from the ontologies (such as ‘home’ and ‘office’). These values are in turn used to group buddies in the ContextWatcher client application [14,60] (see Chapter 7) based on their situation; for example, making it easy to address all buddies who are currently at a soccer stadium in order to send them an SMS. If a set of qualitative context elements characterising a user's context, collected from different CPs, is linked to the same set of decidable ontologies, an automatic reasoner can be used to classify her situation [66,69]. This reasoning process might also include derivations in component ontologies; i.e. reasoning on social relations in an agent component ontology [15]. The determined situation provides a useful criterion for dynamically filtering application contents, such as the list of points of interests in the ContextWatcher [67,68], and to support the matchmaking of semantic services [76,77,87].

The goal in this book has been to define useful ontologies for supporting context-aware computing applications. Vocabulary has been included only if it can be used for demonstrating some aspect of the context-aware computing vision. The approach is less focused on building a comprehensive ontology library for context-aware computing and does not add new concepts just because they could be easily written down. Instead, the number of abstract concepts has been kept to a minimum in order to construct a context ontology that is simple, as well as easy to understand and use. The domains covered are the following:

- *Agent* – covering persons and organisations;
- *Location* – describing global, referenced and named locations, such as buildings, travel points and also moving locations like bus or train;
- *Device* – describing mobile, computer and communication devices and combinations thereof;
- *Time* – describing moments and periods of time, including the named time indication such as day of the week, months, and the more flexible periods (e.g. lunchtime or office hours);
- *Activities* – describing the actions of a person, such as walking, studying, meeting, or relaxing;
- *Schedule* – basically a list of planned activities for specific time periods for a specific person;
- *Situations* – describing complex state compositions of different concepts, like the fact that two colleagues travelling by the same means of transport in the same period during business hours are on a business trip.

As an ontology is referred to here as a logical theory accounting for the intended meaning of a formal vocabulary (i.e. its ontological commitment to a particular conceptualisation), the decidability of the selected ontology language is crucial. The Web Ontology Language (OWL) [89], and more specifically its less expressive variants OWL DL and OWL Lite [10], can be seen as syntactic variants of the Description Logics (DL) SHIF and SHOIN [3], respectively. In this book, the OWL DL fragment of OWL has been chosen to specify the Mobile Context Ontologies, as, besides being the largest decidable fragment of OWL, it is highly expressive and has the potential to become the standard ontology language for the Semantic Web. The good tool support available for OWL DL covering both reasoning (e.g. Racer [44,93] and Pellet [90,103]) and development (e.g. Protégé [56,93] and SWOOP [53,106]) resulted in the construction of high-quality ontologies. The XML serialisation of OWL eased the integration of ontological representations into the overall context representation meta-model. However, there are a number of things that cannot be represented in OWL because of its focus on decidability. Many of these involve property chaining, i.e. the ability to express constraints among multiple properties.

The Mobile Context Ontologies developed for this book have been validated against the language specification using the WonderWeb [33] OWL-Validator and common design flaws have been avoided using the ontology test rules that have been integrated into Protégé. Their consistency has been ensured by the fully automatic OWL DL inference engines Racer and Pellet. Furthermore, the sublanguage (i.e. the DL complexity) used in each component ontology and the number of constructs used (i.e. its size) have been determined using the ontology editor SWOOP. To ensure easy maintainability of the ontologies, particular naming conventions have been strictly followed.

The ontologies have been developed in an incremental way guided by the application scenarios (see the Appendix). At first, the initial baseline Mobile Context Ontologies were defined in the SHIF(D) fragment of OWL DL, consisting of six components covering most of the intended vocabulary. A first (naive) refinement of those ontologies resulted in considerable extensions by a factor of five in size, adding detailed descriptions of the initial concepts, formulating additional vocabulary informed by vCard standard [110] and the FOAF (Friend-Of-A-Friend) vocabulary, linking to standard ontologies like time-entry (a OWL DL variant of Time OWL as used in OWL-S), and adding three additional components. The evaluation of these refined ontologies revealed deficits in reasoning performance, mainly caused by the size of the extensional knowledge formulated and due to the use of modelling constructs known to be computationally expensive (such as nominals in combination with inverse roles) pushing the underlying logic to SHOIF(D). In a second refinement step expensive constructs not strictly necessary were avoided, the logic was restricted to SHIF(D), and all components in their extensional and intentional parts were split resulting in a total number of twelve components with a much improved reasoning performance, although vocabulary covering most of IETF RPID [99] and Microsoft's Mappoint [71] was added. This version of the ontology was then referenced by all CRF components. The experiences gained by their use in the various applications (see the later discussion in Section 4.1.3.2) influenced the development of the final polished version of the ontologies during the maintenance phase.

Ontology-based reasoning is concerned with drawing conclusions from formal concept, role and individual descriptions. It employs the Open World Assumption (OWA) as enforced by the technology underlying OWL. This means that 'what cannot be proven to be true is not believed to be false'. Implicit knowledge about concepts and individuals can be inferred automatically with the help of sound and complete inference algorithms that are known for a wide variety of DLs. In particular, relationships between concepts as well as instance relationships between individuals and concepts play important roles.

The approach used in this book to realise high-level situational reasoning is to apply dynamic assertional classification of situation descriptions represented as concrete individuals. Each 'situation individual' is assembled from a set of entities representing qualitative context information such as the location (e.g. office), the time (e.g. afternoon) and the persons in proximity (e.g. friends). Finally, the direct subsuming concepts of the situation individual determine the user's abstract situation. Classification, too, has a role during the design phase to ensure a high quality as discussed above. In this way, direct subset relationships between the set of objects described by two concepts (i.e. deriving that two concept definitions are implicitly related by subsumption, equivalence or disjointness) can be discovered. This process may result in a restructuring of the concept taxonomy to satisfy the discovered subsumption relations. The approach used in ontology-based situational reasoning is best explained by considering the following example.

Important Business Meeting at Tokyo Station

Two travellers, Dawson Campbell and his boss Fiona Davidson, arrive on a Friday morning at Tokyo's main station. Gordon Green, a project partner, is already waiting for them at the platform. The group is looking for a quick transfer to the airport.

To derive the situation of Dawson automatically, first each piece of context information such as the location (Tokyo station), the time (Friday morning), and all companions (Dawson's boss Fiona and his project partner Gordon) are represented in terms of vocabulary formalised by the context ontologies. This requires the mapping of sensed quantitative data to qualitative representations (e.g. a time-stamp is mapped to an individual representing a Friday morning). The qualitative representations are enriched by the world-knowledge formalised in the component ontologies and are combined to an individual in the situation ontology. A lookup in the knowledge base reveals that Dawson and Fiona are colleagues and that the scene takes place on a weekday's morning.

Computed by the reasoning engine, the direct concept type for the corresponding situation instance is Important Business Meeting. In this case, the location of the scene is a public place (as Tokyo station is an instance of the concept *Station*, which in turn is a subconcept of *Public place*) during office hours (as the individual Friday morning is classified as *Office hours*) and the main actor Dawson is accompanied by his supervisor and a business partner.

The situational reasoning process described above is supported by deductions in all *component ontologies*. For example, the *agent ontology* specifies in detail the semantics of social relations between people. Based on the knowledge encoded within the ontology, it can be inferred that two persons (like Dawson and Fiona) are colleagues, taking into account the transitivity of this relationship in case they have a common colleague.

Related *context ontologies* that are formulated in OWL DL are CONON [111] and SOUPA [21]. CONON is an upper-context ontology for pervasive computing applications defining almost 200 concepts. Rule-reasoning is used to derive high-level context information and to check its consistency. However, rule-reasoning cannot be complete for OWL (not even for OWL Lite [17]) and it easily leads to undecidability [43]. To cope with the observed delay of several seconds caused by the reasoning process, complex reasoning tasks are computed offline in CONON. However, this approach is not feasible in the dynamic setup used here. SOUPA, another ontology designed for ubiquitous applications, is about the same size as the CONON ontology. Its extension CoBra-Ont is used by a *Context Broker* architecture to realise a scenario where people on a university campus come together for a meeting. To limit the reasoning overhead caused by importing standard ontologies, single concepts are mapped to foreign ontology terms. Still, the SOUPA ontology is of a rather high complexity corresponding to SHOIF(D), because it contains nominals.

4.1.3.2 Lessons Learned

Several projects consider the use of ontologies as a key requirement for building context-aware applications. Closely related to the approach described in this book is the work done in the CALI project [55] as it explores the use of Description Logics (DLs) and the associated inferencing. To overcome the limitations of pure DL-based reasoning, a hybrid approach is proposed. However, earlier experiments [68] indicated that the suggested loose coupling of a DL reasoner with an external generic rule engine may lead to severe performance problems.

To achieve completeness, both reasoners have to be applied successively until no new facts have been derived. Furthermore, it remains unclear how consistency can be guaranteed taking both the knowledge base and the rule base into account.

Overall, OWL DL seems to be currently the best option for the representation of context ontologies. Having OWL as a well-defined standard ontology language triggered tremendous progress in the tool support for OWL in recent years, both regarding ontology editors and reasoning engines. Especially, the decidable OWL DL fragment turned out to perfectly support the realisation of a high-level context reasoning engine. The completeness of the logic (along with its decidability) pin-pointed several modelling mistakes during the ontology development for this book, but also identified inconsistencies in the sample dataset used for testing.

Some of the requirements initially defined regarding the Mobile Context Ontology and its development could not be fulfilled completely. The original requirements for modularity (i.e. 'the ontology should be structured such that it supports modularity') and versioning (i.e. 'the ontology should provide a versioning system') are only partly fulfilled, by splitting the ontology into several components and by establishing a different main version, as *owl:import* does not fully support modularity on a semantic level [65] and change management is not supported by the existing standard OWL tools. Furthermore, existing standard ontologies were not referred to for performance reasons detailed above, violating the original requirement to refer to existing standard ontologies.

A recent analysis revealed that 'existing ontologies are either very generalised, very application specific, or inflexible' [26]. According to the analysis of existing standard ontologies built for context representation, one could add that it is currently not an option to link to those ontologies, if decidability and performance of reasoning matters. One either inherits a high DL complexity or introduces undecidability. For example, SOUPA [21] and time-entry, an OWL DL subset of OWL Time [47] used in OWL-S [73], are in SHOIF(D), and the OWL DL variant of the Descriptive Ontology for Linguistic and Cognitive Engineering, DOLCE-lite [41], is listed among the Top10 of the most challenging ontologies [42]. The Suggested Upper Merged Ontology proposed by IEEE SUMO [79] is undecidable and was recently found to be inconsistent [48].

Existing performance results of DL reasoners are often limited to the classification of static ontologies [112]. However, in the case of frequent updates (a KB submission, discarding, and re-submission cycle), the communication overhead introduced on loading the ontology can easily dominate the overall performance. In this respect, the delay caused by ontology-based inferencing can become a major obstacle for its use in context-aware applications. The whole process of determining the situation of a user, including the gathering and transformation of the relevant context data, is limited to about 2 seconds per classification. Retraction can improve the performance for this type of application drastically, since only a small fraction of the ontology changes between two requests. However, the standard DL interface DIG 1.1 [12] does not support the removal of specific axioms, making it necessary to resubmit the complete ontology for each request. For the DIG 2.0 standard [9], it is therefore proposed to have a modular extension to the interface that supports incremental reasoning and retraction [11]. Unfortunately, current reasoners provide only some kind of batch-oriented reasoning procedure. A notable exception is RacerPro, which offers low-level retraction support for most of its statements.

For this book, different retraction strategies implemented in Racer were compared. Reloading of ontologies from a local web server can be accelerated by either loading from an image file (up to three times faster) or by cloning an ontology in memory (up to 70 times faster). For small extensional knowledge bases, cloning the ontology outperformed even the retraction of single axioms with forget statements (usually 80 times faster). However, it turned out that the fastest strategy was to keep situation individuals up to a certain number (about 20 in the tested case) before cloning a fresh pre-loaded extensional knowledge base (keeping individuals and axioms is possible only if they do not influence later classifications). Due to the lack of incremental classification algorithms, RacerPro still initiates a complete reclassification after each change in the ontology. Initial empirical results [45], performed with an experimental version of Pellet, indicate that such algorithms for SHOIN(D) can be quite effective.

Without retraction support, the time needed to compute simple reasoning problems is easily dominated by the communication overhead caused by the reasoner interface. For example, accessing RacerPro via its native API using TCP is about 1.5 times faster then via HTTP/DIG and even 2 times faster than the access realised with the triple-oriented framework Jena2 [20]. The best performance can be achieved by using the Pellet reasoner running in the same Java virtual machine as the application itself, without the need for any external communication.

4.1.4 Context Providers and Context Consumers

Context Providers (CPs) act also as Context Consumers (CCs) for other CPs. In addition, the reference applications in Chapter 7 are naturally Context Consumers.

The following are examples of CPs:

- *Location Provider.* This stores automatically recorded location data from a GPS device or GSM/GPRS/UMTS cell information, enhances that information with geo-location and address, and performs clustering to find personal frequently visited places.
- *Environment Provider.* This delivers environmental information, namely weather and information about the moon and the sun, for a given location.
- *Wellness Provider.* This stores automatically recorded body data from a heart sensor and a foot sensor.
- *Preference Provider.* This offers personal and group-related profile and personalisation data.
- *Experience Context Provider.* This provides context information on the status (absent, away, busy, etc.), mood (bored, desperate, excited, happy, neutral, sad, surprised, tired), activity, drinking, driving activity, eating and safety status of users. In addition to these categories, this CP also provides custom presence information specified by the user. Figure 4.3 shows a ContextWatcher screenshot, where a user added the drink category containing relevant drinks including coffee.
- *C-log Provider.* This composes automatically daily summaries by gathering information from different CPs. These daily summaries can be for private use (diary), for business use (activity reporting) or for public use (towards personal blogs).
- *Photo Provider.* This stores context-tagged photos.
- *Terminal Location Context Provider.* This provides different kinds of location information of a user or a mobile terminal, using different means for determining the location.
- *KML Provider.* This transforms context information like location, photos, experience, etc., into the KML format to visualise it in the Google Earth client.

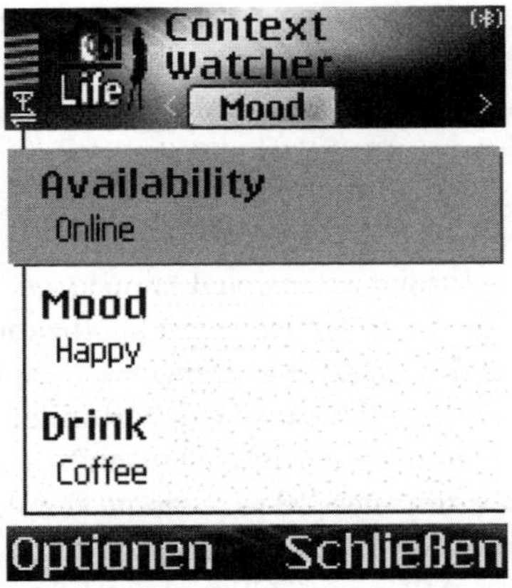

Figure 4.3 Example of experience context information, as visualised in the *ContextWatcher* application.

4.1.5 Context Broker

The *Context Broker* (CB) is a powerful query and repository service component within the CMF's architecture. The CB component was designed to address two functional concerns for proper operation and interaction between Context Provider and Context Consumer software entities within the CMF. First the CB allows CPs to publish context information by registering, updating and deregistering CP advertisements that uniquely describe the functionalities of the Context Provider. Second, it accepts searches as queries from CCs that can be matched against CP advertisements allowing introspection into the advertisements previously published to its repository.

The goal during the design of the CB was to enable registration and discovery of CPs, their interfaces, and the context elements that they can potentially supply to Context Consumers. Additionally, during the design of the CB the need was considered to allow multiple distributed CB instances to co-operatively provide discovery and registration functionalities to external CMF components, while still allowing each CB to remain agnostic as to the technology used for the implementation of other CBs.

To achieve the first goal, general-purpose interfaces and operations that allowed querying and discovery by CCs and CPs were defined. These general-purpose interfaces employ the use of filtering to allow label-based and logic-based predicates to be used by CCs, requiring context information, to query the CB to obtain CP advertisements. In order to achieve matchmaking of queries, CPs were required to register themselves with the CB.

To achieve the second goal, whereby multiple CB instances can co-operatively provide discovery and registration functionalities that could potentially span multiple administrative domains, a federated proxy pattern was chosen. The result of using a proxy-based design for the interoperation of distributed CBs is that CPs and CCs can access the set of CBs through a single *Context Broker Proxy* ingress point. This design has the advantage of being both simplistic and scalable.

A general view of the structure of the Context Broker and its interaction with external CMF components is given in Figure 4.4.

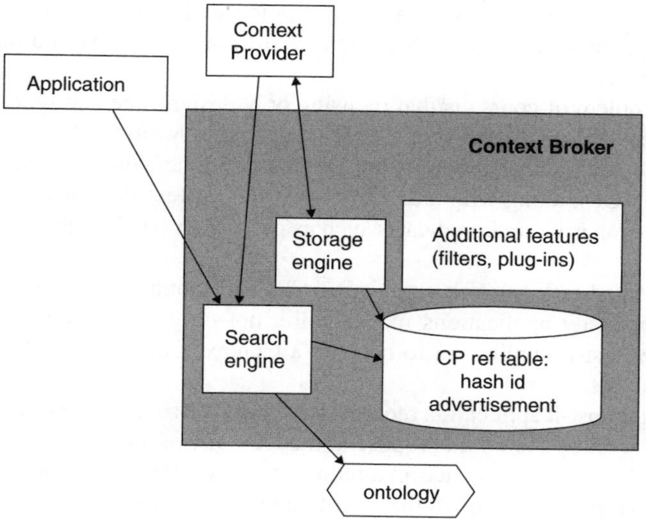

Figure 4.4 Context Broker general view, in which the Context Broker uses a reference table to store Context Provider Advertisements and handle queries to find a Context Provider.

4.1.5.1 Motivating the Context Broker: the Virtual Post-it

Let us illustrate how the Context Broker can be a necessary component in the overall context architecture through the example of a practical application: the virtual Post-it.

In real life, a Post-it is a small piece of paper with an adhesive strip that can contain a handwritten message and may be posted somewhere to be seen. The consequence of this is that the message is public to anyone located in front of it, yet there is no guarantee of the message being delivered to a recipient. For example, a Post-it stuck on a fridge in the kitchen will never be read by anyone who goes straight from her bedroom to the front door. As such, it is a context-based message that is characterised by the situation of the Post-it entity.

A virtual Post-it is an extension of this concept. It may be seen only by readers whose contexts follow certain rules, for example whose position is 'in front of the fridge'. Of course, it is much more powerful. First of all, any context may be checked, not only location; also, it may be sent to a given person or a list of people, not anyone who is there. An example of a rule for delivery of a virtual Post-it could be:

> Show the following message to all my buddies, if they come around my place, and they are in a good mood, and it's before 2 p.m. today: 'Come and see me, my brother is visiting.'

Let's assume that a person has 50 buddies known to the system, most of them far from her place (or in a bad mood), and that there are millions of users of the system. Only those concerned who are buddies of hers and who are close will get the message and only before 2 p.m., so that they would not receive a message tomorrow warning them of an obsolete event. This is, of course, impossible with SMS, email etc., or, on the other hand, a non-virtual paper Post-it.

A simple version of the virtual Post-it application could offer a user interface with location and time data fields, as well as target person and message. It could be developed knowing of one, perhaps two, CPs that offer the location of people and the current time. The first problem will be: what happens when these CPs are removed from service? Would the virtual Post-it application stop working?

The second problem of course is that by using only design-time configuration of CPs and not run-time discovery of CPs, the virtual Post-it application would not be a generic, powerful context-based messaging application. What the Post-it application really needs to know is a list of all existing CPs, along with a description of the context data that they offer, that can be filtered using properties and predicates such as policies or conditions of their use (privacy, precision etc.).

It is a complicated task to track all existing CPs and monitor their availability. It is also functionality that many applications may require, not only the virtual Post-it. Thus, it is desirable for the Post-it application to become a Context Consumer component within the CMF to discover CPs.

What the virtual Post-it application requires is a Context Broker of CPs. A CB could answer a query for all known CPs or could propose to the CC a list of possible filter conditions to set on the message. A CB could also accept a request for any CP that provides a particular piece of context information, and, through the use of an ontology, mediate to return an appropriate CP that semantically provides the correct context information. Without a Context Broker, applications such as the virtual Post-it are unnecessarily limited to using design-time context parameters and CPs.

As an additional advantage of a Context Broker, all a CP has to do in order to be taken into account by the Post-it application is to register with the CB; no modification on the Post-it application is needed.

4.1.5.2 Interfaces

The Context Broker has two interfaces handling *registration* and *querying*, respectively.

Context Provider Registration Interface

The CP *registration interface* allows the CB to provide a repository component for the management of CP advertisements within the *Context Management Framework* (CMF). To this end, the Context Broker's CP registration interface has been designed as a Web Service (WS) that supports the *creation*, *retrieval*, *updating* and *deletion* of four types of CP resources. Each of these typed resources can be published to the repository and enable its operation. The four valid resource types supported by the Context Broker's CP registration interface are:

- *Advertisement resources.* At least one mandatory advertisement resource must be registered for all CPs. Advertisement resources are used to store CP advertisements published by a CP within the CB repository.
- *Collection resources.* Collection resources are optionally used to create named sets of typed resources within the CB repository.
- *Label resources.* Label resources are optionally used to append metadata to any typed resource within the CB repository.
- *Generic function resources.* Generic function resources are optionally used to support predicate logic-based queries of any typed resource within the CB repository.

To register a CP advertisement, a CP can either pass to the Context Broker a representation of the advertisement or simply provide an external introspection interface, which may be accessed by the CB in order for the CB to retrieve the advertisement.

For a Context Provider, the advantage of CB registration is to expose the functions described in its CP advertisement to any Context Consumer. The form of publishing by registration of metadata makes the act of exposure explicitly public. Since publishing is explicitly public, any policy enforcement or security a Context Consumer implements must be done directly by the CC.

Registration in the Context Broker is an optional publication mechanism available for Context Providers. The optional nature of the CB is derived from the fact that the only requirement for a Context Consumer to use the capabilities of a CP is to obtain a CP advertisement.

Context Provider Query

In order for a Context Provider to submit queries to find the CP advertisement of a particular set of CPs, a Context Consumer must initiate a call to the CB's query interface. A CB query may use predicates as filter expressions, corresponding to the contents of a CP's advertisement. The following illustrate some examples:

- name of parameter (e.g. query for 'heart-rate');
- name of entity (e.g. query for 'Bob's' parameters);
- quality of parameter (e.g. query for the location of Bob), but only Context Providers that offer location more precise than 5 metres;
- type of entity (e.g. search for parameter 'schedule' of entity type 'bus');
- identifier of a CP. For example, it may be that after some idle time of non-use, a CP is unavailable. Before looking for another CP, a Context Consumer could query for the same advertisement, so the application first checks whether this CP has perhaps unregistered.

Query filters provide a powerful mechanism to perform projections over the set of resources registered in the CB by Context Providers.

4.1.5.3 Ontology

One innovative feature of the Context Broker is its support for ontologies. An ontology is an organisation of knowledge, including hierarchy and relationships between concepts.

An example would be to define 'heart-rate' not only as a known parameter name, but as a reference to an object in a public ontology. In this ontology, it would be a child of object 'wellness', alongside other child objects such as 'temperature', 'blood pressure', etc.; see Figure 4.5.

If a Context Provider provides this parameter for a group of people (constrained by privacy considerations, of course), it may register to the CB with ontology references being part of its advertisement. Instead of providing 'heart-rate', it is understood to provide 'the parameter whose ontology reference is 'http://www.owl-ontologies.com/unnamed.owl#Heart-rate'. The Context Provider may then be found by a query to the CB using this ontology reference.

The second point of using ontologies within the CB is to take advantage of the semantic level of information. If a 'heart-rate' is a 'wellness' parameter, any CC looking for CPs that provide wellness parameters wants to find the *Heart-Rate CP*. This is impossible, if the CB uses

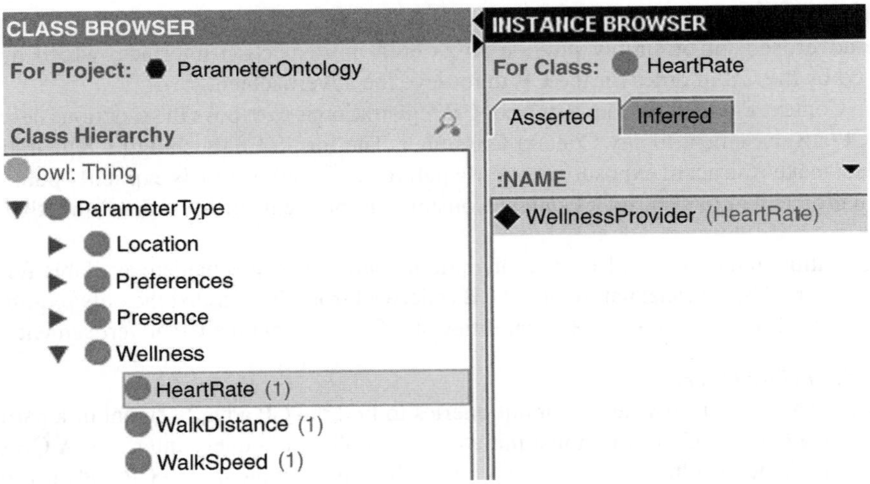

Figure 4.5 Ontology sample, as displayed by the Protégé tool.

only keywords. On the other hand, if the CP was registered using references to a hierarchical ontology, the CC may search for a CP using any term of the ontology, and get back all CPs that provide children of the desired object. In the current example, a CC wishing to publish the general wellness condition of a person may make one request, and get back several CPs, each one providing access to different data concerning the person. The CC does not even need to know beforehand which parameters are available, or what CP is up and running or down.

4.1.5.4 Distribution Using a Context Broker Proxy

One of the main goals of the Context Broker is to provide access to any Context Provider advertisement registered by a CP. In order to be scalable and resilient to fault conditions, the CB is designed to use a proxy-based structure for the interoperation of distributed CBs, with the result that CPs and CCs can access the set of CBs through a single *Context Broker Proxy* ingress point. This tiered design has the advantage of being both simplistic and scalable in an ever-expanding universe of CPs. The Context Broker Proxy is a broker of brokers, acting as an invisible intermediary to any request for a CP.

Just like CPs registering to a Context Broker, the CBs register to the Context Broker Proxy. Any request made to a CB proxy will be transmitted to the first CB that can answer it. This is a federation of Context Brokers to a central ingress point.

4.2 Context Gathering and Simulation

4.2.1 Introduction

Mobile context-aware systems are highly complex systems where the interactions between the users and devices take place in a dynamic and heterogeneous environment. Furthermore, the environment is open in the sense that one cannot specify beforehand all the situations that the users will face. This complexity complicates even the most fundamental tasks, such as data acquisition.

In this section, the focus is on approaches for data acquisition. Obtaining data is one of the most fundamental tasks for research in context-aware computing. In this book, two different data gathering tools are discussed (Section 4.2.2). The first, ContextWatcher [60], has evolved into a fully fledged application and is described in Chapter 7. The second, BeTelGeuse [38], has been designed as a generic tool that supports research in user modelling and in activity recognition. In addition to data gathering, context and behaviour information can be simulated. A simulator for generating data about the context of a single user is described in Section 4.2.3. Additionally, a multi-agent simulator that outputs context and behaviour information from multiple users according to a scenario definition is defined.

4.2.2 Data Gathering Tools

4.2.2.1 Related Work

In terms of data gathering, much work has been conducted on building different kinds of sensing device and tools for data gathering. For example, Muffin [113] integrates an air temperature sensor, a humidity sensor, an alcohol gas sensor, a pulse sensor, a compass and a linear 3D accelerometer to a terminal device. Custom sensor boxes have also been used, for example at Intel Research, Seattle [64], and at Carnegie Mellon University [102].

Also, various tools that gather data from Bluetooth sensors have been proposed. For example, ContextPhone [94] can read GPS information and it provides information about nearby Bluetooth devices. Another similar tool is the ContextWatcher, which is described in Chapter 7.

Another closely related field is wearable computing. Especially in the field of pervasive healthcare, various systems for health monitoring have been proposed; see, for example, the Body Media Sensewear armband PRO [108]. A problem with these systems is that they seldom allow online data gathering, which reduces their applicability for mobile computing.

More generic tools have been proposed. For example, the IBM Mobile Health Toolkit [52] enables online Bluetooth data gathering and it includes a Java tool for gathering the data. However, the Toolkit is restricted to a predefined set of sensors. The Intel Place Lab [104] is a generic, platform-independent tool for data gathering, but it focuses only on location information.

4.2.2.2 BeTelGeuse

While location data can, ideally, serve as a good indicator of the situation of the user, there are many compelling applications where other sources of context data are relevant. Another source of context information that is important for many applications is activity information. Namely, monitoring the physical activity of patients is fundamental for applications of pervasive healthcare (e.g. [61]). Activity information is also important for user modelling and adaptive user interfaces as it can serve as an indicator of the goals and information needs of a user (e.g. [49]). Other important sources of context include the computational context of the user as well as her social context.

Concerning the gathering of activity-related context data, the existing approaches are usually limited to a specific platform (usually Symbian Series 60) or to a specific set of sensors. This

Table 4.1 Main features of BeTelGeuse

Feature	Additional Remarks
Automatic discovery of new devices	BeTelGeuse offers a flexible way of obtaining information about available Bluetooth devices. By default, device discovery is performed periodically, but automated device discovery can be turned off and users can also manually trigger device discovery.
Automatic instantiation of reader components, which handle sensor-specific aspects of communication and data processing	Once the user has specified a mapping between a Bluetooth address and the reader that is used for that Bluetooth address, BeTelGeuse can automatically instantiate the appropriate readers in the current and in future sessions. The mapping does need to use the exact Bluetooth address, but it can also use a free text string that must be contained in the friendly name of the device.
Automatic connection management	If a connection to a Bluetooth device drops, BeTelGeuse attempts to automatically reconnect to the device.
Possibility to read data from the local device	Java APIs have rather restricted access to operating system-specific information. For this reason, BeTelGeuse allows listening to local host sockets for information from the local device.
Possibility to get Bluetooth proximity information	Since the Bluetooth scans are done by default periodically, it is also possible to obtain Bluetooth proximity information from BeTelGeuse.
Plug-in architecture: possibility to extend to tool with new features	BeTelGeuse has a modular structure that offers interfaces to all available information. This makes it possible to extend and to customise the tool.
Not tied to any particular context representation	The core of the tool does not care about how context is represented; and, by building different extensions on top of the tool, it is possible to use representation formats other than the format specified by the Context Representation Framework (CRF).
Runs on many modern devices	Runs on all MIDP2.0 and CLDC1.1 compliant devices to which a JSR-82 compatible Bluetooth stack is available. The tool can be used also on PDAs, laptops, desktops, etc.

clearly hinders the applicability of these approaches for research in context awareness as the used tool defines the sensors and platform that needs to be used for testing. For this reason, BeTelGeuse, a tool for Bluetooth data gathering [38], has been developed. Since BeTelGeuse has been implemented using J2Me, it is platform-independent. In addition, BeTelGeuse can be extended to use any type of sensor. The main features of BeTelGeuse are summarised in Table 4.1. More details about BeTelGeuse can be found at the website [13].

4.2.3 Simulation

Currently there are three main research directions in simulation of context-aware systems. One of the directions is to *simulate physical sensor measurements*. For example, [51] suggests a simulation model for producing sensor measurements in a smart space. Another research direction is to use *multi-agent simulation*. An example of this approach is the generic event simulator [98], which creates context-dependent events that can be used to test, for example, the functionality of a middleware.

The third research direction is to *combine virtual reality simulation with the simulation of physical sensors*. For example, QuakeSim [18,19] uses a multiplayer 3D game (Quake III Arena) as a simulation tool. The game has been modified so that each player is equipped with a virtual GPS that provides the position and altitude of users within the simulated environment. An extension of QuakeSim is the UbiWISE [5,6] simulator, which allows the users also to interact with simulated devices in the virtual reality environment. TATUS [85] uses a 3D game engine for simulation.

One recent and comprehensive approach is the UbiREAL simulator [80]. In UbiREAL, the simulation designer can create a smart space using 3D modelling tools. Within this environment, the designer can place various objects such as sensors, switches and communication devices. Once the environment has been generated, the designer can create an avatar or a virtual actor and define the way the avatar interacts with the devices in the environment. The simulated context variables are based on physical formulas and thus the simulator is capable of providing only environmental context information. As an additional feature, the UbiREAL simulator has support for simulating various ad-hoc and wireless communication protocols and technologies.

4.2.3.1 Physical Sensor Simulation

An important component of context awareness in mobile devices is sensing the user's environment by gathering data that can be used to characterise the user herself, her device or her environment. In context-aware research quite often there is little discussion on the quality of the context data, which may be related to the lack and difficulty of obtaining real measured context data. Examples of real context data show that the data can be noisy and cannot always be interpreted as might be expected. Below, a method and a tool that is used to generate or simulate context data using a probability model are presented.

As well as providing a quick means of testing context-aware applications, simulating context can also contribute to the understanding of context data in a mobile environment. Figure 4.6 shows three approaches that one could take to simulate sensor data. Here, the third approach is used, where one tries to reproduce the statistical characteristics of real sensor data, as opposed to simulating the characteristics of real sensors or simulating the system user-device.

The tool thus models the statistical distribution of the data obtained after the measurement pre-processing and feature extraction steps, for different human-device activities, and simulates sensor data by generating random data following this distribution. The probabilistic method is described in detail in [7].

The simulation tool consists of a script that reads a configuration file and generates corresponding context data, and a graphical interface that helps authoring of the configuration file.

The configuration file is expressed in XML style and contains three important sections: *context variables*, *user contexts* and *scenario*.

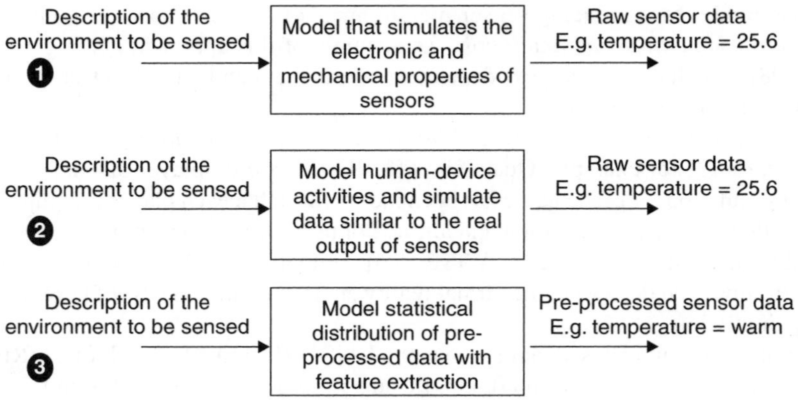

Figure 4.6 Three approaches to simulate sensor data.

- *Context variables* are also commonly referred to as context elements, context features, context pieces or context atoms. A context variable is a column in the table of the generated data table.
  - Examples of context variables: body temperature, GPS location, time, identification of the person, identification of the phone.
- *User contexts* can be understood from a statistical point of view as clusters in the data. In this tool, user contexts are created by selecting several context variables, and giving them statistical properties. User contexts are often referred to simply as context or situation. In other words, user contexts are situations where one expects the context variables to have similar statistical properties, such as similar probability distributions.
  - Examples of user contexts: person is walking, person is running, car is running, people grouped in a room, device is used, device is not used.
- *Scenarios* define how user contexts are grouped together. Scenarios contain all the data that the context simulation tool needs to generate the data.
  - Examples of a scenario: a person is going home (the scenario consists in successive user contexts: person is walking outside, person is sitting in the bus, person is walking outside).

Figure 4.7 shows a screenshot of the graphical interface of the tool. In the left part, the three different categories of the configuration file are shown. In the right part, the information about the context variable 'activity' is displayed. One can see that 'activity' is represented as having three different values: [0=walking, 1=running, 2=stationary].

In Figure 4.8, the configuration of the user context 'Walking outside in Ruoholahti' is displayed. For each of the context variables that one associates to this user context, the statistical distribution that corresponds to it is defined. The probability can either come from prior analysis of real sensor datasets, or one can guesstimate them. In the given example, the context variable 'activity' is given for 'Walking outside in Ruoholahti' with the statistical distribution [walking=80%, running=15%, stationary=5%]. The advantage of this method is that it is easy to include noise, error and accounts for variations in the environment.

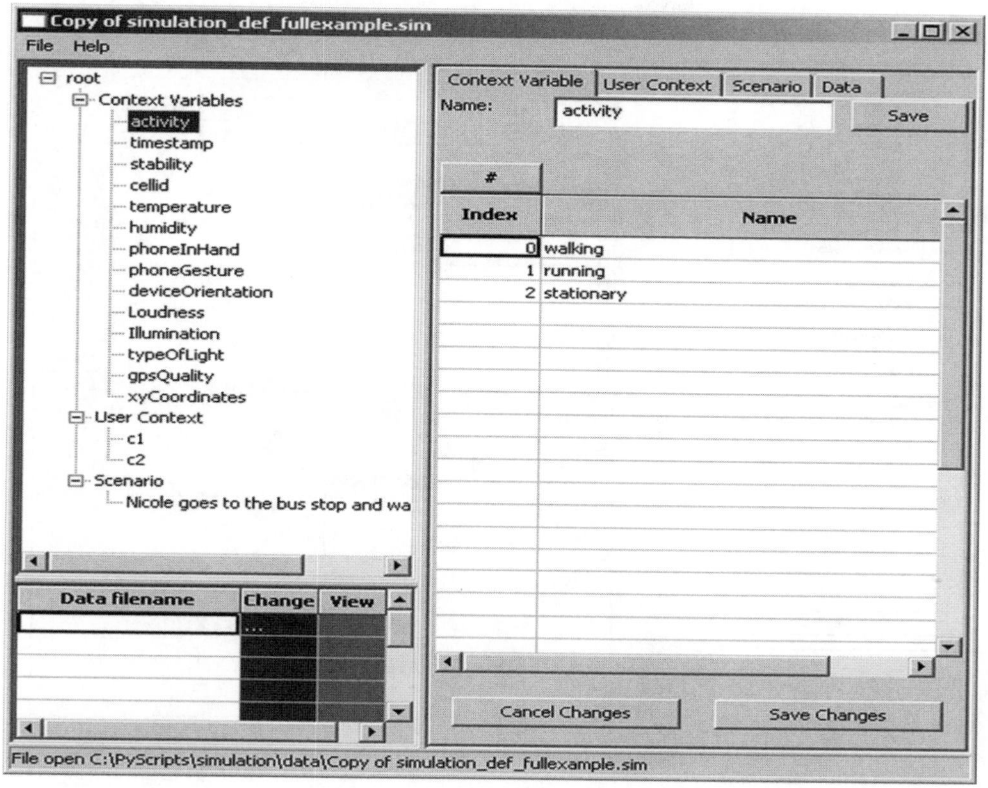

Figure 4.7 Simulation tool – editing of context variable in the graphical interface.

Finally, Figure 4.9 shows the full scenario which is going to be simulated: 'a person who walks from a place A to a place B, then waits around the place B'.

From the scenario, the tool can generate the data. A few lines of it are shown in Table 4.2.

4.2.3.2 Behaviour Simulation: Group Simulator

In order to properly test context-aware applications, one needs to provide the applications with data from multiple users moving around a real environment. However, obtaining this kind of data is currently rather difficult. Unfortunately though, although more and more sensors are available, it is still expensive to equip multiple users with sensors.

Another issue that complicates application testing is that often the applications require context data on a higher abstraction level than what individual sensors produce. Deriving these abstractions, however, is rather complicated and it is difficult to be sure of the correctness and quality of the derived abstractions

To facilitate the testing of context-aware applications, a multi-agent simulation tool for context awareness [74] has been developed. The simulator is called *group simulator*, since it

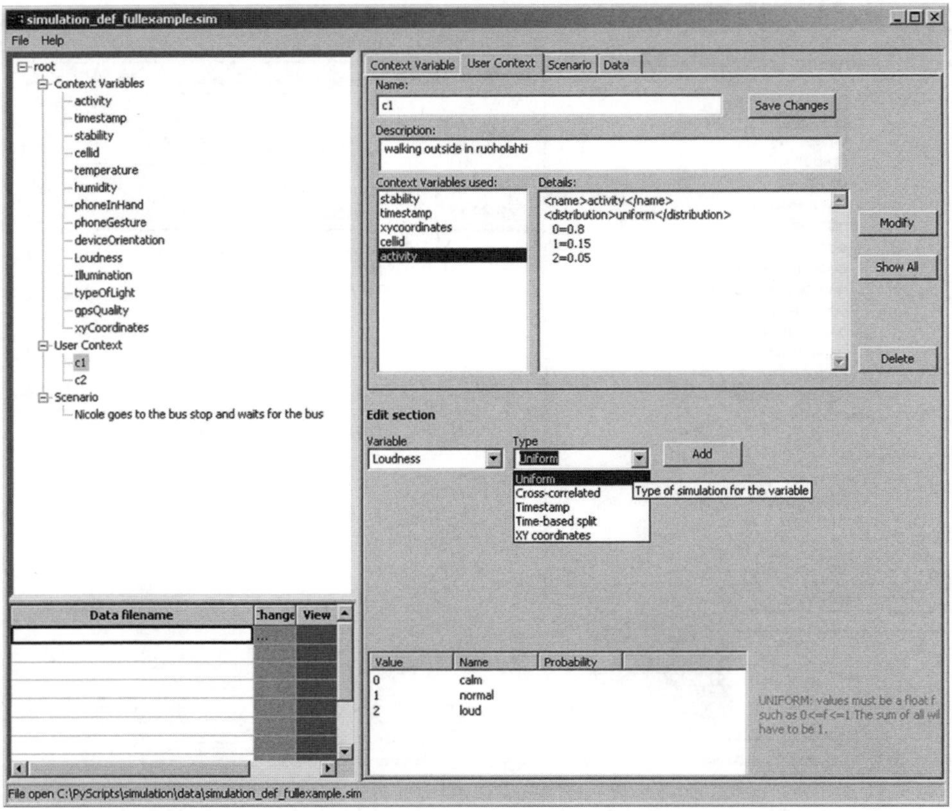

Figure 4.8 Context simulation tool – editing user contexts.

simulates the behaviour of multiple users and since the need to build the simulator originated from the difficulties encountered while attempting to test group awareness-related mobile context-aware applications. In addition to facilitating application testing, the group simulator also serves as a tool for generating data for machine learning and as a tool for aiding demonstrations.

The simulator improves the state of the art in three aspects. First, the simulator is able to generate data from multiple users. Second, the simulator is not confined to smart spaces, but it can be used to simulate more generic scenarios. Third, contrary to the other simulation tools, the group simulator uses extensively techniques from the field of simulation, which, for example, makes it easier to specify new scenarios that are to be simulated.

In order to achieve its goals, the group simulator generates data that contains information about the situation of users and information about how the situation relates to the behaviour of the individuals. To arrive at a simulator that is flexible and that can be easily reconfigured to simulate new scenarios, the three main information sources, actors, environment and scenario, have been separated from each other. An illustration of the simulator is shown in Figure 4.10. In the following, each of the information sources is briefly detailed.

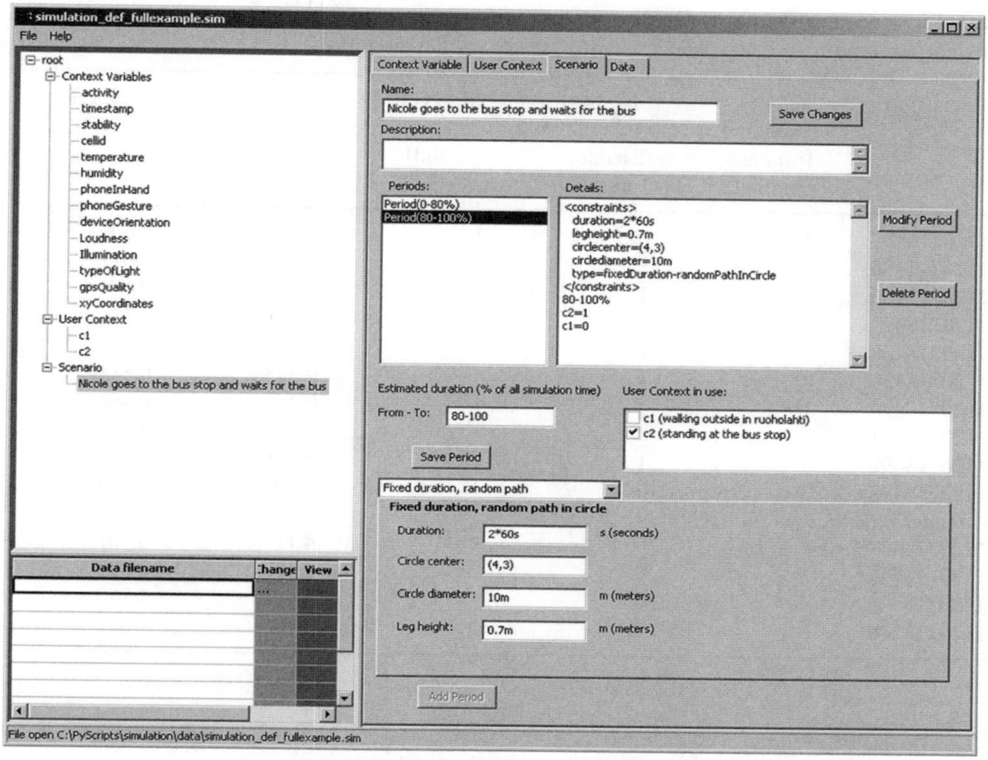

Figure 4.9 Context simulation tool – editing a scenario.

In the group simulator, the actors whose context and behaviour are being simulated are called *agents*. During each simulation step, the agents observe the current state of the environment and decide what to do next. As a simple example, if the agent is a human who is currently at work, the agent could decide, for example, that it needs to go to a meeting, that it should continue working as usual, or that it is time to call it a day. To add randomness to the events, the behaviour can also include stochastic elements. For example, one can say that, with some given probability, the agent needs to go to bathroom on the next simulation step. The agent model is also responsible for modifying the so-called *external state* of the agent. For example, if the agent changes its mood from happy to sad, then the agent model is responsible for making this change visible to the environment.

In addition to the actors moving in the environment, an important part of a simulated scenario are the places and objects that are part of the environment. In the group simulator, these are specified by the world model. For example, in a city scenario, relevant places could include work places, restaurants, homes, etc.

The main output of the simulator is the context of individual agents, which arises from two information sources. The main source of context is the location and activity of a user, which is specified by the agent model. The second source of context is the environmental parameters, which are specified by the context model. Examples of environmental context include WLAN

Table 4.2 Simulated data. The first rows represent a person walking outside between places A and B, and the last rows represent a person standing outside around place B

Timestamp	Activity 0:walking, 1:running, 2:stationary	Stability 0:stable, 1:unstable	GSM cell-ID 1:cellid1, 2:cellid2	X coordinate	Y coordinate
1120718484	2	0	1	0.000	0.000
1120718485	0	1	1	0.077	0.077
1120718486	0	1	1	0.154	0.154
1120718487	2	0	1	0.231	0.231
1120718488	1	1	1	0.308	0.308
1120718489	0	1	1	0.385	0.385
1120718490	1	1	1	0.462	0.462
1120718491	0	1	1	0.538	0.538
1120718492	0	1	1	0.615	0.615
1120718493	0	0	1	0.692	0.692
..... *data removed for clarity*					
1120718543	2	0	2	4.113	3.607
1120718544	2	0	2	4.113	3.607
1120718545	2	0	2	3.692	2.859
1120718546	2	1	2	3.692	2.859

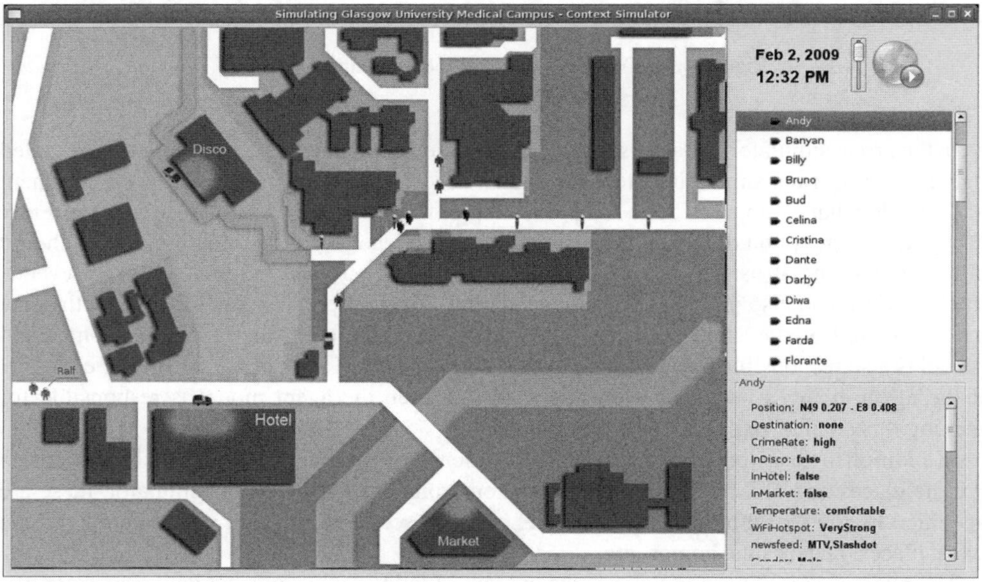

Figure 4.10 Screenshot of the group simulator.

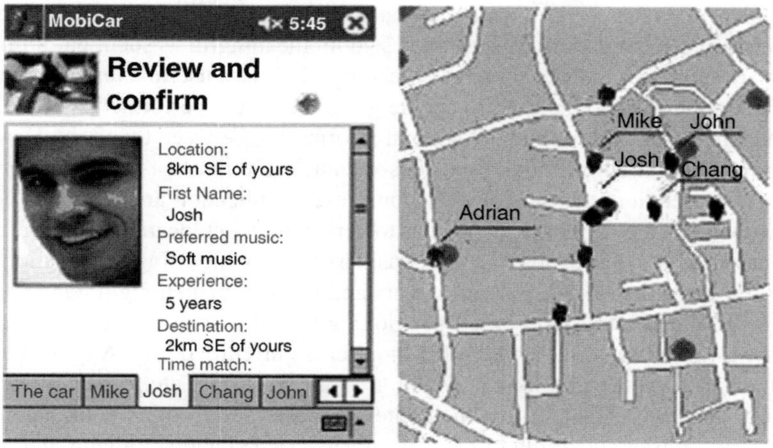

Figure 4.11 *MobiCar*, an example of using the group simulator for demonstrating context-aware applications.

hot-spot coverage, temperature and other physical variables, and area-related parameters such as crime rate.

As an example of how the group simulator has been used, consider the *MobiCar* application (see Chapter 7). MobiCar is a prototype car sharing application that has been designed for studying the effect context information has on group formation.

In MobiCar, users post requests about their need to travel to a specific destination around a given timeframe. The users can also specify additional parameters, such as music type, smoking preferences, etc. Based on the requests, the application attempts to find cars that best match the requests of individual users.

Performing the allocation of users into groups (cars) optimally, however, is a difficult task and finding the best solution for the end application requires evaluating various algorithms and their performance. In addition, the algorithms probably need fine-tuning before they can be used in real-life situations. To this end, it is important that one can test the algorithms in an environment, which does not include real users. Naturally this would not replace actual user tests, but serve only as an intermediate step before the user tests. An example of how the group simulator is used with MobiCar is shown in Figure 4.11. As the figure clearly indicates, the simulator also provides an intuitive illustration of the results of the allocation.

4.3 Machine Learning for Context Awareness

4.3.1 Introduction

Machine learning is defined as the study of algorithms that are able to improve automatically through experience [75]. This definition is rather wide and thus techniques ranging from data analysis and data mining to reinforcement learning and Kernel-based methods can be seen as machine learning approaches.

Mobile, context-aware environments offer ample opportunities for machine learning approaches. First of all, the data from sensors is seldom meaningful as such, but it must be first mapped into descriptions that are somehow meaningful. This is useful to explain the results of potential adaptation rules to the user, to enable generalising the performance of the system to new, unseen situations, and to share situational information between the users. Here, the task of deriving meaningful descriptions from sensor data is referred to as *context interpretation*. In terms of interpretation, the most widely considered application areas are *positioning* and *activity recognition*. In positioning, the goal is to derive the exact location of a user from noisy, real-life measurements. For example, [23] considers positioning using information about the estimated GPS coordinates of GSM cell towers and GSM signal strength. Other approaches include WiFi [24] and ultrasound-based positioning [91]. In activity recognition, the goal is to use sensor information to recognise what the user is currently doing. A typical example of this is recognising whether the user is walking, running, sitting, etc., based on acceleration data; see, for example, [4] and [82]. The context interpretation in this book has focused on (1) *symbol string-based clustering of context information* [8] and (2) *identifying meaningful locations from GSM and GPS data* [83]. These topics are further discussed in Section 4.3.2.

Another domain for machine learning in context-aware environments is *personalisation*. A widely used approach for personalisation is *user modelling* [57]. In user modelling, the aim is to construct models that capture the beliefs, intentions, goals and needs of a user [49]. In context-aware environments, one also needs to associate the captured interests (needs, beliefs, etc.) with the situation of the user. The main techniques for user modelling are knowledge representation (KR) methods and predictive user models [116]. The former covers traditional expert systems and logic-based systems, whereas the latter includes statistical machine learning techniques, such as rule induction, neural networks and Bayesian networks. In terms of user modelling, one of the most widely used application areas is location-based services, especially map-based mobile tourist guides; see [100] for a review. In mobile tourist guides, user modelling is used, for example, to learn the preferences of the user and to filter out or to highlight information that is considered relevant to the user. However, most of the solution techniques that are used are rather heuristic and domain-specific, which makes it difficult to apply user modelling on a wider application scale. The approach in this book is based on generic user modelling systems in context-aware systems and, instead of resorting to domain-specific heuristics, the focus is on statistical user modelling methods, such as Bayesian networks and rule induction. The approach for user modelling is described in Section 4.3.3.

4.3.2 Interpretation

4.3.2.1 Location Clustering

In context-aware mobile computing, location information has been, without doubt, the most widely studied source of contextual information [1,22,96]. The main reason for the situation is that current terminal devices can readily access location-related information, whereas other types of contextual information are harder to gather and process. For example, mobile phones can access the GSM cell identifier and PDAs can use information about WiFi access points. In addition, prices of GPS devices with Bluetooth capabilities have decreased significantly and the number of PDAs that have GPS modules integrated to them has increased rapidly, which makes using GPS a feasible option.

Different sources of location information have their peculiarities. GSM cell tower identifiers give coarse estimates of the location; the numbering of the identifiers is seemingly random and, even though operators allow the obtaining of the current cell identifier, they normally do not offer services to convert cell identifiers into geographic locations. On the other hand, GSM cell tower identifier information is available also indoors, whereas GPS measurements are not. GPS signals may also be blocked in an outdoor environment by buildings and trees. Finally, WiFi access point information can be used for positioning when the locations of access points are known [63].

Regardless of the source of location information, the raw measurements are usually meaningless to the user. As a consequence, much work has been conducted on identifying significant locations from the raw data. A *significant location* is defined to be a location that is meaningful to the user and to which the user can attach some (meaningful) semantics. For example, 'home', 'work' and 'airport' are significant locations whereas 'SomeStreet 52', '(60.42, 42.36)' or 'Cell 4287' are not.

Existing approaches for location clustering can be categorised based on the nature of the used data or based on the type of information that is used to identify the significant locations. In the following, the first approach is used.

In terms of data, the most popular approach has been to use periodically gathered streams of GPS coordinate data. In bounded areas such as office buildings, campuses, research laboratories or individual cities, background information about the physical location of landmark beacons, such as WiFi access points, may be available. Hence, the second type of data that has been considered is WiFi access point information (access point identifier, signal strength, etc.). Finally, GSM cell tower identifiers have been used to identify significant locations [62].

The algorithms for GPS coordinate data typically employ a heuristic approach that is based either on signal loss or on duration. For example, Marmasse *et al.* [72] first use the geometric distance of succeeding measurements and loss of GPS signal to identify buildings. More specifically, let $l(t)$ be the location of the user at time t and assume that a signal loss occurs. If the next location $l(t+1)$ lies within distance r from the location $l(t)$, the significant location is inferred to be a building. After the buildings have been identified, the number of visits is used to identify significant locations.

Ashbrook and Starner [2] use a cut-off parameter to determine whether the user stays long enough within an area of a preset radius r. If the duration of stay exceeds the value of the cut-off parameter, the location is determined to be a significant location. A variation of this work is presented by Toyama *et al.* [109], who use multiple values for the radius parameter r. Initially they use the approach of Ashbrook and Starner to identify locations, after which the value of the radius parameter is decreased and the same procedure is used to identify sub-locations within the previously identified locations. This process is then iterated until no more sub-locations are found.

Kang *et al.* [54] use background knowledge about the physical location of WiFi access points and their MAC addresses to identify significant locations. Their algorithm first builds clusters using a customised DBScan algorithm [32], after which a cluster is marked as a significant location if the user stays long enough within the cluster. A similar approach has also been adopted by Zhou *et al.* [115], who use a modified DBScan algorithm together with temporal pre-processing for inferring significant locations. The temporal pre-processing ensures that the significant locations are really visited frequently enough, and the modification to the DBScan algorithm is needed to cope with signal errors.

The approach by Laasonen *et al.* [62] works with GSM cell tower identifiers. The used algorithm calculates statistics such as total and average duration of the visits for recently visited cells. In addition to the statistics, constraints on the graph induced by the GSM cell transitions are used to build cell clusters. Next a duration test is used to check whether the average stay in a location is significant. If this is the case, the cell cluster is considered to be a significant location.

It should be noted that all of the existing approaches first apply a customised version of a density-based clustering algorithm to identify regions of interest, after which temporal constraints are used to detect whether or not the regions are significant for the user.

A problem in existing work is that most approaches are not well suited to large-scale mobile environments. Namely, even if the GPS equipment and GPS-enabled devices are becoming more common, users seldom have access to GPS information and GSM cell identifiers do not allow separating significant locations that are near each other (i.e. that are mainly covered by the same cells). In this book, a setting where GSM cell transitions are logged to enrich the information using the GPS coordinates of the transition point whenever a GPS device is available is considered.

The ContextWatcher application (see Chapter 7) has been used to monitor cell changes. Whenever a cell change occurs, the available location information is sent to the Location Provider (see Section 4.1.4). If there is a GPS device connected to ContextWatcher, the GPS coordinates at the transition point are sent to the Location Provider. If no GPS coordinates are provided, the Location Provider can use GPS measurements of other users to give an estimate for the GPS coordinates of the cell. All in all, the setting considered results in partially incomplete data: The GSM cell identifier is almost always present, as well as country information. Latitude and longitude information is present in approximately 70% of the cases, city information in 60% of the cases and street information in 50% of the cases. In the data gathering effort performed, relatively many people had GPS devices available, but commonly the situation would not be as good.

The Location Provider is also responsible for identifying significant locations from the available location information. In [83], four algorithms have been presented for identifying significant locations in the described setting. The first two algorithms are based on graph clustering whereas the two last are based on duration and cell transition information. Also, these algorithms are variants of spatial clustering and the two last use temporal processing.

4.3.2.2 Symbol String Clustering

In [8], a set of recorded data is analysed using exploratory data analysis and visualisation techniques, such as the Self-organizing Map (SOM) [59]. The information sources used range from time and location to sensors such as air pressure, temperature and acceleration. The main conclusion drawn from [8] is that high-level user contexts can be defined as clusters in the data generated from the combination of multiple information sources. For several reasons, methods like the SOM cannot be directly applied to clustering data or context awareness in a mobile device environment. The reasons, referred to in [8], are related to the statistical distribution of the data and the limited computational and memory resources of a mobile device. Another problem is related to data representation where the data comes from many different sources. For example, temperature is represented as a real number and location by a discrete index (GSM cell ID) or a two dimensional vector (GPS). The question is how to combine temperature and location as an input to a classical clustering algorithm based on a Euclidean distance measure.

To address some of these problems, the *K-SCM* (K-Symbol String Clustering Map), an unsupervised clustering algorithm for symbol strings, has been introduced [35,36]. The functional concept of the K-SCM is quite similar to that of the well known K-Means algorithm [70]. The internal structure consists of a set of nodes. For each node there is a symbol string and a weight vector. Each weight of the weight vector is associated with a symbol from the symbol string. At any given time, the K-SCM processes an input symbol string. The best matching node for the input is identified and the node is updated such that it better matches the input sample. The matching and learning phases are described in more detail in [35]. The result of this learning is each node representing a cluster in the symbol string data in much the same way as in the K-means algorithm.

As the K-SCM is an unsupervised clustering algorithm, one has a means of carrying out a personalised form of context awareness that requires no user interaction. The learned contexts are entirely specific to the user and the history of the user. However, because of the unsupervised nature of the learning, there is no indication beforehand what the learned contexts will be. In the following, a very simple learning mechanism is described, which associates each of the learned contexts with different user requirements in that context. This mechanism allows the learned contexts to be exploited in a way that requires no user interaction.

4.3.2.3 Context Labelling

One general approach to the context labelling problem is to use a second learning mechanism to label the contexts in a quasi-supervised manner. In [35], one such mechanism is used as follows. Consider the illustration in Figure 4.12, which shows on the left a set of recognisable contexts and on the right a set of applications. Note that the contexts can be recognised in any manner and this does not apply specifically to the K-SCM. The applications are those available on a user's mobile device, but they could also represent any information or service available on the device. In the middle there is a set of weights ω_{ij}, which represent an association between context i and application j. In practice, if the user is identified as being in context i and uses application j, then the weight is increased. If application j is not used in context i, then the

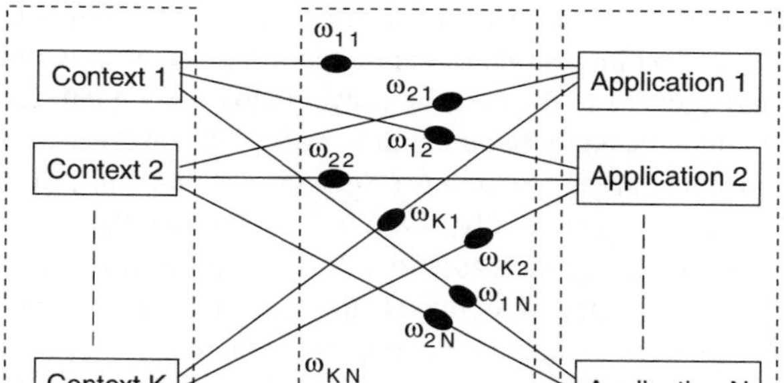

Figure 4.12 Quasi-supervised labelling of contexts.

weight is decreased. It is possible to show from this learning mechanism that the weight ω_{ij} converges to the probability of the user using application *j* in context *i*. It is shown in [35] how this mechanism can be used with the K-SCM to generate a context-sensitive shortcut menu for a mobile device.

4.3.2.4 Communicating Context Information

In *unsupervised learning*, the interpretation of context is derived from the user and the environment, but without any global framework as a reference. Essentially, the problem is that the learning for different users is independent and isolated and thus there is no common language in which to communicate context information. A paradigm described in detail in [37] is based on short-range, wireless communication that allows a set of users' devices to arrive at a common labelling scheme for different contexts. From now on, each user's context-aware device is referred to as an agent. This paradigm is based on a simple learning process and is independent of the manner in which each agent recognises contexts.

Consider a set of mobile agents that are regularly in close physical proximity. Assume each agent has a short-range wireless ability and agents can transfer data between each other. Furthermore, each agent has some means of recognising contexts. Each agent maintains a list of names for each context and associated with each name is a weight in the range [0,1]. The names can be ordered with respect to the values of the weights. The name an agent uses for a context is the name with the highest weight. In a recognised context, each agent communicates its name for the context to each of the other agents. If a majority of the agents have the same name for a context, then each agent increases the weight associated with that name. If an agent does not have that name in its list, then the name and a weight is added to the list. All the weights associated with names which are not a majority name are decreased in each of the agent's lists. In the case where there is no majority name, the weight each agent has associated with their name is used to decide on a majority name. This type of update of the names and the weights means that the current majority name is more likely to be the name used by the agents to label this context. The agents are free to move between different contexts and communicate with different agents in different contexts. Initially, the name each agent has for all the different possible contexts is chosen randomly.

Figure 4.13 shows the results of running the simulation for different numbers of agents and contexts. The average number of times an agent must visit a context before all agents have the same name for the context is plotted. The average is calculated over the number of times each agent visits each context in a single run and then the simulation is run ten times for each number of contexts and agents and the average is averaged over the ten runs. Each curve is for a different number of contexts. Note that the number of agents on the *x*-axis is the $\log_{10}$ value.

The result is straight line curves – which is a very important result. This means that, as the number of agents increases exponentially, the number of times they need to visit a context before reaching common labelling increases in a linear fashion. It is clear that as the number of agents falls below the number of contexts, the time increases exponentially. This can be understood, as the chances of agents meeting to exchange labels in a context decreases considerably as the number of contexts is greater than the number of agents.

4.3.3 Context-dependent User Modelling

Previous work on context-dependent user modelling has mainly focused on knowledge representation (KR) methods. The most common approach has been the use of static rules, which

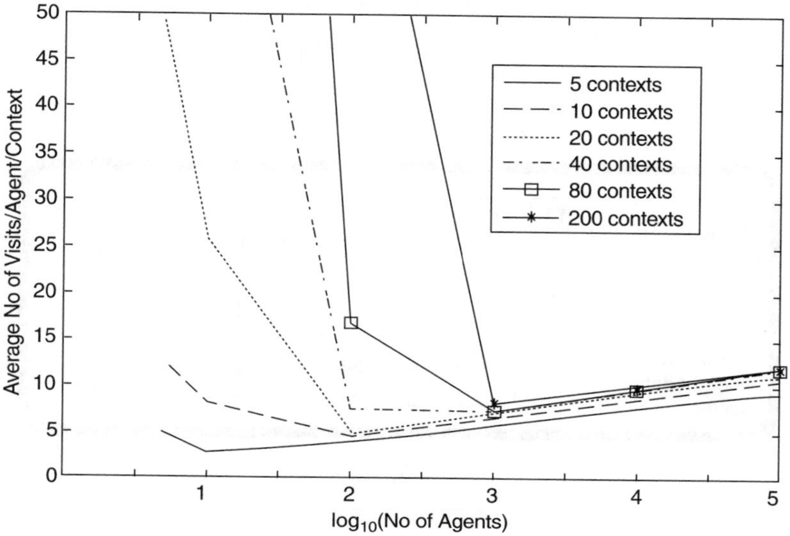

Figure 4.13 The average number of times an agent must visit a context before all agents have the same name for each context for different numbers of agents in the simulation. Each curve is for a different number of contexts. The average is taken over ten simulation runs for each number of contexts and each number of agents.

are usually handcrafted and provided by the application designer (e.g. [101,102]). The so-called preference approach (e.g. [46]), where users can specify application-specific rules, falls under the scope of KR-based methods. In addition to rules, ontology reasoning (e.g. [97]) and case-based reasoning (e.g. [58]) have been suggested. Finally, many context-aware middleware provide system-level support for using KR methods (e.g. [95,114]).

The KR methods suffer from two major problems. First, the techniques do not have a way to cope with uncertainty. Second, KR methods are not usually able to generalise their performance; i.e. to work well in previously unseen situations. However, in context-aware environments, there are various sources of uncertainty (e.g. the uncertainty about the goals of a user and inaccurate sensor signals) and the number of different situations that are relevant to a user might be very large. As a consequence, predictive user modelling seems a promising option. At the moment, however, work on using predictive user models in context-aware settings has been rather limited and all the uses are confined to a single application (e.g. [50]) or to a well-defined spatial area such as a smart home (e.g. [107]) or a smart office (e.g. [81]).

4.3.3.1 Generic Context-dependent User Modelling System

In this section, the generic user modelling system [84] used in the book is described. The system consists of three components: a *Usage Record Provider*, a *Recommender* and a *Profile Manager* (Figure 4.14). In the following each of these components is described in detail.

The Usage Record Provider is a repository that stores information about the behaviour of a user and the context in which the behaviour takes place. For example, the Infotainer (see Chapter 7) stores the selected modality and the corresponding situation of the user. The

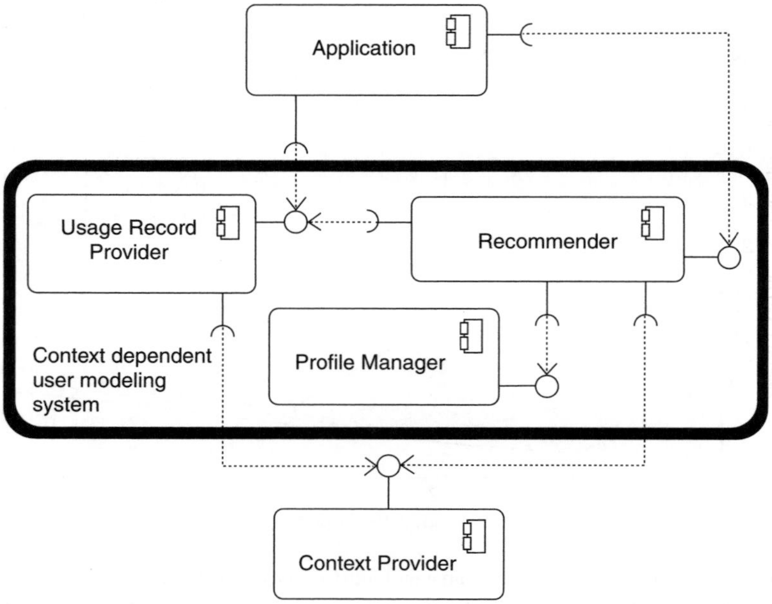

Figure 4.14 Component diagram of the generic user modelling system.

information that is stored in the Usage Record Provider is then used by the Recommender to learn new models and to adapt existing user models.

The entries that the system logs are called Usage Records; an example is shown in Figure 4.15. All context information is contained within the part of the XML that is labelled with the tag *contextElement*. For privacy reasons, some of the context parameters in the example have been removed and obfuscated.

The Usage Record Provider also acts as an entry point for applications and services into the system. Namely, to get useful information from the system, applications must send usage information to the Usage Record Provider. The sent information should contain at least unique identifiers for the user and for the action the user performs. When the Usage Record Provider receives usage information, it checks whether it can enrich the context associated with the usage information. This is done by contacting those Context Providers that are known to contain parameters that are not yet part of the received usage information. This step is especially useful for mobile applications, as it reduces the communication cost of the client; both the Usage Record Provider and the Context Providers typically reside on servers.

The Recommender is responsible for performing all tasks related to user modelling. Thus, the Recommender is responsible for learning and updating user models and making inferences with the learned models. Within the component, the functionalities have been divided into three kinds of modules: *Finders*, *Mappers* and *Reasoners*.

The Finders are responsible for finding components and for the interactions that take place between the Recommender and other components. There is a Finder for each of the components with which the Recommender needs to interact. The reason for separating the

```
<usageRecord
        action="update"
        actor="355023003598706"
        application="Buddy"
        feedback="0.0"
        initiationType="manual"
        recommendationType="action"
        timestamp="2005-12-13T07:51:34.000Z">
  <contextElement>
    <parameter name=”location”>
      <parameter name=”cluster” value=”1”>
        <parameter name="mcc" value="204"/>
        <parameter name="mnc" value="815"/>
        <parameter name="cellid" value="45960"/>
        <parameter name="latitude" value="32.232845916875"/>
        <parameter name="longitude" value="9.8896758573835"/>
      </parameter>
    </parameter>
  </contextElement>
</usageRecord>
```

Figure 4.15 Example of a Usage Record.

interactions with external components is that it offers more flexibility as the Recommender is not tied to a particular implementation of the external components. Thus, one can modify the external components at any stage and the only thing one needs to change in the Recommender is the corresponding Finder.

The Mappers are responsible for mapping structured information into a non-structured (flat) form, and vice versa. Context and behaviour information are often structured (e.g. nested context parameters, as illustrated in Figure 4.15, or references to a tree-like or graph-like ontology of user behaviour), whereas common machine learning algorithms can handle only flat data (e.g. numeric vectors or name–value pairs).

The requirements of the Mappers are two-fold. First, the data formats must be made compatible. Second, the translated data should contain information that makes it as efficient as possible to learn and apply the context-dependent user model. While the first requirement could be handled in an application-independent manner, this is not the case for the second requirement. A good mapping may depend on the application. For example, an application-specific ontology of service categories can be used to map fine-grained identifiers of individual services into higher-level identifiers of service categories; this way the system can make useful predictions with considerably less training data and the predictions generalise to new situations. Furthermore, a good mapping may also depend on the specific machine learning algorithm.

Examples of possible mappings include the following:

- Flattening structured information; for example, translating the record of Figure 4.15 into (cluster=1; mcc=204; . . .; longitude=9.89).
- Feature selection (choosing a subset of input variables); for example, translating the record of Figure 4.15 into (cluster=1).

- Extracting higher-level features in an application-specific manner; for example, using an external ontology of context or behaviour information.
- Technical low-level translations as required by the specific machine learning algorithms; for example, discretising real-valued data to integral values, or mapping class-valued data to Boolean vectors.

The Reasoners are responsible for encapsulating implementations of individual machine learning algorithms into the Recommender. Thus, the Reasoners are the part of the architecture where the actual learning and inference are done. The Reasoners interact closely with the Mappers. Before usage records are used for learning, the Mappers map the context and behaviour information into vectors that are given to the Reasoners. Similarly, once the Reasoners are used for inference, the Mappers take the results of the inference and map them so that the applications can understand the results.

To have a clear separation between learning and inference, separate interfaces for components that offer learning functionalities and for components that offer inference functionalities have been specified. The separation of interfaces is crucial as in a mobile device one seldom has enough resources to run the learning phase. However, once one has learned a model, it may be possible to run the inference stage even on the phone. Another advantage of the clearly separated interfaces is that this further improves the reusability of the implementations of individual algorithms.

Once a Reasoner (that implements the learning interface) has learned a new model, the models are stored into a Profile Manager. The motivation for this is improved scalability and better distribution of functionalities. Furthermore, by storing the models in a Profile Manager, the user models become part of the user profile, and thus all relevant preference-related data is stored in a single place.

In the current reference implementation of the Recommender, *tree augmented naive Bayesian classifiers* (TAN) (e.g. [40]) and *rule-based reasoning* are supported. The algorithm that is used for learning the TAN classifiers is described in [40]. The so-called quasi-Bayesian algorithm is used as the inference method.

Rule-learning uses the Ripper algorithm [27] to express the user behaviour patterns in rules. The output of the algorithm is then represented in RuleML [16], which is a rule markup language designed for the Semantic Web. Each learned rule is assigned a weight; by using the weights of the rules, the Recommender can produce a list of recommendations and their scores. Rule-based inference uses a general interface called *Rule-based Context Reasoning Interface* (RuBaCRI). The RuBaCRI provides a common interface for inference. Behind this interface one can integrate and include appropriate rule-engines. It serves as a dispatcher and selects the corresponding supported rule engine to carry out inference based on the given rule base. Currently, a specific rule engine is chosen and is improved to support negation that is needed to process the learned rules. The rule engine OO jDREW [86] is an open-source deductive rule engine that supports RuleML reasoning. However, the bottom-up rule engine, provided by OO jDREW, does not support usage of negation in inference. Therefore the RuBaCRI-integrated OO jDREW has been modified to support usage of negation-as-failure in the rules generated and stored in the user model.

The third component of the system is the Profile Manager, and a detailed description of this can be found in [28] and in Chapter 5.

Table 4.3 Summary of the used datasets and results of experiments

Dataset	Records	Attributes	Classes	Accuracy	Avg. log-loss
Vote	232 (435)	16	2	94.4%	0.165
Zoo	101	16(17)	7	98.0%	0.121

In order to evaluate the quality of the recommendations that the system defined above produces, and to ensure that the system works properly, two datasets from the UCI machine learning repository [78] were selected. The data was exported into a test system by mapping the data records into usage records. After this, a leave-one-out cross-validation was used to test the classification accuracy of the system. In the experiments, the TAN was used and since the system is able to make probabilistic predictions, also the logarithmic loss $-\log p_i$ was measured, where p_i is the probability assigned to the correct class.

The datasets and the experimental results are summarised in Table 4.3. The first dataset, Vote, consisted of 435 records. The dataset contains many records with missing values. In this experiment, only the part of the dataset with complete information was considered, resulting in 232 records. The classification task is a binary task and the system used was able to achieve 94.4% accuracy on the dataset. The second data, Zoo, consisted of 101 records, each of which had 17 attributes. However, since one of the attributes was a unique identifier, it was removed. On this dataset, the system achieved an accuracy of 98%.

4.4 Proactive Service Provisioning

4.4.1 Introduction

As mobile terminals are becoming increasingly powerful, and the availability of data connections increases, mobile phones can become the terminals on which networked services are delivered. This is not a new trend. Mobile phones are already used as terminals to deliver all kinds of different data. For example, it is possible to use mobile phones to book train rides in Germany [30], or to receive SMS notifications about the delay of flights and to check for both train and flight schedules.

Given that the number of services increases, a crucial problem that is emerging is how to find these services. Right now the services are advertised through written material such as brochures and web pages that are somehow delivered to *Service Requesters*. For instance, the German railway system advertises on its web page the ability to purchase tickets using a mobile phone. However, with the increase of services to be delivered, a better discovery mechanism is required.

In the Mobile Service Architecture, *Proactive Service Provisioning* (PSP) tackles this problem; it is a general architecture for service discovery and provisioning that provides the starting point on which a platform for service delivery can be built.

The approach adopted is based on the *Service Oriented Architecture* (SOA) [88], which provides a general framework for service delivery and invocation. Essentially, SOA identifies three parties that are required in the delivery of a service. The first party is the Service Provider

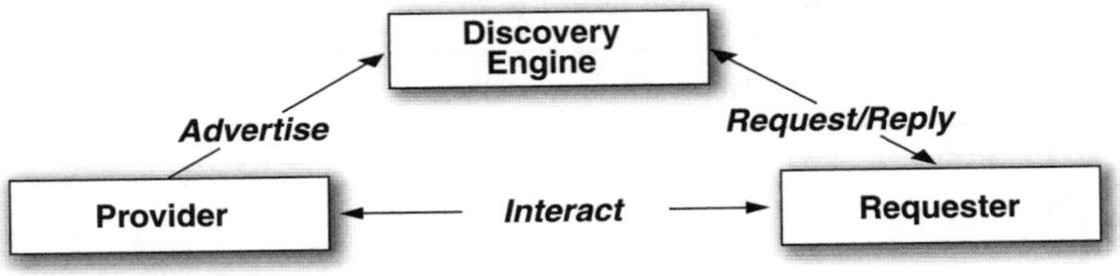

Figure 4.16 Structure of the Service Oriented Architecture.

(SP) that delivers the service, the second party is the Service Requester that looks for the service, and the third party is the *service discovery* mechanism that matches the needs of the requester with the appropriate SP. Following SOA, the structure of a transaction is displayed in Figure 4.16, where the SP advertises its services in the Discovery Engine, and the Requester decides which services it needs and searches for them by means of the Discovery Engine. The Discovery Engine replies with a list of services that fit the requirements of the Requester, and finally the Requester selects the service of a preferred SP and interacts with it.

SOA is widely used to deliver services in business-to-business (B2B) environments, where the Service Provider and the Service Requester correspond to services of well-structured organisations. Furthermore, this usage is based on a growing number of web services standards that implement different aspects of SOA. The result is that the programmers' efforts in delivering services under the SOA framework are greatly facilitated, and a number of interoperability issues are automatically resolved.

Because of SOA's impact on B2B systems, it was an obvious starting point on which to build a service delivery mechanism for mobile environments. Indeed, one could naively claim that Service Providers are already available (the German railroad system was mentioned above as an example), mobile users are of course the Service Requesters, and discovery engines are the only thing that is missing to deliver services in mobile environments. However, this picture proves to be quite simplistic. First, Requesters are not skilled programmers that have the ability to look for services in existing Discovery Engines, and the user may not even be expecting any service to be available at a given time. Second, services for mobile environments may be delivered under very different conditions. Some may be delivered via Bluetooth and therefore be limited to a small physical space. The delivery of other services may be triggered by passive technologies such as Radio Frequency Identification (RFID) or Near Field Communication (NFC) that require a direct action of the user.

As a consequence, the service provisioning architecture presented here extends SOA in different ways. One needs to distinguish the different delivery mechanisms and realise that they require different discovery mechanisms; therefore a distinction between *Registered Services* and *Self-promoting Services* is needed. The Registered Services are typically delivered through the Internet and discovered through a registry mechanism. The discovery of these services is based on the SOA model. In contrast, Self-promoting Services do not use a registry, but the proximity of the user, to advertise themselves. For example, a Bluetooth-based service may promote itself to any Bluetooth device that passes within its range.

Another change produced within the service provisioning architecture presented here is the origin of the service request. Current implementations of the SOA discovery mechanism, such as UDDI, require Requesters to express their needs consistently with very formal specifications.

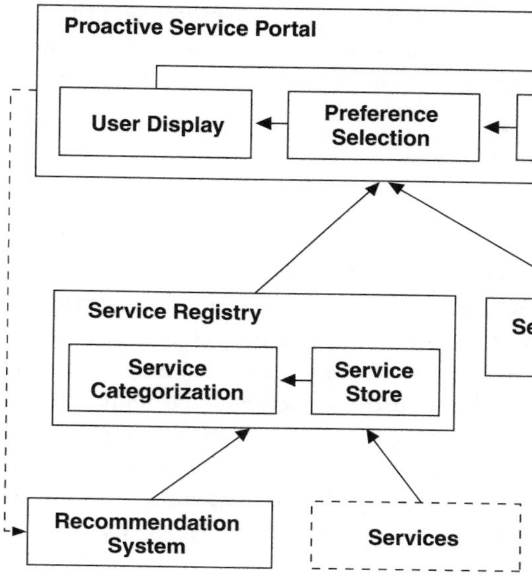

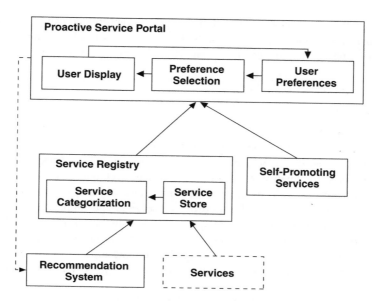

Figure 4.17 Architecture of the Proactive Service Portal.

Whereas this requirement is easily satisfied by hard-coded programs that look for services, it does not hold true in a mobile environment, since one cannot expect the user to query a registry to find the desired service. Rather, the infrastructure needs to 'guess' the needs of the user and propose the most appropriate services for delivery.

The resulting architecture is displayed in Figure 4.17. The user is placed in direct contact with a *Proactive Service Portal*, which displays services that are gauged to be of interest. The discovery of these services follows different paths. One of them is based on the SOA proposal in which services register with the Service Registry as they would do in SOA; but, whereas in SOA the queries come directly from the Requester, in the Mobile Services Architecture they come from the Recommender: a recommendation system that exploits the history of the user's service usage to find the best service to be delivered at a given time. Alternatively, discovery is based on self-promotion: services promote themselves directly to the service portal so that the user can be notified of their availability.

The following sections discuss how the different components of the Proactive Service Provisioning mechanism work. Specifically, in Section 4.4.2, the service categorisation mechanism, which includes the representation of services and the matching mechanism to verify whether a request for a service is satisfied by a given service advertisement, is discussed. The service recommendation mechanism discussions in Section 4.4.3 and the Self-promoting Services discussions in Section 4.4.4 complete the section.

4.4.2 Service Categorisation

The Service Categorisation function of the PSP provides the ability to retrieve services from a repository on the basis of a given classification schema. Specifically, the service categorisation process requires two functionalities: first, the services are represented using the classification schema; second, the service retrieval exploits the representation schema and its underlying logics to find the best match to a given request. These issues are covered in the following two subsections.

4.4.2.1 Representation

The representation schema should specify the capabilities of the service. More precisely, the capabilities of a service are defined as the function that it computes; for example, providing weather forecast or selling books. In addition, capabilities are qualified by a number of properties that assess the quality of the provided service. Such quality properties include the speed of response and the media quality and resolution, and they may extend to computational and bandwidth limitations, the media used by the service, and more; e.g. whether data is received as video streaming or text messaging, and whether the service is delivered on a per-use model or with a flat fee. Finally, restrictions may be imposed on the use of the service by specifying what properties the user and the device need to have to use the service. For example, the user may need to be a member of a club, or the device screen should have a required resolution.

Since services have many different aspects, there is the need for a flexible representation schema that allows the Service Providers to express all the important features of the services, and the Service Requesters to express exactly what kind of services they expect. To support such flexibility, any service representation schema should be based on an inference mechanism that allows the extraction of knowledge that is implicitly stated in the description, but not explicitly exposed, and the inference should allow the Service Requesters to 'reason' about the service specifications that they receive. On the other side, any computational mechanism should also be effectively computable. In practice, any representation schema should be based on the DL fragment of OWL [10], because OWL DL is a very expressive language that also guarantees a decidable and sound inference mechanism. In addition, OWL DL inference engines are highly optimised and can be applied in many practical cases.

There are essentially two ways to express services using OWL. The first is to provide a taxonomical representation of services in which each class corresponds to a service type.[1] The second way is to use the OWL-S service specification, and in this case the OWL-S Profile [73]. The two representation schemata strike different trade-offs that can be exploited to create a very powerful service representation paradigm.

Use of a taxonomy of services allows the enumeration and the naming of all possible service types, and their organisation in a class/subclass taxonomy. For the domain of Mobile Services Architecture, a fragment of the ontology is shown in Figure 4.18. This type of representation is very convenient, because it is easy to precisely pinpoint what type of service is provided or required. Furthermore, this type of representation schema greatly facilitates the discovery process, because for a given type of service that is desired, the only task of the discovery process is to collect all the instances of services of that class. If more complex queries are

[1] The term 'taxonomy' usually refers to impoverished ontologies in which only subclass relations are used. Here it is used to indicate a representation that results in a tree of service categories. However, in addition to the subclass relation, here it is allowed the use of OWL properties that are usually not represented in a taxonomy. Therefore the representation results in an ontology in the sense of a conceptualisation of a domain expressed as a set of logic axioms.

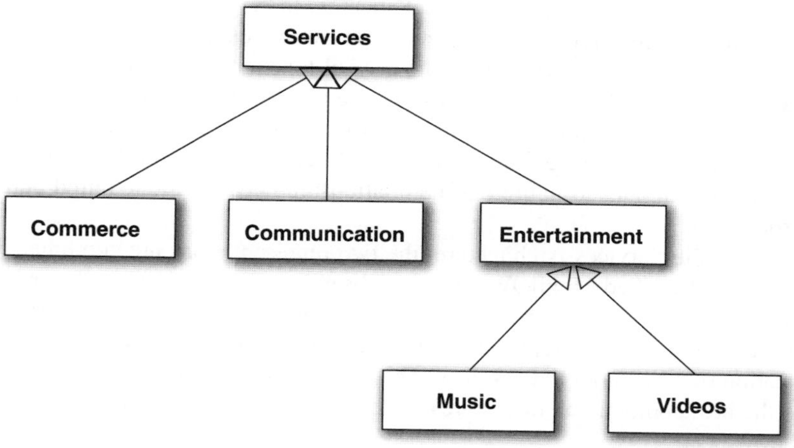

Figure 4.18 A fragment of a taxonomy of services.

needed, it is possible to exploit the subsumption reasoning that is provided by OWL inference engines. For example, a Service Provider may specify the general features of a service, such as that it provides music outputs of a given format and so on, and this service will be automatically classified within the music class. There is no need for the Service Provider to specify what the classification should be.

However, the taxonomical representation has two drawbacks. The first one is some serious representation issues with OWL. Typically, it is easy to push the representation of services to OWL-Full, which is very problematic since the required inference becomes very expensive. Second, it is very difficult to support goal-directed reasoning: the Requester cannot say that what she wants are services that provide a given type of information, such as music streaming.

The OWL-S Profile provides a canonical representation of capabilities of services. The viewpoint of the OWL-S Profile is that a service can be described at two levels of abstraction: at the *functional level* the service can be described by the transformation it produces; at the *non-functional level*, a number of parameters may be used to specify the quality of the service provided, and additional restrictions on the service. The functional level is described by the information transformation from the inputs that the service requires, the outputs that it produces, the state transformation that is defined by a set of preconditions that should be satisfied when the service is executed, and the set of effects that result when the service completes.

The OWL-S service description has advantages and disadvantages that are complementary to the taxonomy representation discussed above. The first advantage is that the functional level of OWL-S does not depend on any taxonomical representation; therefore any service can be described even if there is no corresponding service type in the taxonomy. Another advantage of the OWL-S Profile is that it represents explicitly the data that the service manipulates. This greatly facilitates the representation of the requirements of the Requester, since the Requester can simply specify the data that it expects from the service. There is no need to abstract it to the type of service.

The disadvantages of OWL-S are also complementary to the disadvantages of the taxonomical representation. Since OWL-S does not rely on taxonomies to describe services, it is not possible to name the type of service, as in the case of taxonomical representations. Furthermore, the automatic classification algorithms that are used for OWL do not apply. Therefore, the discovery algorithms need to define a similarity measure to recognise when the service provided is similar to the service requested. Often, to compute this similarity measure, they use OWL inference engines to figure out the relationship between the required output and a service output.

Within the Mobile Services Architecture, the two representations are combined. For this, a version of OWL-S, called MobilOWL-S, specialised to represent services for mobile computing, is defined. MobilOWL-S is based on OWL-S, but extends it with a representation of the type of service and with a representation of a wide range of parameter types, and with the specification of a taxonomy of services. The Service Provider and the Service Requester in turn use the best combination of features to describe what services they provide and what services they expect.

4.4.2.2 Inference

The service description as discussed above can be used with two goals in mind: first, to represent the services that are provided to mobile users; and second, to represent the services that are requested by them. In the latter case the mobile user can represent the 'ideal' service that she would like to use and then ask a registry to find similar services.

The responsibility of the registry is to select the advertised services that match the required services. An additional confounding factor is that there may not be any service that matches exactly the required service, but there are a number of similar (though somewhat different) services. The algorithm adopted here attempts to perform the abstraction that is required to deal with finding similar enough services. Essentially, the algorithm first attempts to classify the requested service using OWL classifiers. This classification allows one to know exactly the type of service requested. Exploiting this knowledge, one can therefore retrieve all services of the same type, but also services that are more general, and more specific. Furthermore, one can retrieve services that are somewhat different, but share some features with the requested service. In addition to exploiting the OWL representation, one can also exploit the OWL-S representation looking for all the services that provide the information required by the requester. Here again one can look for more general, more specific and similar services.

The result of the discovery algorithm used here is that one can find services that are similar to what was requested, even though they are not exactly the same. Furthermore, one can support very efficient indexing mechanisms to find the candidate services very quickly, making the service discovery a very efficient process in practice.

4.4.3 Recommendation-based Service Discovery

Service categorisation makes it possible to organise the vast number of services that may be available on the Internet, but finding the relevant services still requires knowledge of the ontologies used to describe the services, the ability to express the request in a very formal way, and a considerable amount of user interaction. These requirements are not satisfied in mobile computing, where the user has neither the knowledge nor the time to formally describe the services that she needs. The goal of recommendation-based service discovery is to make it easier to find and activate the services that are relevant in the current situation.

Ideally, the user interface should display only a brief list of relevant services from which she can pick up the ones that she needs. However, the relevance of a given service depends on both the particular user and the current situation. Thus, to be able to recommend relevant services, a user model that captures the interests of each user is needed.

To construct the user models, an assumption is made that the recent behaviour of a user in a similar context reflects the user's current needs. Therefore, in recommendation-based service discovery, the generic framework of context-dependent user modelling, as described in Section 4.3.3, is used. This framework makes it possible to (1) learn context-dependent user models based on the user's past behaviour, (2) store the user models, and (3) use the models to provide recommendations.

As the recommendations are based on the previous behaviour of the user, care should be taken as the number of different contexts and the number of different services may be huge, so one cannot assume that enough past data on any given service in any given situation is available. Furthermore, recommendations are needed also in completely new situations (e.g. travelling in an unfamiliar country) and in the case of completely new services (e.g. a new competing service).

To make the user models generalise to new contexts and new services, the higher-level structure of the data is exploited. First, one should make use of higher-level context clusters instead of raw context data. This way one can reason on types of context rather than single instances eliminating irrelevant details. Second, one should make use of types of services instead of individual services, and then use the service categorisation mechanism described in Section 4.4.2 to map from the type of service to the concrete service that will be proposed to the user.

4.4.3.1 Components and Data Flows

In recommendation-based service discovery, the following generic components that were introduced in the contextual user-modelling framework (Section 4.3.3) are used: the *Usage Record Provider*, the *Profile Manager* and the *Recommender*. Here, the Usage Record Provider is sometimes referred to as the Basic Usage Record Provider (BURP). The overall architecture of the recommendation-based discovery is shown in Figure 4.19.

In the following, the process of recommendation-based service discovery is presented, starting from the components that collect data on the behaviour of the user and moving towards the components that provide recommendations using those models. Here, the Proactive Service Portal (see Chapter 7 for more information) is used as an example of a user interface that both gathers behavioural information and provides the user with recommendations.

When the user activates a service in the user interface, the Portal constructs a usage record (see Section 4.3.3) and sends it to the Usage Record Provider. In this case, the usage record contains the category of the service. If needed, the Usage Record Provider queries the relevant Context Providers in order to complete the usage record with contextual information, and stores it in a database. The Recommender queries the Usage Record Provider, fetches the latest behavioural information, updates the user model, and stores the new model in the Profile Manager. See Section 4.3.3 for more information on these parts of the process.

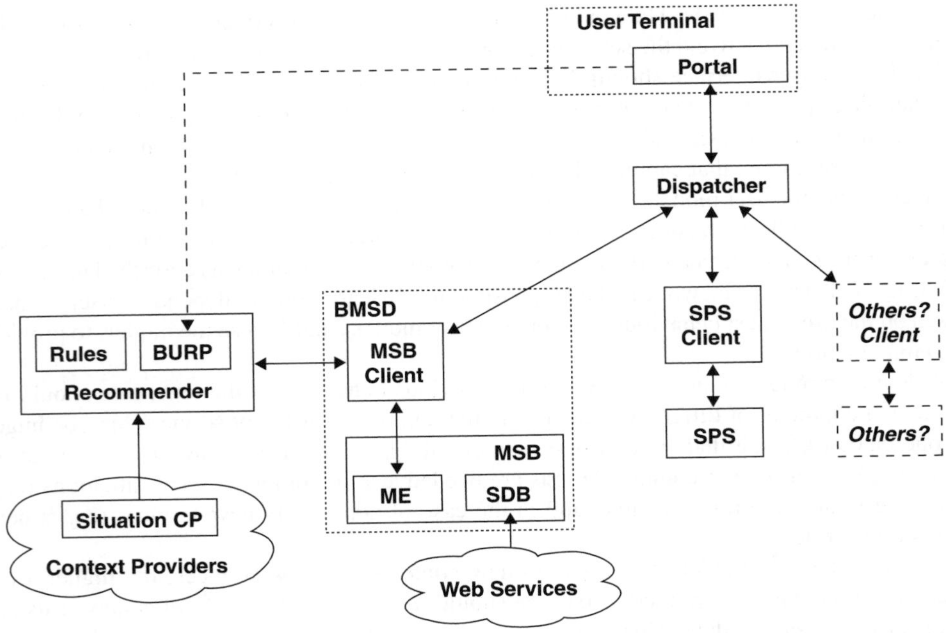

Figure 4.19 Architecture of the recommendation-based service discovery.

To construct a list of recommended services, the Portal contacts the *Dispatcher*, which provides a unified view of service discovery mechanisms. In the case of recommendation-based service discovery, the Dispatcher queries the *Best Matching Service Discovery* (BMSD) component. The BMSD fetches a list of recommended service categories from the Recommender, matches the service categories against the database of available services, and returns a list of relevant services. In the end, the list of services is presented to the user.

The rest of this section details the functionality and interactions of the Dispatcher and BMSD.

4.4.3.2 BMSD Architecture

The role of the BMSD is to match the service requests that come from the Dispatcher with the service advertisements that come from services available in the system. The BMSD is composed of the *Mobile Service Broker* (MSB) and the *MSB Client*. The MSB is responsible for collecting the advertisements and matching them against the service recommendations that are received from the Recommender. The MSB Client provides a level of abstraction interfacing the MSB to the rest of the system.

The MSB provides essentially two functionalities. First, it receives the advertisements of the services that are available in the system and stores them in the *Services DB* (SDB). These advertisements are described using the MobilOWL-S service description as discussed in Section 4.4.2. The second component of the MSB is the *Matching Engine* (ME), which selects the advertisements from the SDB that match the types of service recommended by the

Recommender. Such matching is performed using the *Mobile Context Ontologies* and *Service Ontologies* to classify the services first and to retrieve them later. This process is described in more detail in the following subsection. Only services that are available are matched during the discovery process; when a service is no longer available, its advertisement is removed from the SBD and it will not be found by the ME.

4.4.3.3 Control Flow within the BMSD

Upon receiving a request of service from the Dispatcher, the service discovery process starts. The Dispatcher calls the MSB Client and provides user identifier information. The client uses this information to create an XML compliant file and then registers a subscription of the user with the MSB. At this point the MSB generates a new thread dedicated to the subscribing user and returns a subscription identifier to the Dispatcher. The thread starts a cycle that periodically (e.g. each minute) interrogates the Recommender for services that may be useful to the user. This query results in a list of categories of services that are likely to be useful to the user. Upon receiving the recommendation, if the list of categories is changed with respect to the one advertised in the previous cycle, the MSB transmits the list to the Matching Engine, which selects new services to propose to the user. If the user's context has not changed and the output of the Recommender remains the same, the cycle is restarted, avoiding the forwarding of redundant service lists.

The Matching Engine monitors the list of available services in the SDB and it loads all the context and service ontologies that describe them. The context and service ontologies are used as background knowledge of the Matching Engine. Specifically, they are loaded in a *Description Logics classifier*,[2] which first derives the relations between the concepts specified in the ontology. Such inference, technically named *TBox classification*, results in the final taxonomy tree of all concepts. Once the concept inferences are completed, the descriptions of the services in the system are classified using the concepts in the ontology. During such classification, each service description is added as an instance of the service type to which it belongs. For example, a service that reports weather information may be classified as an *Environment Context Provider*. The process of identifying the type of services in the system is technically referenced as *ABox classification*. As a consequence of this process, the Matching Engine stores an up-to-date indexing of the available services, which reduces the service matching to little more than a table lookup.

The matching process is performed whenever the Matching Engine receives a list of service categories from the Recommender. This list is ordered by importance: the most relevant category first, followed by the less relevant categories depending on the user's current situation.

For each category of the ordered list of categories above, the Matching Engine performs the following three operations:

1. Look for all the services that are indexed as instances of the requested service.
2. Look for all the services that are instances of a category that is more generic than the category requested.
3. Look for all the services that are instances of a category that is either more specific, or intersects the category requested.

[2]In the current reference implementation of the MSB, the Description Logics classifier used is Pellet.

Each one of these three steps results in an ordered list of services, which mimics the ordered list provided by the Recommender. When the list is ready, the Matching Engine returns it to the Dispatcher. When the user or application is no longer interested in receiving updated lists of services, the Dispatcher calls the unsubscribe method of the BMSD Client. The database is contacted using the subscription identifier as a parameter and the related XML document is deleted, causing the end of the MSB activities in the service discovery process.

4.4.4 Self-promoting Services

Self-promotion of Services (SPS) is a newly introduced paradigm for service discovery in which services are explicitly promoted to the user via local radio networks or Personal Area Networks (PANs). From the discovery point of view, the specific characteristic of these services is that they cannot be covered by broker-based discovery mechanisms. The main reason behind the need of SPS is that services with such limited special scope are simply not able to reach a service register such as the MSB. Furthermore, even when they can be registered, the registration has little or no use, since the use of the service requires the user to be in a specific location. Discovering the service anywhere else would be useless.

There are different characteristics which make a service suitable for a self-promotion based discovery, rather than for a brokered one. For example, if the service is available only through local networks in a peer-to-peer manner, SPS could be used for its discovery. Another case would be for the discovery of services that make sense only in the direct proximity of the user – for example, if the service is bound to a specific hardware (such as a printer or a vending machine) or if it directly relates to the local environment (such as a local attraction or a building). If a service has a significantly short lifecycle or is consumable just one time, it is inefficient to publish it in global service repositories, so SPS is appropriate for the discovery process. Other services that are suitable to be discovered via SPS are services that should be detectable only locally because of privacy or security concerns. Moreover, the service (Service Provider) which is dynamically changing its location could also be advertised via SPS. For services provided by normal users (e.g. an online game), for whom it is too much effort to register it globally, SPS provides a very simple means to make the service anyway known to the people close by.

Beyond those application realms, SPS can of course also serve as a mechanism to augment global service repository and at the end to make the services be registered and detectable by other service discovery technologies.

4.4.4.1 Realisation and Architecture

From the architecture point of view, the SPS consists of three major components: (1) the service promoter, (2) service detection and republishing, and (3) SPS management (Figure 4.20). Typically, the service promoter is either another active node like an embedded device or just a passive tag like an RFID or a (2D) barcode. Whereas the service detection together with the republishing resides in the user's terminal, the SPS management is instead realised as a network-hosted service. For discovery, SPS makes use of a modified UPnP discovery protocol[3] based on unicast messages; furthermore, since services are transient and a service

[3]UPnP is designed for multicast traffic. Here, it has been modified to be unicast enabled, which allows to re-publish the service advertisement across network domain boundaries.

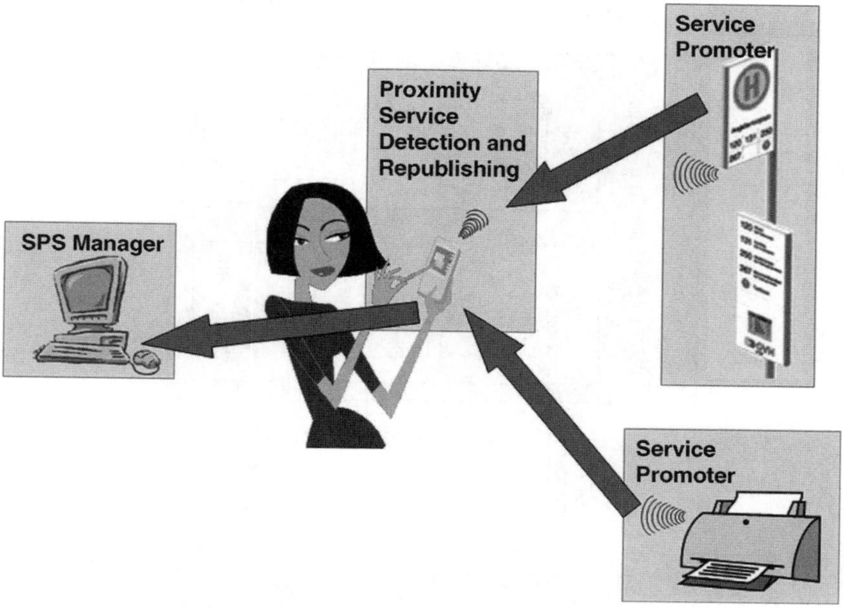

Figure 4.20 SPS components.

discovered at a given time may not be available at a later time, SPS includes a refresh process that maintains the list of advertisements up to date.

Figure 4.21 depicts the architecture of the SPS framework. It describes the three device components and its modules for the case of an active promoter. Details are as follows.

- *Service promoters*. The service promoters can be active or passive. The *active* promoters are devices which permanently promote services by setting up local radio network connections with the service detection entity, which is hosted in the user's end device. Normally a service promoter is hosted by an embedded device, but normal stationary or mobile computers, or even mobile devices, can host a service promoter. The communication between the promoter and the user device is based on the applied UPnP-based SPS protocol. The *passive* promoters are just information sources which have to be scanned actively by the user's device. As passive promoters, RFID tags, barcodes etc. can be used. The user's device requires specific hardware (e.g. an RFID reader or a digital camera and barcode recognition software); the SPS protocol cannot be used. The user needs to specifically trigger the detection of each promoter.
- *SPS detection and republishing*. The SPS detection and republishing system is hosted by the user's device. The detection is, as already described, based on the local radio network interface or on specific hardware for passive promoters. Once a service promotion is detected, the information about the service is processed and handed over to the republishing component. This provides the information via UPnP to the SPS management. The specific characteristic of the republisher is that it acts as a UPnP sink ('Control Point') towards the promoter and as a UPnP source ('Device') towards the SPS management.

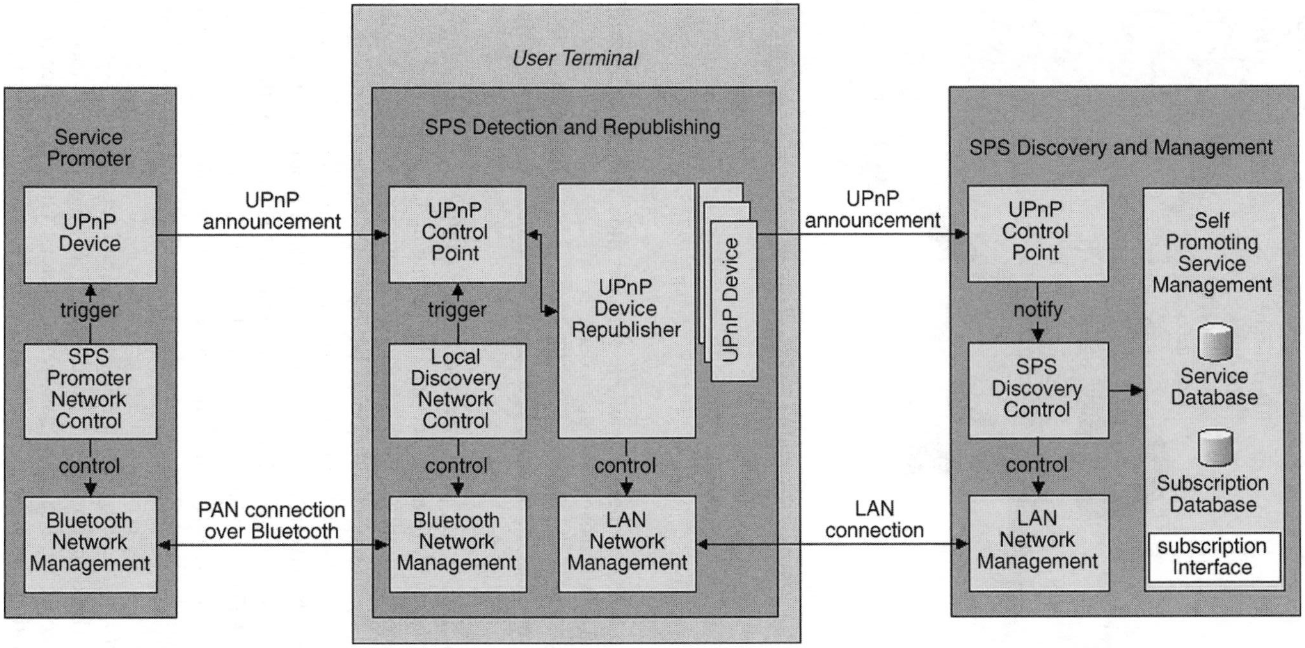

Figure 4.21 SPS architecture.

- *SPS discovery and management*. The SPS management is implemented as a network-based service, which provides information about discovered SPS services. In addition to that, it runs a standard UPnP discovery engine, to which the SPS republisher connects. The SPS Manager also hosts sophisticated filter mechanisms, which considers restrictions from user preferences or service requirements.

4.4.4.2 SPS Protocol

To be compliant with existing discovery frameworks, SPS implements the discovery-related aspects of the UPnP protocol. Here the detected services are treated as 'devices' in UPnP notation. In SPS, the UPnP mechanism experienced the following modifications:

- *Unicast UPnP*. Since the SPS republishing is typically performed across network domain boundaries, the multicast model of UPnP had to be modified towards a unicast-enabled UPnP communication.
- *Lifecycle management*. With respect to the dynamic connectivity of the service promoters, a new lifecycle management replaced the traditional one, which foresees a time-controlled frequent 'alive' notification. In SPS, the refresh notification is performed every time when a connection is established. A timeout mechanism was applied to the UPnP sink to handle in an intelligent manner a missing refresh message.

With UPnP, the SPS service discovery can be embedded in an external service discovery framework.

4.5 Conclusions

This chapter has concentrated on context management and proactive service provisioning technologies to enable fully context-aware services with a certain proactive behaviour. The central results of the presented work are the *Context Management Framework* (CMF) that allows the component-based development of such applications, a recommender framework with algorithms for reasoning on context information, as well as a framework for proactive service provisioning. The work includes novel methods and paradigms for context representation, context ontologies, as well as methods for context analysis and reasoning. The presented *Proactive Service Provisioning* framework builds on CMF in adding rich service models grounded in common context representations to facilitate smart and proactive services. Services and applications that build on the presented technologies are included in Chapter 7. Applications that fully exploit the presented foundational technologies include the Proactive Service Portal, ContextWatcher as well as the Personal Context Monitor.

4.6 Acknowledgements

The following individuals contributed to this chapter: Agathe Battestini (Nokia, Finland), Adrian Flanagan (Nokia, Finland), Patrik Floréen (University of Helsinki, Finland), Stefan Gessler (NEC, Germany), Johan Koolwaaij (Telematica Instituut, The Netherlands), Eemil Lagerspetz (University of Helsinki, Finland), Sian Lun Lau (University of Kassel, Germany),

Marko Luther (DoCoMo Euro-Labs, Germany), Miquel Martin (NEC, Germany), Jean Millerat (Motorola SAS, France), Bernd Mrohs (Fraunhofer FOKUS, Germany), Petteri Nurmi (University of Helsinki, Finland), Massimo Paolucci (DoCoMo Euro-Labs, Germany), Julien Robinson (Alcatel-CIT, France), Jukka Suomela (University of Helsinki, Finland), Claudia Villalonga (NEC, Germany), and Matthias Wagner (DoCoMo Euro-Labs, Germany).

References

[1] Abowd G.D., Atkeson C.G., Hong J., Long S., Kooper R. and Pinkerton M.: 'Cyberguide: a mobile context-aware tour guide'. *Wireless Networks* Vol. 3, No. 5, pp. 421–433, 1997.

[2] Ashbrook D. and Starner T.: 'Using GPS to Learn Significant Locations and Predict Movements across Multiple Users'. *Personal and Ubiquitous Computing*, Vol. 7, No. 5, pp. 275–286, 2003.

[3] Baader F., Calvanese D., McGuinness D., Nardi D. and Patel-Schneider P.: *The Description Logic Handbook: Theory, implementation and applications*. Cambridge University Press, Cambridge, 2003.

[4] Bao L. and Intille S.S.: 'Activity Recognition from User-annotated Acceleration Data'. Proceedings of the 2nd International Conference on Pervasive Computing (PERVASIVE 2004, Vienna, Austria, April 2004), *Lecture Notes in Computer Science 3001*, Springer-Verlag, Berlin, 2004, pp. 1–17.

[5] Barton J.J. and Vijayaraghavan V.: 'UBIWISE, a Ubiquitous Wireless Infrastructure Simulation Environment'. Technical Report, HPL-2002-303, Hewlett-Packard Laboratories, Palo Alto, 2002.

[6] Barton J.J. and Vijayaraghavan V.: 'UBIWISE, a Simulator for Ubiquitous Computing Systems Design'. Technical Report, HPL-2003-93, Hewlett-Packard Laboratories, Palo Alto, 2003.

[7] Battestini A. and Flanagan J.A.: 'Modelling and Simulating Context Data in a Mobile Environment'. Proceedings of the Workshop on Context Awareness for Proactive Systems (CAPS 2005, Helsinki, Finland, June 2005), Helsinki University Press, 2005, pp. 127–136.

[8] Battestini A. and Flanagan J.A.: 'Analysis and Cluster-based Modeling and Recognition of Context in a Mobile Environment'. In 2nd International Workshop on Modeling and Retrieval of Context (MRC2005, IJCAI Workshop, Edinburgh, Scotland, Aug 2005).

[9] Bechhofer S.: 'DIG 2.0: The DIG Description Logic Interface'. "September 9, 2006 'DIG 2.0: The DIG Description Logic Interface'. September 9, 2006. Online: dig.cs.manchester.ac.uk.

[10] Bechhofer S., van Harmelen F., Hendler J., Horrocks I., McGuinness D., Patel-Schneider P. and Stein L.A.: 'OWL Web Ontology Language Reference'. W3C Recommendation, The Worldwide Web Consortium, February 2004.

[11] Bechhofer S., Liebig T., Luther M., Noppens O., Patel-Schneider P., Suntisrivaraporn B., Turhan A.-Y. and Weithöner T.: 'DIG2.0: Towards a flexible interface for description logic reasoners'. Proceedings of the OWL: Experiences and Directions Workshop at the 5th International Semantic Web Conference ISWC'06, (Athens, Georgia, November 2006).

[12] Bechhofer S., Möller R. and Crowther P.: 'The DIG Description Logic Interface: DIG/1.1'. Proceedings of the 2003 International Workshop on Description Logics (DL'03, Rome, Italy, June 2003).

[13] BeTelGeuse homepage. Online: www.cs.helsinki.fi/group/acs/betelgeuse/.

[14] Böhm S., Luther M., Koolwaaij J. and Wagner M.: 'ContextWatcher: Connecting to places, people, and the world', Proceedings of the Demo Track of the 5th International Semantic Web Conference (ISWC'06, Athens, Georgia, November 2006).

[15] Böhm S., Luther M. and Wagner M.: 'Smarter Groups: Reasoning on qualitative information from your desktop'. Proceedings of the 1st Workshop on the Semantic Desktop at the 4th International Semantic Web Conference ISWC'05 (Galway, Ireland, November 2005), CEUR-WS Publication, Vol. 175, pp. 276–280.

[16] Boley H., Tabet S. and Wagner G.: 'Design Rationale of RuleML: a markup language for semantic web rules'. Proceedings of the 1st International Semantic Web Working Symposium (SWWS-1, Stanford University, California, July/August 2001), 2001, pp. 105–113.

[17] de Bruin J. and Fensel D.: '*Owl-*'. WSML Deliverable D20.1, 2005.

[18] Bylund M. and Espinoza F.: 'Using Quake III Arena to Simulate Sensors and Actuators when Evaluating and Testing Mobile Services'. Proceedings of the CHI 2001 Conference on Human Factors in Computing Systems (CHI 2001. Seattle, Washington, March/April 2001), ACM Press, 2001, pp. 241–243.

[19] Bylund M. and Espinoza F.: 'Testing and Demonstrating Services with Quake III Arena'. *Communications of the ACM*, Vol. 45, No. 1, pp. 46–48, 2002.
[20] Carroll J., Dickinson I., Dollin C., Reynolds D., Seaborne A. and Wilkinson K.: 'Jena: Implementing the semantic web recommendations'. Proceedings of the 13th International World Wide Web Conference, (WWW2004, New York, New York, May 2004), ACM Press, 2004, pp. 74–83.
[21] Chen H., Finin T. and Joshi A.: 'The SOUPA Ontology for Pervasive Computing'. In Tamma V., Cranefield S., Finin T.W. and Willmott S. (eds), *Ontologies for Agents: Theory and experiences*. Springer-Verlag, Berlin, 2005.
[22] Chen G. and Kotz D.: 'A Survey of Context-aware Mobile Computing Research'.Dartmouth Computer Science Technical Report TR2000-381, Hanover, New Hampshire, 2000.
[23] Chen M.Y., Sohn T., Chmelev D., Haehnel D., Hightower J., Hughes J., LaMarca A., Potter F., Smith I. and Varshavsky A.: 'Practical Metropolitan-scale Positioning for GSM Phones'. Proceedings of the 8th International Conference on Ubiquitous Computing (UbiComp 2006, Orange County, California, September 2006), *Lecture Notes in Computer Science 2406*, Springer-Verlag, Berlin, 2006, pp. 225–242.
[24] Cheng Y.-C., Chawathe Y., LaMarca A. and Krumm J.: 'Accuracy Characterization for Metropolitan-scale Wi-Fi Localization'. Proceedings of the 3rd International Conference on Mobile Systems, Applications, and Services (MobiSys, Seattle, Washington, June 2005), ACM Press, 2005, pp. 233–245.
[25] Christensen E., Curbera F., Meredith G. and Weerawarana S.: 'Web Services Description Language (WSDL) 1.1'. W3C Note, 15 March 2001. Online: www.w3.org/TR/wsdl.
[26] Clear A., Knox S., Ye J., Coyle L., Dobson S. and Nixon P.: 'Integrating Multiple Contexts and Ontologies in a Pervasive Computing Framework'. Proceedings of the ECAI'06 Workshop on Contexts and Ontologies: Theory, Practice and Applications (Riva del Garda, Italy, August 2006), pp. 37–41.
[27] Cohen W.W.: 'Fast Effective Rule Induction'. Proceedings of the 12th International Conference on Machine Learning (ICML 1995, Tahoe City, California, July 2005), Morgan Kaufmann, 1995, pp. 115–123.
[28] Coutand O., Sutterer M., Lun Lau S., Droegehorn O. and David K.: 'User Profile Management for Personalizing Services in Pervasive Computing'. Proceedings of the 6th International Workshop on Applications and Services in Wireless Networks (ASWN 2006, Berlin, May 2006), 2006, pp. 3–11.
[29] Cozman F.: 'A Derivation of Quasi-Bayesian Theory'. Technical Report CMU-RITR-97-37, Robotics Institute, Carnegie Mellon University, Pittsburgh, Pennsylvania, 1997.
[30] Deutsche Bahn online booking system. Online: www.db.de/site/bahn/en/travelling/tickets/ticket_booking/ticket_booking.html.
[31] Dey A.K. and Abowd G.D.: 'Towards a Better Understanding of Context and Context-awareness'. Technical Report GIT-GVU-99-22, Georgia Institute of Technology, College of Computing, Atlanta, Georgia, June 1999.
[32] Ester M., Kriegel H.-P., Sander J. and Xu X.: 'A Density Algorithm for Discovering Clusters in Large Spatial Databases with Noise'. Proceedings of the 2nd International Conference on Knowledge Discovery and Data Mining (KDD-96, Portland, Oregon, August 1996), AAAI Press, 1996, pp. 226–231.
[33] EU IST Project 2001-33052 WonderWeb: Ontology Infrastructure for the Semantic Web, project homepage. Online: wonderweb.semanticweb.org.
[34] EU IST Project FP6-IST-511607 MobiLife: Mobile Life, project homepage. Online: www.ist-mobilife.org.
[35] Flanagan J.A.: 'Unsupervised Clustering of Context Data and Learning User Requirements for a Mobile Device'. Proceedings of the 5th International and Interdisciplinary Conference on Modeling and Using Context (CONTEXT-05, Paris, France, July 2005), pp. 155–168.
[36] Flanagan J.A.: 'Clustering of Context Data using k-means with an Integrate and Fire Type Neuron Model'. Proceedings of the 5th Workshop on Self-Organizing Maps (WSOM2005, Paris, France, September 2005), pp. 17–24.
[37] Flanagan J.A.: 'An Unsupervised Learning Paradigm for Peer-to-Peer Labeling and Naming of Locations and Contexts'. Proceedings of the 2nd International Workshop on Location- and Context-Awareness (LoCA 2006, Dublin, Ireland, May 2006), *Lecture Notes in Computer Science 3987*, Springer-Verlag, Berlin, pp. 204–221.
[38] Floréen P., Kukkonen J., Lagerspetz E., Nurmi P. and Suomela J.: 'BeTelGeuse: Tool for context data gathering via Bluetooth'. Proceedings of the 2nd Workshop on Context Awareness for Proactive Systems (CAPS 2006, Kassel, Germany, June 2006), Kassel University Press, 2006, pp. 137–139.
[39] Floréen P., Przybilski M., Nurmi P., Koolwaaij J., Tarlano A., Wagner M., Luther M., Bataille F., Boussard M., Mrohs B. and Lun Lau S.: 'Towards a Context Management Framework for MobiLife'. Proceedings of the IST Mobile & Communications Summit 2005, Dresden, Germany, June 2005.

[40] Friedman N., Geiger D. and Goldszmidt M.: 'Bayesian Network Classifiers'. *Machine Learning* Vol. 29, No. 2/3, 1997, pp. 131–163.

[41] Gangemi A., Guarino N., Masolo C., Oltramari A. and Schneider L.: 'Sweetening Ontologies with DOLCE'. Proceedings of the 13th International Conference on Knowledge Engineering and Knowledge Management. Ontologies and the Semantic Web, (EKAW 2002, Siguenza, Spain, October 2002), *Lecture Notes in Computer Science 2473*, Springer-Verlag, Berlin, 2002, pp. 166–181.

[42] Gardiner T., Horrocks I. and Tsarkov D.: 'Automated Benchmarking of Description Logic Reasoners'. Proceedings of the 2006 International Workshop on Description Logics (DL'06, Lake District, UK, May/June 2006), CEUR-WS Publication, Vol. 189, pp. 167–174.

[43] Grosof B., Horrocks I., Volz R.and Decker S.: 'Description Logic Programs: Combining logic programs with description logic'. Proceedings of the 12th International World Wide Web Conference (WWW2003, Budapest, Hungary, May 2003), ACM Press, 2003.

[44] Haarslev V. and Möller R.: 'Racer: an OWL reasoning agent for the Semantic Web'. Proceedings of the International Workshop on Applications, Products and Services of Web-based Support Systems, in conjunction with the 2003 IEEE/WIC International Conference on Web Intelligence (WI 2003, Halifax, Canada, October 2003), pp. 91–95.

[45] Halaschek-Wiener C., Parsia B., Sirin E. and Kalyanpur A.: 'Description Logic Reasoning for Dynamic ABoxes'. In Horrocks *et al.* (eds), Proceedings of the 2006 International Workshop on Description Logics (DL'06, Lake District, UK, May/June 2006), CEUR-WS Publication, Vol. 189.

[46] Henricksen K. and Indulska J.: 'Personalizing Context-aware Applications'. Proceedings of the OTM 2005 Workshop on Context-Aware Mobile Systems (CAMS 2005, Agia Napa, Cyprus, October/November 2005), *Lecture Notes in Computer Science 3762,* Springer-Verlag, Berlin, 2005, pp. 122–131.

[47] Hobbs J.R. and Pan F.: 'An Ontology of Time for the Semantic Web'. *ACM Transactions on Asian Language Information Processing* (TALIP), Vol. 3, No. 1, pp. 66–85, 2004.

[48] Horrocks I. and Voronkov A.: 'Reasoning Support for Expressive Ontology Languages using a Theorem Prover'. Proceedings of the 4th International Symposium on Foundations of Information and Knowledge Systems (FoIKS, Salzau Castle, Germany, February 2006), *Lecture Notes in Computer Science 3861*, Springer-Verlag, Berlin, 2006, pp. 201–218.

[49] Horvitz E., Breese J., Heckerman D., Hovel D. and Rommelse K.: 'The Lumière Project: Bayesian user modelling for inferring the goals and needs of software users'. Proceedings of the 14th Conference on Uncertainty in Artificial Intelligence (UAI, Madison; Wisconsin, July 1998), Morgan-Kaufmann, 1998, pp. 256–265.

[50] Horvitz E., Koch P., Sarin R., Apacible J. and Subramani M.: 'BayesPhone: Precomputation of context-sensitive policies for inquiry and action in mobile devices'. Proceedings of the 10th International Conference on User Modeling (UM'05, Edinburgh, Scotland, July 2005), *Lecture Notes in Computer Science 3538*, Springer-Verlag, Berlin, 2005, pp. 251–260.

[51] Huebscher M.C. and McCann J.A.: 'Simulation Model for Self-adaptive Applications in Pervasive Computing'. Proceedings of the 15th International Workshop on Database and Expert Systems Applications (DEXA'04, University of Zaragoza, Spain, August/September 2004), IEEE Computer Society, 2004, pp 694–698.

[52] Husemann D., Narayanaswa C. and Nidd M.: 'Personal Mobile Hub'. Proceedings of the 8th International Symposium Wearable Computers (ISWC'04, Arlington, Virginia, October/November 2004), IEEE Computer Society, 2004, pp. 85–91.

[53] Kalyanpur A., Parsia B. and Hendler J.: 'A Tool for Working with Web Ontologies'. Proceedings of the International Journal on Semantic Web and Information Systems, Vol. 1 (2005), pp. 36–49.

[54] Kang J.H., Welbourne W., Stewart B. and Borriello G.: 'Extracting Places from Traces of Locations'. Proceedings of the 2nd ACM International Workshop on Wireless Mobile Applications and Services on WLAN Hotspots (WMASH'04, Philadelphia, Pennsylvania, October 2004), ACM Press, 2004, pp. 110–118.

[55] Khushraj D. and Lassila O.: 'CALI: Context awareness via logical inference'. ISWC 2004 Workshop on Semantic Web Technology for Mobile and Ubiquitous Applications, Hiroshima, Japan, November 2004.

[56] Knublauch H., Fergerson R., Noy N. and Musen M.: 'The Protégé OWL Plugin: an open development environment for Semantic Web applications'. Proceedings of the 3rd International Semantic Web Conference (ISWC2004, Hiroshima, Japan, November 2004), *Lecture Notes in Computer Science 3298*, Springer-Verlag, Berlin, 2004, pp. 229–243.

[57] Kobsa A.: 'Generic User Modeling Systems'. *User Modeling and User-Adapted Interaction*, Vol. 11, No. 1/2, 2001, pp. 49–63.

[58] Kofod-Petersen A. and Aamodt A.: 'Case-based Situation Assessment in a Mobile Context-aware System'. Artificial Intelligence in Mobile Systems 2003 (AIMS 2003), Seattle, Washington, October 2003.
[59] Kohonen T.: *Self-organizing Maps*. 3rd edition. Springer-Verlag, Berlin, 2001.
[60] Koolwaaij J., Tarlano A., Luther M., Nurmi P., Mrohs B., Battestini A. and Vaidya R.: 'ContextWatcher: Sharing context information in everyday life'. Proceedings of the IASTED conference on Web Technologies, Applications and Services (WTAS, Calgary, Alberta, Canada, July 2006), IASTED, 2006.
[61] Korhonen I., Pärkkä J. and van Gils M.: 'Health Monitoring in the Home of the Future'. *IEEE Engineering in Medicine and Biology Magazine*, Vol. 22, No. 3, 2003, pp. 66–73.
[62] Laasonen K., Raento M. and Toivonen H.: 'Adaptive On-device Location Recognition'. Proceedings of the 2nd International Conference on Pervasive Computing (PERVASIVE 2004, Vienna, Austria, April 2004), *Lecture Notes in Computer Science 3001*, Springer-Verlag, Berlin, 2004, pp. 287–304.
[63] LaMarca A., Chawathe Y., Consolvo S., Hightower J., Smith I., Scott J., Sohn T., Howard J., Hughes J., Potter F., Tabert J., Powledge P., Borriello G. and Schilit B.: 'Place Lab: Device positioning using radio beacons in the wild'. Proceedings of the 3rd International Conference on Pervasive Computing (PERVASIVE 2005, Munich, Germany, May 2005), *Lecture Notes in Computer Science 3468*, Springer-Verlag, Berlin, 2005, pp. 116–133.
[64] Lester J., Choudhury T. and Borriello G.: 'A Practical Approach to Recognizing Physical Activities'. Proceedings of the 4th International Conference on Pervasive Computing (PERVASIVE 2006, Dublin, Ireland, May 2006), *Lecture Notes in Computer Science 3968*, Springer-Verlag, Berlin, 2006, pp. 1–16.
[65] Liebig T., Luther M., Noppens O., Paolucci M., Wagner M. and von Henke F.: 'Building Applications and Tools for OWL: Experiences and suggestions'. Proceedings of the International Workshop on OWL: Experiences and Directions (OWLED, Galway, Ireland, November 2005), pp. 190–201.
[66] Luther M., Böhm S., Wagner M. and Koolwaaij J.: 'Enhanced Presence Tracking for Mobile Applications'. Proceedings of the 4th International Semantic Web Conference (Demo Track, ISWC2005, Galway, Ireland, November 2005), pp. 105–108.
[67] Luther M., Fukazawa Y., Souville B., Fujii K., Naganuma T., Wagner M. and Kurakake S.: 'Classification-based Situational Reasoning for Task-oriented Mobile Service Recommendation'. Proceedings of the ECAI'06 Workshop on Contexts and Ontologies: Theory, Practice and Applications (Riva del Garda, Italy, August 2006), pp. 31–36.
[68] Luther M., Mrohs B., Vaidya R. and Wagner M.: 'OWL-SF: Distributed OWL-based reasoning on objects in the real world'. Proceedings of the 4th International Semantic Web Conference (Demo Track, ISWC2005, Galway, Ireland, November 2005), pp. 112–114.
[69] Luther M., Mrohs B., Wagner M., Steglich S. and Kellerer W.: 'Situational Reasoning: a practical OWL use case'. Proceedings of the 7th International Symposium on Autonomous Decentralized Systems (ISADS'05, Chengdu, China, April 2005), pp. 96–103.
[70] MacQueen J.B.: 'Some Methods for Classification and Analysis of Multivariate Observations'. Proceedings of the 5th Berkeley Symposium on Mathematical Statistics and Probability, Vol. 1, University of California, 1967, pp. 281–297.
[71] Microsoft's MapPoint Web Service. Online: support.microsoft.com/default.aspx?scid=kb;en-us;884771.
[72] Marmasse N. and Schmandt C.: 'A User-centered Location Model'. *Personal and Ubiquitous Computing*, Vol. 6, No. 5/6, 2002, pp. 318–321.
[73] Martin D., Burstein M., Hobbs J., Lassila O., McDermott D., McIlraith S., Narayanan S., Paolucci M., Parsia B., Payne T., Sirin E., Srinivasan N. and Sycara K.: 'OWL-S: Semantic markup for web services'. W3C Member Submission, November 2004. Online: www.w3.org/Submission/2004/SUBM-OWL-S-20041122/.
[74] Martin M. and Nurmi P.: 'A Generic Large-scale Simulator for Ubiquitous Computing'. Proceedings of the 3rd Annual International Conference on Mobile and Ubiquitous Systems: Networks and Services (MobiQuitous, San Jose, July 2006), IEEE Computer Society, 2006.
[75] Mitchell T.M.: *Machine Learning*. McGraw Hill, New York, 1997.
[76] Mrohs B., Luther M. and Vaidya R.: 'Context-aware Presence Management'. Proceedings of the Workshop on Context Awareness for Proactive Systems (CAPS 2005, Helsinki, Finland, June 2005), Helsinki University Press, 2005, pp. 177–180.
[77] Mrohs B., Luther M., Vaidya R., Wagner M., Steglich S., Kellerer W. and Arbanowski S.: 'OWL-SF: a distributed semantic service framework'. Proceedings of the Workshop on Context Awareness for Proactive Systems (CAPS 2005, Helsinki, Finland, June 2005), Helsinki University Press, 2005, pp. 67–78.

[78] Newman D., Hettich S., Blake C. and Merz C.: 'UCI Repository of Machine Learning Databases'. University of California, Irvine, Dept. of Information and Computer Sciences, 1998. Online: www.ics.uci.edu/~mlearn/MLRepository.html.

[79] Niles I. and Pease A.: 'Towards a Standard Upper Ontology'. Proceedings of the 2nd International Conference on Formal Ontology in Information Systems (FOIS-2001, Ogunquit, Maine, October 2001), pp. 2–9.

[80] Nishikawa H., Yamamoto S., Tamai M., Nishigaki K., Kitani T., Shibata N., Yasumoto K. and Ito M.: 'UbiREAL: Realistic smartspace simulator for systematic testing'. Proceedings of the Conference on Ubiquitous Computing (UbiComp 2006, Orange County, California, September 2006), Lecture Notes in Computer Science 4206, Springer-Verlag, Berlin, 2006, pp. 459–476.

[81] Oliver N., Horvitz E. and Garg A.: 'Layered Representations for Recognizing Office Activity'. Proceedings of the 4th IEEE International Conference on Multimodal Interaction (ICMI'02, Pittsburgh, Pennsylvania, October 2002), IEEE Press, 2002, pp. 3–8.

[82] Nurmi P., Floréen P., Przybilski M. and Lindén G.: 'A Framework for Distributed Activity Recognition in Ubiquitous Systems'. Proceedings of the International Conference on Artificial Intelligence (ICAI 2005, Las Vegas, Nevada, June 2005), pp. 650–655.

[83] Nurmi P. and Koolwaaij J.: 'Identifying Meaningful Locations'. Proceedings of the 3rd Annual International Conference on Mobile and Ubiquitous Systems: Networks and Services (MobiQuitous, Sun José, July 2006), IEEE Computer Society, 2006.

[84] Nurmi P., Salden A., Lun Lau S., Suomela J., Sutterer M., Millerat J., Martin M., Lagerspetz E. and Poortinga R.: 'A System for Context Dependent User Modeling'. Proceedings of the OTM 2nd International Workshop on Context-Aware Mobile Systems (CAMS 2006, Montpellier, France, October/November 2006), *Lecture Notes in Computer Science 4278*, Springer-Verlag, Berlin, 2006, pp. 1894–1903.

[85] O'Neill E., Klepal M., Lewis D., O'Donnell T., O'Sullivan D. and Pesch D.: 'A Testbed for Evaluating Human Interaction with Ubiquitous Computing Environments'. Proceedings of the 1st International Conference on Testbeds & Research Infrastructures for the DEvelopment of NeTworks and COMmunities (TRIDENTCOM 2005, Trento, Italy, February 2005), pp. 60–69.

[86] OO jDREW deductive reasoning engine for the RuleML web rule language. Online: www.jdrew.org/oojdrew.

[87] Paolucci M., Goix W., Andreetto A., Luther M. and Wagner M.: 'Representing Services for Mobile Computing using OWL and OWL-S: an initial investigation'. Proceedings of the WI 2005 workshop on WWW Service Composition with Semantic Web Services (wscomp05, Compiègne, France, September 2005), pp. 44–53.

[88] Papazoglou M.P. and Georgakopoulos D.: 'Service-oriented Computing'. *Communications of the ACM*, Vol. 46, No. 10, 2003, pp. 24–28.

[89] Patel-Schneider P.F., Hayes P. and Horrocks I.: 'OWL Web Ontology Language Semantics and Abstract Syntax'. W3C recommendation, The Worldwide Web Consortium, 2004.

[90] Pellet: 'An OWL DL Reasoner'. Online: pellet.owldl.com.

[91] Priyantha N.B., Chakraborty A. and Balakrishnan H.: 'The Cricket Location–Support System'. Proceedings of the 6th Annual International Conference on Mobile Computing and Networking (MobiCom 2000, Boston Massachusetts, August 2000), ACM Press, 2000, pp. 32–43.

[92] Protégé: Online: protege.stanford.edu.

[93] Racer: Online: www.racer-systems.com.

[94] Raento M., Oulasvirta A., Petit R. and Toivonen H.: 'ContextPhone: a prototyping platform for context-aware mobile applications'. *IEEE Pervasive Computing*, Vol. 4, No. 2, 2005, pp. 51–59.

[95] Ranganathan A., Al-Muhtadi J. and Campbell R.H.: 'Reasoning About Uncertain Contexts in Pervasive Computing Environments'. *IEEE Pervasive Computing*, Vol. 3, No. 2, 2004, pp. 62–70.

[96] Rao B. and Minakakis L.: 'Evolution of Mobile Location-based Services'. *Communications of the ACM*, Vol. 46, No. 2, 2003, pp. 61–65.

[97] Sadeh N.M., Gandon F.L. and Kwon, O.B.: 'Ambient Intelligence: the MyCampus experience'.Technical report CMU-ISRI-05-123, Carnegie Mellon University, Mellon University, Pittsburgh, Pennsylvania, 2005.

[98] Sanmugalingam K. and Coulouris G.: 'A Generic Location Event Simulator'. Proceedings of the 4th International Conference on Ubiquitous Computing (UbiComp 2002, Gothenburg, Sweden, September/October 2002), *Lecture Notes in Computer Science 2498*, Springer-Verlag, Berlin, 2002, pp. 308–315.

[99] Schulzrinne H., Gurbani V., Kyzivat P. and Rosenberg J.: 'RPID: rich presence extensions to the Presence Information Data Format (PIDF)'. Internet draft, IETF, December 20, 2005. Online: www3.ietf.org/proceedings/06mar/IDs/draft-ietf-simple-rpid-10.txt.

[100] Schwinger W., Grün Ch., Pröll B., Retschitzegger W. and Schaurhuber A.: 'Context-awareness in Mobile Tourism Guides: a comprehensive survey'. Technical report, Johannes Kepler Universität Linz, Austria, July 2005.

[101] van Setten M., Pokraev S. and Koolwaaij J.: 'Context-aware Recommendations in the Mobile Tourist Application COMPASS'. Proceedings of the 3rd International Conference on Adaptive Hypermedia and Adaptive Web-based Systems (AH 2004, Eindhoven, the Netherlands, Aug 2004), *Lecture Notes in Computer Science 3173* Springer-Verlag, Berlin, 2004, pp. 122–131.

[102] Siewiorek D., Smailagic A., Furukawa J., Krause A., Moraveji N., Reiger K., Shaffer J. and Wong F.L.: 'SenSay: A context-aware mobile phone'. Proceedings of the 7th International Symposium on Wearable Computers (ISWC'03, White Plains, New York, October 2003), IEEE Computer Society, 2003, pp. 248–249.

[103] Sirin E. and Parsia B.: 'Pellet: an OWL DL reasoner'. Proceedings of the 2004 International Workshop on Description Logics (DL'04, Whistler, British Columbia, Canada, June 2004), pp. 212–213.

[104] Sohn T., Grisworld W.G., Scott J., LaMarca A., Chawathe Y., Smith I. and Chen M.Y.: 'Experiences with Place Lab: an open source toolkit for location-aware computing'. Proceedings of the 28th International Conference on Software Engineering (ICSE'06, Shanghai, China, May 2006), ACM Press, 2006, pp. 462–471.

[105] Staab S. and Studer R. (eds): *Handbook on Ontologies*. Springer-Verlag, Berlin, 2004.

[106] SWOOP semantic web ontology editor. Online: code.google.com/p/swoop/.

[107] Tapia E.M., Intille S.S. and Larson K.: 'Activity Recognition in the Home Setting using Simple and Ubiquitous Sensors'. Proceedings of the 2nd International Conference on Pervasive Computing (PERVASIVE 2004, Vienna, Austria, April 2004), Springer-Verlag, Berlin, 2004, pp. 158–175.

[108] Teller A. and Stivoric J.: 'The BodyMedia platform: continuous body intelligence'. Proceedings of the 1st ACM Workshop on Continuous Archival and Retrieval of Personal Experiences (CARPE 2004, New York, New York, October 2004), ACM Press, 2004, pp. 114–115.

[109] Toyama N., Ota T., Kato F., Toyota Y., Hattori T. and Hagino T.: 'Exploiting Multiple Radii to Learn Significant Locations'. Proceedings of the International Workshop on Location- and Context-Awareness (LoCA 2005, Oberpfaffenhofen, Germany, May 2005), *Lecture Notes in Computer Science 3479*, Springer-Verlag, 2005, pp. 157–168.

[110] vCard: The Electronic Business Card, Version 2.1 Specification, September 18, 1996. Online: www.imc.org/pdi/vcard-21.txt.

[111] Wang X., Zhang D.Q., Gu T. and Pung H.K.: 'Ontology-based Context Modeling and Reasoning using OWL'. Proceedings of the PerCom Workshop on Context Modeling and Reasoning (CoMoRea 2004, Orlando, Florida, March 2004), pp. 18–22.

[112] Weithöner T., Liebig T., Luther M. and Böhm S.: 'What's Wrong with OWL Benchmarks?'. Proceedings of the 2nd International Workshop on Scalable Semantic Web Knowledge Base Systems (SSWS 2006, Athens, Georgia, November 2006).

[113] Yamabe T., Takagi A. and Nakajima T.: 'Citron: a context information acquisition framework for personal devices'. Proceedings of the 11th IEEE International Conference on Embedded and Real-time Computing Systems and Applications (RTCSA'05, Hong Kong, August 2005), IEEE Computer Society, 2005, pp. 489–495.

[114] Yau S.S. and Karim F.: 'An Adaptive Middleware for Context-sensitive Communications for Real-time Applications in Ubiquitous Computing Environments'. *Real-Time Systems*, Vol. 26, No. 1, 2004, pp. 29–61.

[115] Zhou C., Frankowski D., Ludford P., Shekhar S. and Terveen L.: 'Discovering Personal Gazetteers: an interactive clustering approach'. Proceedings of the 12th Annual ACM International Workshop on Geographic Information Systems (GIS'04, Washington DC, November 2004), ACM Press, 2004, pp. 266–273.

[116] Zukerman I. and Albrecht D.W.: 'Predictive Statistical Models for User Modeling'. *User Modeling and User-Adapted Interaction*, Vol. 11, No. 1/2, 2001, pp. 5–18.

5

Multimodality and Personalisation

Edited by David Bonnefoy (Motorola SAS, France),
Olaf Drögehorn (University of Kassel, Germany)
and Ralf Kernchen (University of Surrey, UK)

This chapter focuses on the following aspects of the mobile user experience:

- How to make the best use of available devices, those that the user carries with her (such as a cell phone) or those that become available while she moves (such as an audio car kit); and for each of those devices, how to make the best use of the various modalities (such as sound, video and text) they offer.
- How to ensure that the applications offered to the user are tailored to her specific needs and her specific situation.

The end goal of the multimodality and personalisation research presented here is to serve the end-user. For this to happen, it is important that any application can easily access the capabilities of these technologies. Multimodality and personalisation are thus seen here as services that are made available to any application. This chapter shows how a decoupling can be ensured between the applications and the management of devices, modality, profiles, etc., making it easier for application developers to add multimodality and personalisation to their applications. Another benefit of such an approach is that multimodality and personalisation are handled in a consistent way across devices and applications, maintaining a consistent experience for the user.

Section 5.1 introduces a multimodality framework that allows an application to easily receive user input and to provide content to the user on the most suitable device and using the most suitable modality. Section 5.2 presents a personalisation framework that manages the user's personal information and helps applications to adapt themselves to the user needs. Section 5.3 is a short conclusion.

Enabling Technologies for Mobile Services: The MobiLife Book Edited by Mika Klemettinen

5.1 Multimodal Interfaces in Mobile Environments

Human social interaction undoubtedly is multimodal in essence. Precise ways in which humans use talk, gesture, gaze and aspects of the material surroundings have been researched as described in [16]. Research on human interactions has been carried out in research fields such as psychology and very recently in semiotics. Nevertheless, the human–computer interaction (HCI) community has pinpointed the research field of multimodal computer interaction to overcome the hurdles of current human–computer interaction. Findings from these allied fields are therefore applied to the human–computer interaction.

The ultimate goal of multimodal user interfaces is the enrichment of human–machine interaction, promising the empowerment of the user to interact with a computer system as naturally as possible. Human beings therefore are exchanging information with a machine through various information channels assigned as *computer input vs. human output channels* and *computer output vs. human input channels* [15]. From the perspective of multimodal systems, human output channels are referred to as 'modality inputs', and computer output channels can be seen as 'media output' representations. A widely accepted definition of multimodal interfaces is:

> Multimodality is the capacity of the system to communicate with a user along different types of communication channels and to extract and convey meaning automatically. [12]

Several multimodal applications have been created over recent years. Very popular are applications which are integrating modality input such as speech and pen gestures in map navigation applications [7,14]. Navigation in 3D applications for construction [13] or in games [3] is a further example.

In the above cases, multimodality is implemented specifically for each application. There are many drawbacks of such approaches. First, in most cases systems are bimodal and concentrate on one device providing the modalities. Only recent activities, as for example in [5, 6,11,19], are proposing frameworks for multimodal interaction. Furthermore, for commercialisation, widespread acceptance and more general use, robust and scalable architectures are needed [12]. Controlling sophisticated visualisation and multimedia output is demanding and provides the potential for further research. For *mobile* multimodal interfaces, several new challenges arise: '. . . future multimodal interfaces, especially mobile ones, will require active adaptation to the user, task, ongoing dialogue, and environmental context' [12]. In addition, there is the notion of personal preferences of users. Preferences are usually either entered manually by the user or they can be learned automatically as described later in this chapter.

Both handset manufacturers and network operators could offer a better service experience for users by applying multimodal interface technologies to their handsets or platforms. Nevertheless, multimodality should not be constrained to terminal devices themselves, but rather incorporated in devices in the user's neighbouring environment. Either multimodal input devices (e.g. specialised devices such as keyboards, microphones, and gesture input pads) or media output presentation devices (e.g. video on a large screen or voice output for emails and news) discovered in the vicinity of the user should be integrated based on the current user's *context*. Context awareness, media content transformations, user input integration, content and device selection, and device and modality service discovery are aspects which need to be researched and integrated in order to achieve a flexible and modular overall software architecture allowing for a better user experience. Finally, the functionalities of the software architecture should be provided in a developer-friendly way via a well-defined Application

Programming Interface (API) to enable a widespread reuse of the components for mobile services and applications.

5.1.1 Introduction

Multimodal interfaces in mobile environments, as described in this chapter, aim at providing an overall framework for integrating human–computer interaction advances in the mobile applications world. Presentation of mobile applications media content and interpretation of user input can be improved by taking into account the relevant and available context information. The generic provision of multimodal interface technology for mobile applications is a research challenge which can be resolved only by pursuing an integrated approach, taking into account relevant aspects from multimodal interface research, discovery mechanisms, device independence work, context awareness and mobile device platforms. The framework of the Mobile Services Architecture has led to the design of a model called the *User Interface Adaptation Function* (UIAF). The model is the functional realisation of user-context and preferences-dependent multimodal input and output in changing environments. Device independence supporting multiple devices and modalities is achieved by using a device-independent model which is complemented by a device discovery mechanism. In addition, the approach aims at being application-agnostic, providing its functionalities through a defined Application Programming Interface (API).

In a mobile environment, people face very frequent changes in their environment, which imposes a very heavy cognitive load on users. While moving, different devices may appear or disappear, preferences may change, user input and output may be understood or delivered differently.

Therefore multimodality research on mobile systems should fulfil at least the following requirements:

- *Mobility.* Multimodal interfaces in mobile environments have to prove to be useful in different situations mobile users could be in. A mobile situation by nature has some consequences on available devices and therefore modalities available to a user to interact with her mobile applications. A mobile multimodal interface model needs to be developed and the definition has to target multi-device user interfaces that best match the changing environments.
- *Reference architecture.* In order to realise a functional system, a software architecture needs to be defined, as indicated earlier. It should cover all defined functional requirements on a high level by defining a component structure and their interactions.
- *Application interface.* A well-defined application interface (API) for easy development of multimodal applications is necessary. It needs to define how applications can be supported for multimodal output and multimodal input interpretation, and how to receive sensor information.
- *Device and modality descriptions.* In order to support mobility, devices need to have a description of their capabilities and offered modalities. Device descriptions offer the flexibility to handle different devices in a consistent manner.
- *Mechanism for input interpretation and selection.* Definition and specification of mechanisms to analyse the user input is a research subject on its own. Interpretation ensures the selection of the right input channel for the user's situation. Inputs should be captured in an input model and forwarded to the application for further interpretation in the current user context.

- *Mechanisms for output selection and adaptation.* Media data provided by the applications needs to be delivered to the best media output channel. Considering the user's situation, and therefore context information, the best selection of these channels (audio, video, etc.) should be ensured. Selection algorithms providing the expected behaviour need to be defined, taking into account the available devices. The output selection mechanism needs to be able to handle the provided device descriptions.
- *Uniform device interfaces.* Heterogeneity of devices makes uniform interfacing with devices a difficult task. By defining uniform software components, the exchange of device and modality information and also the request for modality input and output is handled in a consistent way. The definition of so-called device agents can accomplish this task.
- *Discovery mechanisms.* Mobility itself presumes the available devices in the user's environment change. It is thus necessary to continuously monitor the environment to discover devices that become available in the user's vicinity. Discovery mechanisms, based on technologies such as Bluetooth, Wireless LAN or UPnP, can provide these functionalities, but they need to be incorporated into a discovery framework in order to allow consistent access to the information, whatever the underlying protocols.
- *Mechanisms to integrate context information.* Context information is essential to achieve the targeted adaptation results. Therefore, mechanisms to take user preferences into account and interpret context information need to be elaborated. For the multimodal output decision, modality recommendations which provide the means of integrating context information in a unified way into the decision process are used. Modality recommendations define which device or which modality should be used or omitted in the current user situation based on context interpretation and context inference. Recommendations are provided by the personalisation function, described later in this chapter.

In addition to these essential research requirements, the setting-up of a multimodal interface model in mobile environments and software architecture, in order to achieve a reference implementation for these requirements, is an important step in order to achieve true multimodality for mobile applications. The next sections introduce possible general solutions and some specific details on them.

5.1.2 General Approach

5.1.2.1 Model for Multimodal Interfaces in Mobile Environments

The physical multimodal user interface model considered here consists of a user equipped with a *portal device* – the main device the user carries at all times – and, possibly, several other user interface (UI) devices (displays, speakers, cameras, etc.) providing various modalities for input and output, an approach described in [8]. In general, the user is interacting with the portal device and the available user interface devices, together forming the current multimodal user interface, as shown in the upper part of Figure 5.1. The user interface combines user input modalities and serves available user output modalities with application requested media output.

In a simple scenario the applications are executed on the portal device. Limitations in the currently available devices that could act as such a portal device (e.g. PDA, smart phones) could make it necessary to consider distribution of parts of the applications and the interface management in the network to provide more computing capacity.

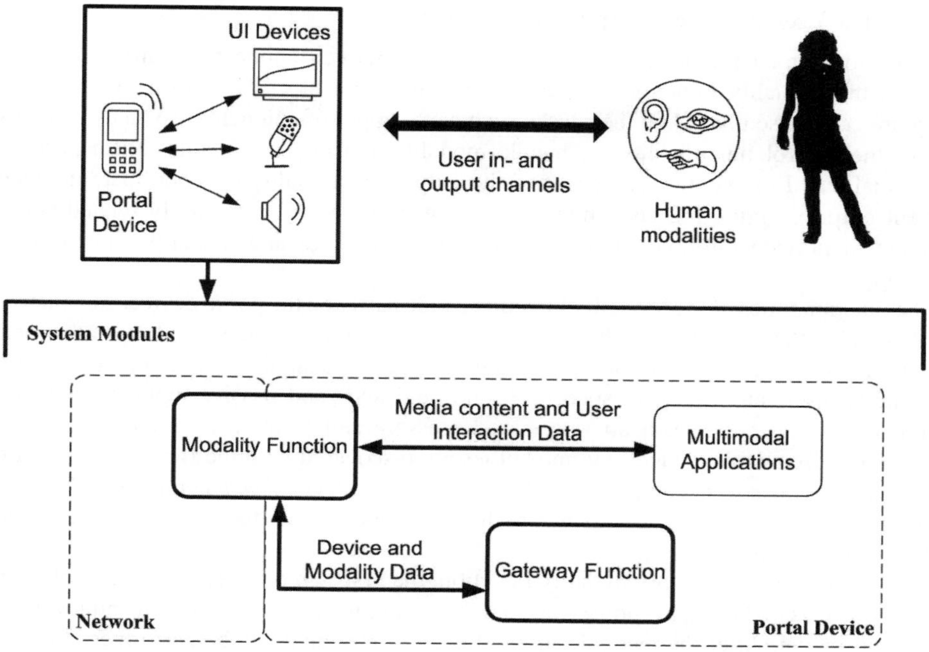

Figure 5.1 Physical component distribution and system modules.

As shown on the lower part of Figure 5.1, the system managing the interface is composed of two main modules:

- The *Gateway Function* discovers and directly interacts with the surrounding interface devices. Due to the direct connection with other devices, it runs entirely on the portal device.
- The *Modality Function* manages the interface: decisions are made on modality output, integration of input, content transformation (e.g. text to speech), etc. It can be distributed between the portal device and the network.

The interface devices provide support for the user interaction facilitating one or more modalities. They have the ability to be discovered by the Gateway Function and are described by modality capabilities. The very high-level objective addressed by the Modality Function is to handle the available modalities as discovered by the Gateway Function. It also processes the application media content as sent by the applications, analyses the user input, coordinates the media output and decides which devices are to be included in the actual user interface.

The decision on which output devices are bound optimally into the current multimodal user interface is in particular based on contextual information, such as the user's situation, personal preferences, available modality and device information, etc. Based on this, the media content provided by the applications is presented in the most efficient way by the available output modalities.

5.1.2.2 The User Interface Adaptation Function

In order to abstract the actual physical interaction between applications and the physical devices, the Modality Function defines several functional blocks as shown in Figure 5.2. They are categorised based on their tasks such as the input functionality (to analyse the user input), the control functionality (to handle modality and media content, collect necessary contextual data from context sources) and the output functionality (to coordinate the media content output). Figure 5.2 also shows the Gateway Function as a closely related support function to provide information of the discovery of devices and modalities in the user's proximity.

The functionalities of the Modality Function reside partly on the portal device, user interface devices and in the network (e.g. the recognition part could be on the user interface device).

A user gives active input to an application through the available active modalities in the multimodal user interface (e.g. speech, gesture and text) and passive input through other user interface devices (e.g. heart rate, blood pressure and location information through a GPS device). Available devices and modalities are discovered with the functionalities of the *Device/Service Discovery* component and managed by the *Device Agent* component. The user's active and passive input data is delivered to the input functionality blocks through the *Content Routing* component.

Within the input part of the Modality Function, the first step is the recognition of the input modalities. This includes data processing and the generation of a synchronised and formalised user input model. After the recognition, the fusion of the input can be performed. Fusion is the process of combining certain input modalities in a meaningful way to create and provide a semantically consistent interpretation of the user input for the running applications. In addition, the *Fusion* component filters relevant context information from the user input as raw input for

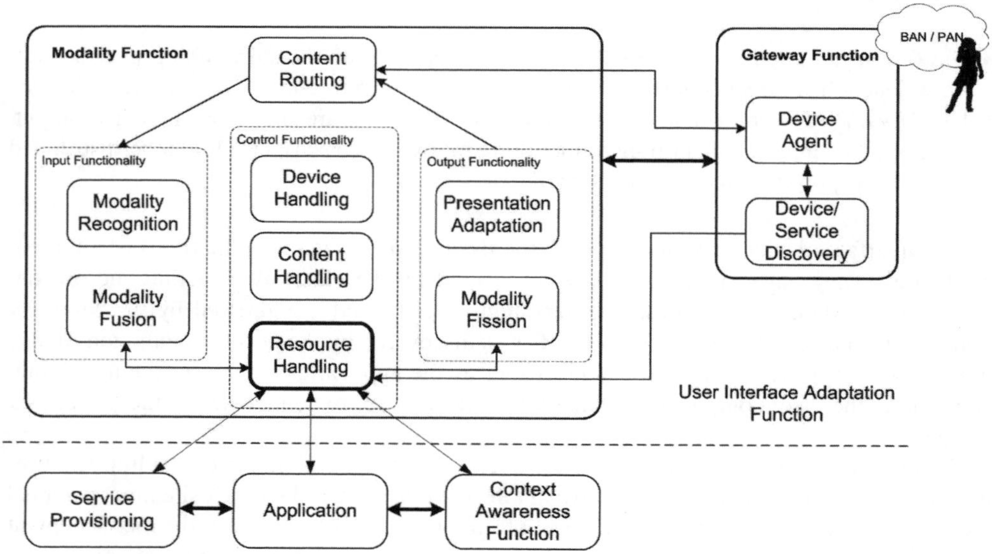

Figure 5.2 System architecture.

the system. All generated information included in the user interaction will go to the *Resource Handling* block.

The Resource Handling component controls the Modality Function data flow, requests data from the external context sources and routes the application output to the output components. The behaviour of the Resource Handling component is based on the user interaction model generated by the Fusion component and the requests of the application. The application will provide the content that should be delivered through the output part of the Modality Function. Output requests will be routed towards the *Fission* component. The decisions made by the Fission component leads to the choice of which device should be used for media output; that, in turn, could depend on certain user preferences, the user's situation, available devices and the desirable format of the content.

The Fission component implements the logic and the appropriate algorithms for the decision on how media output requests from a potential multimodal application will be delivered to the available output modalities. Information about available devices, content transformers as managed from the device and content handling components, is taken into account along with context information requested from the *Personal Context Function*. The last part of the decision process is the transformation of media data from the source format into a requested target format (e.g. text to speech), as well as splitting of the media data if required, adaptation of the media presentation and delivery of this media content to the *Content Routing* component.

As the last step, the Content Routing component does the physical coupling of adapted output towards the corresponding device agents.

5.1.3 Encapsulation of Physical Devices

The Gateway Function defines an infrastructure provided to enable plug-in support of physical devices and their access. Physical devices are encapsulated by so called *Device Agents*.

Device Agents are used to provide the software with a standard interface at programming level, which will encapsulate all the device-specific physical interfacing functionality. This ensures that the infrastructure will always use the same interface for initialising, sending requests, receiving notifications, etc., regardless of the device being used. Obviously some devices, based on their intrinsic capabilities, will be able to support only a certain set of requests; for example, input devices support input operations and notifications, while output devices typically support only output requests.

Possible *UI Devices* are:

- display (portal device, small display, external large display);
- sound (portal device speaker, Bluetooth earphones, external amplifier);
- keyboard;
- voice (internal microphone, external microphone);
- GPS;
- heart rate, speed and distance sensors;
- RFID reader: sensor input used both for location discovery and data input.

The Gateway Function framework allows an easy integration of additional device agents, simply by providing the agent code and adding specific descriptions for configuration files.

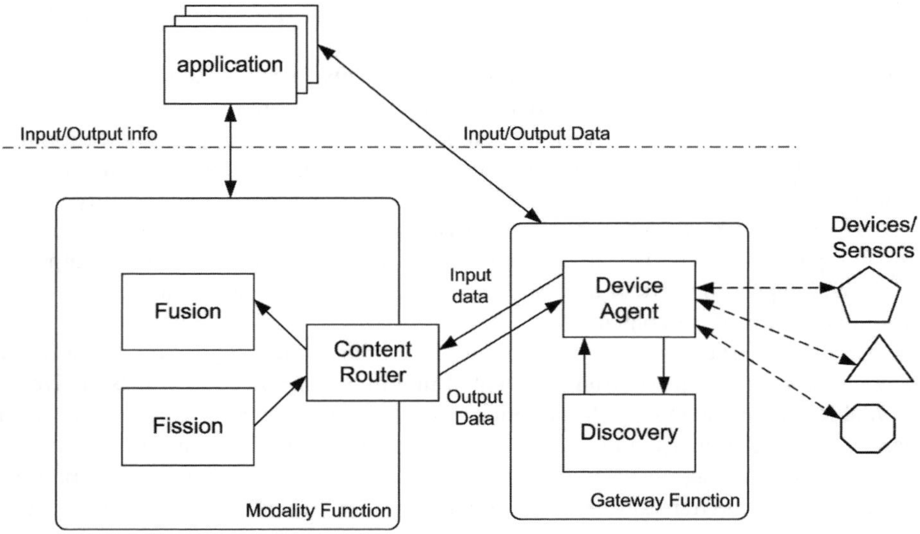

Figure 5.3 Gateway Function interfaces.

This allows the *Device Discovery* module to dynamically discover the presence/availability of physical devices and automatically include them in the list of usable devices.

Figure 5.3 highlights the canonical interfaces of the Device Agent module, the ones typically used by other Modality Function modules through the *Content Router*; these interfaces are available also to other components, service providers and applications, although this is not the typical scenario.

The Device Agent, on the one hand, ensures the establishment, maintenance and termination of physical short-range network connections between system input and output channels dependent on available device services for serving the users modalities. On the other hand, it connects back into the system, using specific communication protocols, with its counterpart, the Content Router module of the Modality Function.

In this case, the Device Agent component of the Gateway Function module acts as a gateway between the local modality channels (both input and output) accessible by the physical portal device and its logical proxy (the respective content router instance) in the Modality Function module.

This component works in close connection with the Device/Service Discovery module, as it has open channels to all available local resources, and therefore it needs to communicate any disconnection or unusual event. Furthermore, it will receive indications of the availability of additional resources, and will be ready to open new channels on implicit or explicit request from the Modality Function components. Moreover, it will take care of the physical resources of the portal device itself (the display, the input devices, etc.) in terms of both content delivery and retrieval. This module will host agents for specific interpreters/viewers to handle specific high-level requests sent by the *Modality Function Content Router* on behalf of the Fission and Fusion modules.

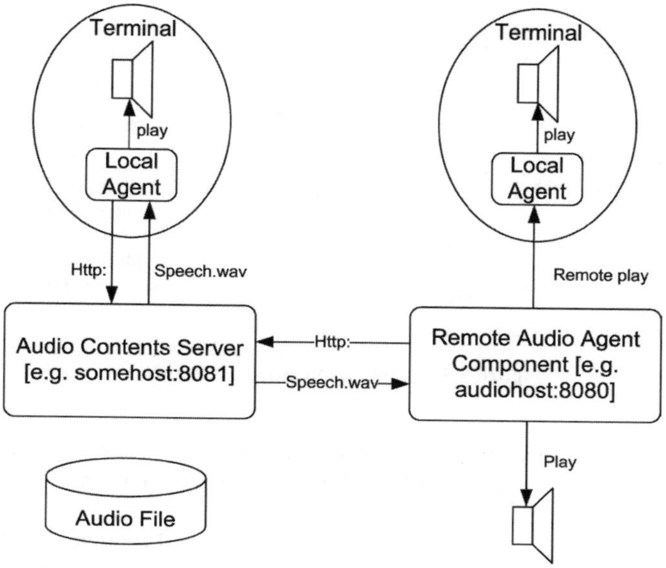

Figure 5.4 Local and remote audio agents.

Inside the Gateway Function module, the Device Agent interacts with the Device Discovery Module to exchange information on device availability and device communication failures – as possible indication of device unavailability. The complete encapsulation will also support the utilisation of the Device Agent code outside the main framework on the portal device, for instance inside the 'remote agent' application used to access devices which have a physical connections on specific sever hosts.

Figure 5.4 shows the difference between a 'local' device agent and a 'remote' one: as an example an audio agent has been used in the figure. The local agent will play the audio file, retrieved from the audio content server through the network, using local resources of the portal device (i.e. the local speaker). When a remote audio device is discovered and is chosen by the user (automatically or explicitly, depending on the context), the play request is sent by the portal device to the remote agent application, supplying the URL of the audio file (not the contents). The remote agent will retrieve the audio file through the network and then play it using the available resource (e.g. the speakers of a car or the media-centre at home).

5.1.4 Application Interface

5.1.4.1 Overview

The Modality Function and the Gateway Function have been designed with the assumption that both multimodality functionalities and the user mobile applications are executed on the terminal side. Although the applications are therefore physically close to the multimodal interface implementation, decoupling of multimodal functionalities and the mobile applications is one of the target requirements. In this way, mobile applications can use multimodality

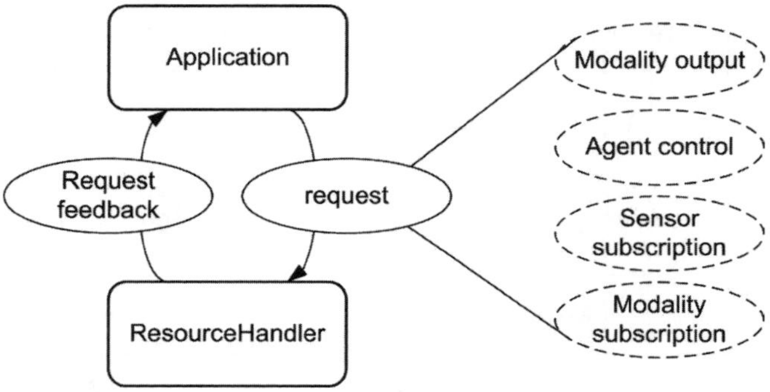

Figure 5.5 Application request overview.

functionalities without implementing multimodality features themselves, but rather reuse provided functionalities from the modular Modality Function and Gateway Function implementations.

Therefore, the provision and definition of an Application Programming Interface (API) to mobile applications to ease support of the development of multimodal applications is an important aspect of the Modality Function.

Through this interface, applications are able to request modality input and modality output, control media presentation and subscribe to sensor data as depicted in Figure 5.5. Feedback for every request is given according to the requested functionality.

5.1.4.2 Application Interface Mechanism

Any kind of application request towards the Modality Function has to invoke one request interface method provided by the Resource Handler, which acts as the interface component of the Modality Function. This accepts parameters defined in an XML-based format. The description format has been defined by using the W3C XML schema description syntax. A collapsed visualisation of the schema is shown in Figure 5.6.

The request is processed inside the Resource Handling component using a number of helper classes, and according to the application request the different functionalities inside the Modality Function are triggered.

An application has the possibility to request a modality output (*app_output_request*) defined by its media content (*content*). For example, an application could request a modality output of type *text* (e.g. news text and web page content). The choice for describing content types is provided by SMIL media elements [18]. SMIL has two major advantages: it provides a standardised way to express a comprehensive choice of media elements, and certain browsers can actually handle this format natively. An XML example of an application request for modality output is shown in Figure 5.7. It includes content tags in text and audio formats using the SMIL statement. The *src* attributes of the SMIL elements are showing placeholders for the required URL, where to find the media content in the referenced format.

In Figure 5.7, the application receives feedback to its request. For backward reference, this contains a unique identifier for every request an application sends to the Modality Function. This unique identifier can be used by the application to control the executed multimedia

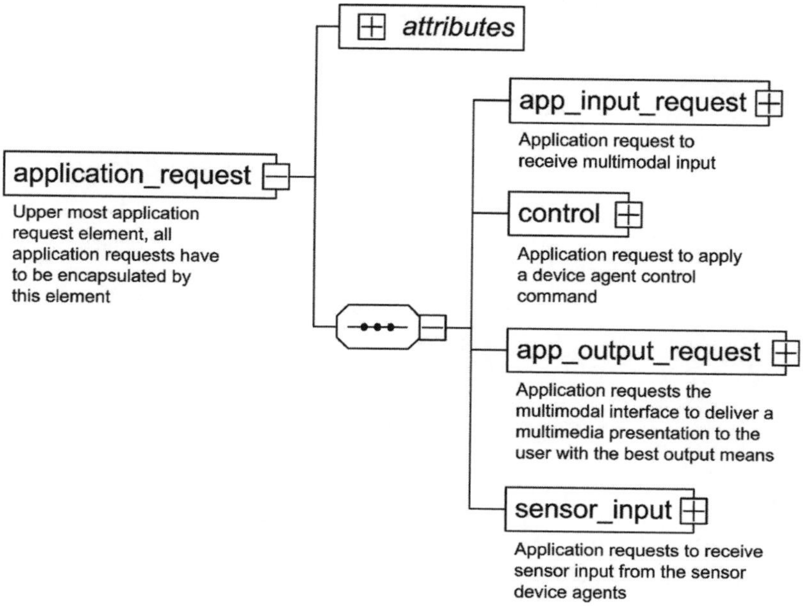

Figure 5.6 Application interface XML schema.

presentations. When, for example, an application requests the playback of an audio file, the request will trigger the Modality Function to stream the audio to a selected audio agent. During run-time, the application can now decide to stop or pause the streaming audio presentation. Therefore, the application can use the output feedback request ID and wrap it into a control request to influence the presentation state without having knowledge about the actual device. Everything will be handled by the Modality Function internally.

The Modality Function also provides a request type for subscription to certain sensor input (*sensor_input*) that is forwarded through the Modality Function (e.g. a GPS sensor is connected via Bluetooth to the portal device and the application or other components might be interested in this GPS information).

A similar mechanism is used for subscription to certain modality input devices (*app_input_request*) as keyboards providing text input, etc. More details on how the multimodality fusion and the corresponding request work together are given in the modality input section of this chapter (Section 5.1.7).

5.1.5 Discovery of the User Environment: Device and Modality Description

5.1.5.1 Gateway Function Discovery Mechanisms

The Gateway Function includes a *Device/Service Discovery* component which relies on the underlying capabilities of the *Portal Device* through the use of modules named *adapters*. Each adapter implements an interface towards a particular discovery mechanism (Bluetooth, WiFi, RFID, UPnP, etc.). Regularly, adapters launch a discovery procedure and report changes in the user environment to the framework.

```
<?xml version="1.0" encoding="UTF-8"?>
<application_request port="10000" ipaddress="192.168.0.4">
          <app_output_request startTime="20:15:06">
                    <content>
                              <smil20:text type="text/xhtml" src="XHTML-URL" system-language="en"/>
                              <smil20:text type="text/ascii" src="ASCII-URL" system-language="en"/>
                              <smil20:audio type="audio/wav" src="WAV-URL"/>
                              <smil20:video type="video/wmv" src="WMV-URL" system-bitrate="128" system-screen-size="1024x768"/>
                    </content>
          </app_output_request>
</application request>
```

Figure 5.7 Example application output request.

These changes are notified to the *Personal Context Function*, and used internally by the UIAF, which maintains the list of available devices and their related modality services. Based on this information, the system is able to make binding decisions when receiving an application request.

For example, the following discovery adapters are available in the current reference implementation:

- Bluetooth adapter: discovers surrounding Bluetooth devices, based on a JSR82 implementation;
- WiFi adapter: discovers surrounding access points, based on a native library;
- RFID adapter: discovers RFID tags using an external reading software;
- UPnP adapter: discovers UPnP devices and services;
- local adapter: pre-configures the user terminal with related modality agents.

5.1.5.2 Device and Modality Description

To enable a context-driven user interface realisation, there is a need for a complete and accurate device and modality description framework. The huge differences between the different available discovery mechanisms do not allow discovering a rich and uniform set of information about devices in the vicinity. For example, while the Bluetooth protocol allows the discovery of devices and their Bluetooth profile services, WiFi discovery stops at the device level.

Therefore, a framework to describe uniformly the discovered devices and the modality services that they support has to be developed. This work, detailed in [8], relies on the elements shown in Figure 5.8.

Host Devices are defined as self-contained piece of hardware with which the user interacts. They are physically composed of a set of *physical modalities*, which are the different pieces of hardware that render a given modality. For example, a laptop is a Host Device, composed of a screen, speakers, keyboard and microphone, all physical modalities. Furthermore, it is considered that these physical modalities are used through software components called

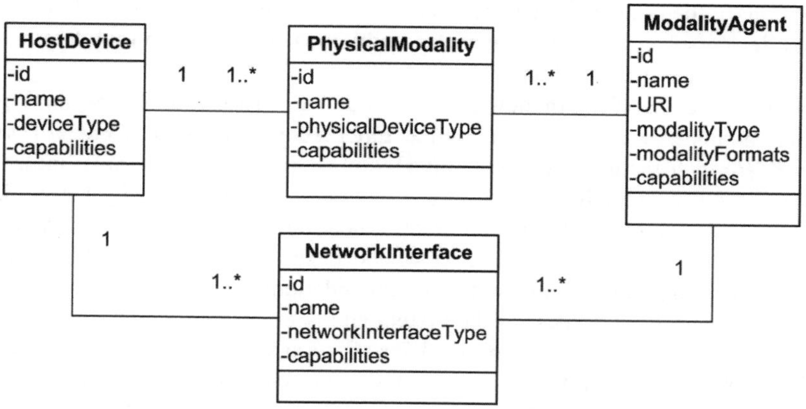

Figure 5.8 Device description main categories and relationships.

Modality Agents. Finally, these software agents are accessed using one of the available network interfaces that are supported by the Host Device.

From this model, the *User Interface Adaptation Function* (UIAF) is mostly interested in Modality Agent descriptions. Typically, information on a Modality Agent includes its modality class (audio, video, text, etc.), the means to access it, and the supported data format as well as other capabilities information. This is used by the Fission and Fusion modules of the Modality Function to decide how service output and input should be realised, respectively.

As most devices cannot provide a comprehensive set of information about themselves and the modality services they support, the description might be retrieved from a distributed database, as it is done in the UAProf approach. In this case, the complete discovery procedure takes two steps:

1. the actual 'physical' discovery of the device in the vicinity and its reporting to the system;
2. the lookup by the Device Handler for information about the device and the service it may host.

For more details of this description framework and associated mechanisms, refer to [9].

5.1.6 Multimodal Output

One important aspect of the work presented here is to take advantage of the interaction between context awareness (including personalisation) and multimodality, in particular for finding the best way to present multimedia information, through the best modality channel. For example, when a user is driving her car and she receives a message, the best modality for output could be considered to be audio output, so as to avoid distracting the user from driving the car. Such behaviour can be achieved by taking into account the media elements coming from the application, the user context and the available devices.

Multimodal applications send their media output to the Modality Function as an application request. In this application request, the application can provide multiple media elements for output. It can provide alternative media element representations to assist the Modality Function in supporting multiple output modalities. For example, it is possible to provide a rich text representation (including images and audio) using XHTML and an alternative plain text representation for the same information. This section will give an indication about how media output and device selection is envisioned for multimodal interfaces in mobile environments based on the work described in [10].

The reference message flow, on how output requests are processed, is shown in Figure 5.9.

Information about the context is received as input from the *Personal Context Function* by the *Resource Handler* in the form of modality recommendations. This includes information on whether or not to use a certain device or more generally a certain modality for media output. Through personal preferences, users can define whenever they would like the system to take automatic modality and device decisions or whenever the system should ask before taking a decision. This is also part of the modality recommendations.

During the initialisation process of the system and later, when the user is moving, devices available in the user vicinity and their characteristics are stored and updated into the device tables of the system. Device information is stored as presented in the previous section.

In addition, the Modality Function is also able to perform transformations on content to better suit the user's context and preferences. These transformations include modality transformation (for instance text-to-speech) and language translation.

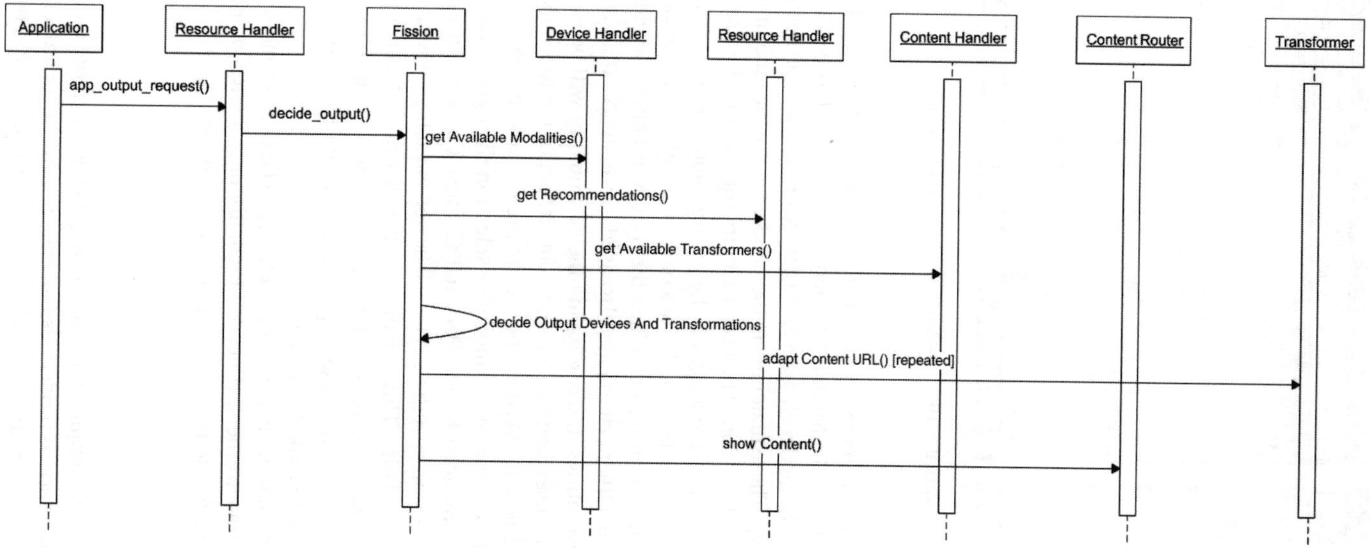

Figure 5.9 Output request processing.

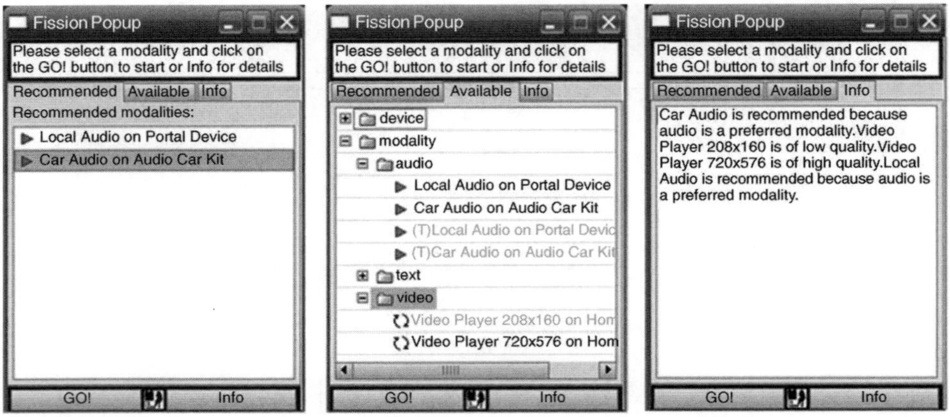

Figure 5.10 Modality selection dialogue.

Considering the media elements received from the application, the available devices and the available content transformers, the Modality Function is able to determine possible ‘matches’. A match is a combination of a media element and a device on which it can be displayed, possibly using content transformers. These matches are then ranked using the recommendations coming from the Personal Context Function. The major factors for ranking the matches are modality and device recommendations. Depending on the context and user preferences, the Personal Context Function can suggest the best modalities or devices to use. For instance, when the user is in her car, audio modality is preferred. When the user is at home, the Home Theatre could be a recommended device. Other criteria are also used: for instance, if a video media element is proposed in several resolutions, the one giving the best quality will be ranked higher.

Then, depending on the user preferences and context, either the top-ranked match is automatically selected, or the user is asked to choose among the possible matches.

Figure 5.10 shows an example of the modality selection dialogue. The first tab shows the modalities that are recommended by the Personal Context Function in the current context. It shows a list of modalities on specific devices. The second tab shows all possible matches, ordered by device and by modality. Through this tab, the user can select her preferred modality even if it is not actually recommended. The third tab gives some information as to why specific modalities were selected. These explanations are derived from the process where recommendations are used to rank the matches.

After this stage, an output modality has been selected. The corresponding media element is then sent to the content transformers, if needed (actually only a URL pointing to the content is sent), and finally the media element is sent to the output device.

5.1.7 Multimodal Input

Multimodal fusion or input integration aims at allowing the user to more naturally interact with computer systems through various modalities (speech, gesture, keyboard, etc.). The user experience can be enhanced by integrating several user input modalities into one consistent input model. The input model then would contain not only the knowledge about single modalities, but also extracted meaning from the user interaction. Although there is some advanced work in this area, multimodal fusion described in this work allows the user to interact with a mobile application through various input channels by selecting one single input channel at a given time. Nevertheless, the framework for fusion described here also aims at being compatible in integrating more then one input modality, which depends on the algorithm implemented in the fusion itself.

Therefore the fusion process consists of the following three main steps:

1. The mobile application registers its request to receive multimodal input giving an application input request based on a defined XML structure to the Modality Function as described in the application interface section (Section 5.1.4).
2. The user's multimodal input (enriched with timing information) is received by the Fusion component coming from recognisers in an EMMA [17] format.
3. The application is informed when an recogniser input matches the given application input request.

5.1.7.1 Mobile Application: Requesting for Modality Input

Mobile applications are enabled to register their input requests using the application input request mechanism. Figure 5.11 shows the sequence when registering for modality input.

Applications send their requests for input subscription to the Resource Handling component in an XML format similar to the one described above for output. These requests are forwarded to the Fusion component, which stores their parameters in an internal registry. When the Fusion component receives user input relevant for the application, this input will be sent back to the application. This continues until the application requests its input subscription to be deleted.

The Fusion component handles elementary input. Management of the dialogue remains the responsibility of each application.

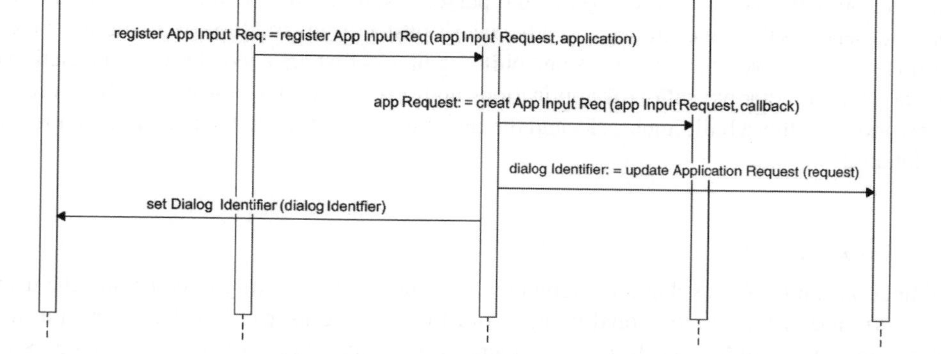

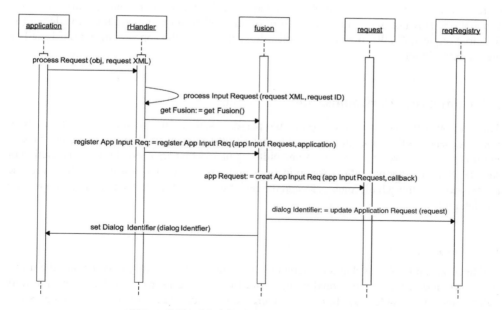

Figure 5.11 Modality input request registration.

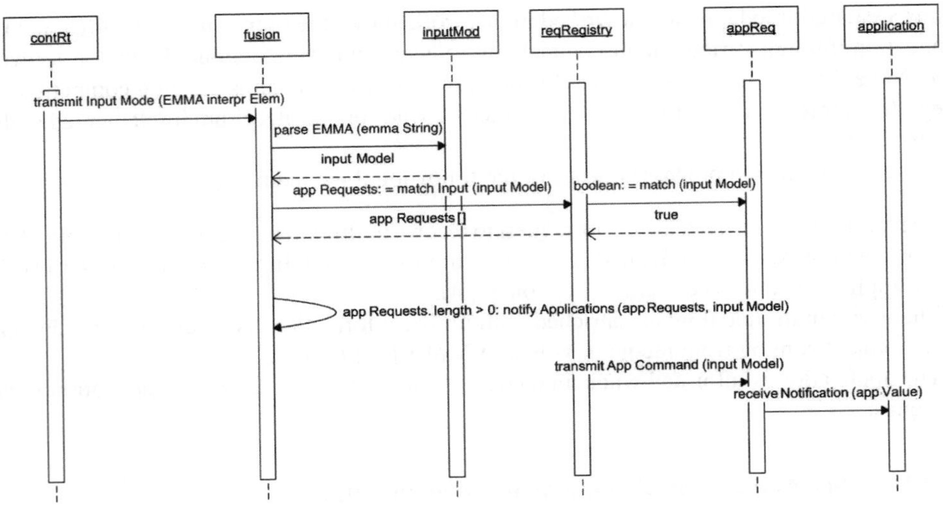

Figure 5.12 Modality input notification.

5.1.7.2 Modality Input Processing and Mobile Application Notification

User input is received constantly through the device agents and their recognisers. The recognisers use the W3C EMMA [17] to notify the Fusion component through the content router of a new modality input token (e.g. voice command 'OK'), as depicted in Figure 5.12.

When the fusion component receives the EMMA string, it parses it into an input model understandable for the fusion process. Previously stored application input models in the request registry will be checked against the retrieved modality input token. If the modality input token matches one of the registered application requests, the Fusion component will inform the application. This notification takes the form of an application command that was defined in the input request. Finally, the mobile application can execute the appropriate actions based on the modality input.

5.2 Contextual Personalisation

This section presents research on contextual personalisation – the personalisation of applications and services that takes into account the user's past, current and future situational context. For this purpose, several mechanisms and building blocks that are needed for the realisation of such contextual personalisation of applications and services are introduced. Furthermore, the interworking of these building blocks and the resulting contextual personalisation architecture are detailed.

5.2.1 Introduction

Applications and services that are augmented and enhanced by taking into account the user's past, current and future situational context aim to assist the user in her daily life in a more natural and pleasurable way. In order to increase user acceptance of these applications and services, user needs and expectations have to be considered in detail. The user should not have to configure and manage devices, local connectivity multimodal interfaces, applications and services in her Personal Area Networks or Body Area Networks by herself.

One key issue in the personalisation of applications and services is the diversity of user contexts. A context may be described in terms of many user-related data elements that represent in turn her geographical coordinates and velocity, activity, mood, time schedule, technical aspects like bandwidth of the network and capabilities of the terminal, sensor input, etc. Contextual information is defined as any information that can be used to characterise the situation of an entity, where an entity can be a person, place, physical or computational object [4].

Another more challenging issue in contextual personalisation concerns the dynamics and emergence of contexts of a user. Preferably, applications and services should automatically, in real time, in the appropriate way adapt themselves to changes in the user's context whenever the user has trusted and allowed them to do so.

Current applications and services offer various means to adapt user interface presentation and operation to individual users' and user groups' needs. However, in existing systems, the support for contextual personalisation is often proprietary and application-specific. For example, Amazon [1] performs personalisation for books, while restaurant guides offer personalisation for user's favourite meals. There is no commonly accepted application-independent architecture which brings together different existing personalisation approaches.

Application-independent context-awareness support is needed to support the personalisation architecture. This contextual personalisation support allows an application or service developer to focus on the creation of context-dependent personalised applications by utilising the functionality provided by the personalisation architecture, instead of having to deal with the complexity of context gathering, processing and reasoning issues on the application side.

User preferences, profiles and profile management are also usually application-specific. Every application uses its own profile representation and maintains its own user profile. Such an application-specific setting does not support the simultaneous re-use and adaptation of user profile data for different applications and does not support the user in easy management of all her data. Hence, the creation and modification of the user's preferences have to be carried out separately for each application, leading to the modification of user data being a tedious and time-consuming task for the user.

Existing profile management approaches only partly support the management and automatic delivery of context-dependent preferences as envisioned in context-aware environments. For instance, the profile management in [2] needs manual activation of the appropriate user profile in order to meet the user's usage condition in a particular situation.

In order to design an architecture for generic contextual personalisation, several personalisation technologies are needed. The first is technologies concerning the representation, management and delivery of context-dependent user profiles. The second concerns the adaptation of the user preferences through learning. The third is about personalised recommendations for the adaptation of applications and services based on the user's current context and corresponding preferences.

This contextual personalisation of mobile applications and services for individuals is achieved in three ways:

- Research and develop a user profile management that supports the management, query and provision of context-dependent user profiles and preferences for various applications and services. In addition, such a heterogeneous application environment may require the need to deal with different application semantics.

- Perform contextual personalisation of applications and services for an individual user by collecting and storing user context data streams, and robustly processing and interpreting the intrinsically inaccurate stored context data streams. In this way, personalisation architecture components can perform sensible contextual personalisation of applications and services.
- Adapt the user profile and corresponding user preferences according to her behaviour by learning user preferences and rules. The user data can be contextualised and also automatically updated for her if she wishes. This learning process is applied in order to distil rules or Bayesian schemes used in application recommendations, including the multimodal interface described in the previous section.

5.2.2 General Approach

The main elements of this contextual personalisation architecture are depicted in Figures 5.13 and 5.14. This approach brings the means for providing context-dependent recommendations and control actions; i.e. the process of yielding recommendations that takes into account the user's current context. The developed approach also supports the learning of new or updates of existing context-dependent user preferences; i.e. adaptation of the user's profile to the user's needs and habits. For this purpose, the learning process utilises the user's usage behaviour in correlation with the user's usage context.

Realisation of these contextual personalisation tasks into concrete building blocks is depicted in Figure 5.15. The Contextual Personalisation Architecture consists of *Context Gathering and Interpretation* (CGI), *Usage Record Provider* (URP), *Individual Profile Management* (IPM), the *Profile Editor* and the *Recommender*.

The Contextual Personalisation Architecture is a receiver of raw sensor data – data about available services, devices and modalities; and application requests from the Modality Function and other services and applications. Subsequently, the architecture processes this data in several steps yielding recommendations and control actions as output to any interested service and application needing this kind of information.

In order to obtain the recommendations and control actions, the above input data has to pass through different functional blocks. The first one is the Context Gathering and Interpretation

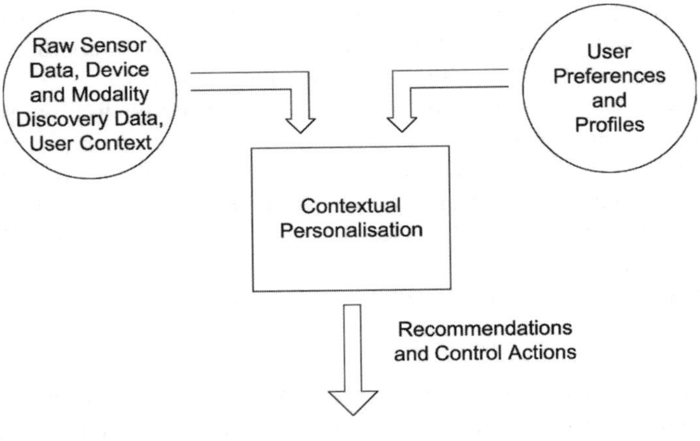

Figure 5.13 Provision of context-dependent recommendations.

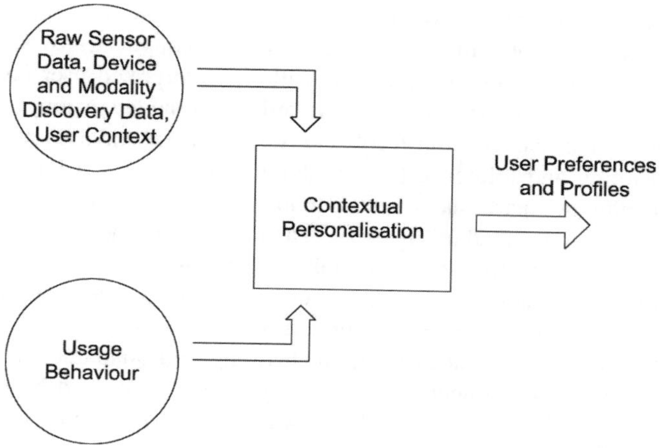

Figure 5.14 Learning and updating context-dependent user preferences and profiles.

component that receives raw sensor data, device and service discovery data, etc. It transforms and grounds all data streams described in context representations and links them to the open upper ontology (introduced in Chapter 4). It produces low-layer and high-layer interpretations of the user's context and situation. For instance, combining GPS input data of the user's car and similar location data provided by the user's mobile device can be sufficient for concluding that a user is driving or sitting in a car.

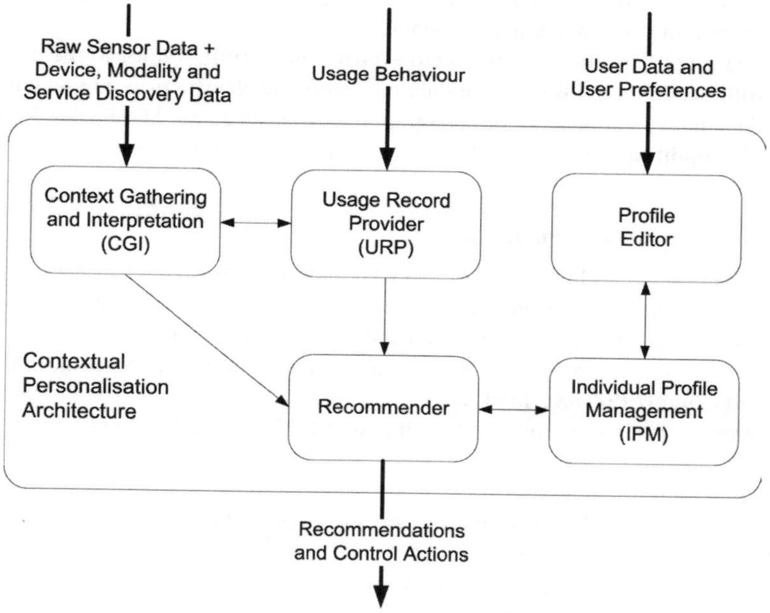

Figure 5.15 Contextual Personalisation Architecture.

The Usage Record Provider receives feedback in terms of the user's usage behaviour or direct user feedback via the Modality Function, services and applications. The Usage Record Provider correlates the user's usage behaviour with the interpreted user context data from the Context Gathering and Interpretation block in order to provide meaningful usage records, which can be described as pairs of user feedback/user behaviour and user context.

Because the learning and inference (e.g. applying rules) tasks are intrinsically coupled, the learning and inference parts have been put into closely linked subcomponents of the Recommender, which is described in Chapter 4. The inference of the Recommender produces the recommendations and control actions by taking user contexts and user models as input. In deriving the recommendations and control actions, the inference part employs different contextual reasoning schemes stored in user models by *Individual Profile Management* (IPM). Sensible contextual reasoning schemes require learning and adaptation to the domain of interest. The learning subcomponent takes the usage records provided by the Basic Usage Record Provider and distils new user models from those usage records and available user models provided by IPM.

Thus, the Contextual Personalisation Architecture provides a means to store all these models into the IPM, from which the Recommender can fetch context-relevant user models including contextual reasoning schemes. In addition, the IPM manages context-dependent user data for various services and applications, and provides interfaces for the Profile Editor and the learning part of the Recommender. It enables creation, update, deletion and inquiry of context-dependent user data and user profiles. The stored user data can be enriched with conditions in linking the user data to a particular user situation such as the time of day, the user's location or the availability of a user's device. It can also be linked to a logical conjunction of different context parameters such as the user's location at a specific time of day. Last but not least, the IPM takes into account the diversity of services and applications in not restricting the structure and semantics of application-specific user data.

Besides means for learning, the architecture provides a Profile Editor to the user for managing her profile and context-dependent data and preferences. The user can add, modify and delete user data and preferences through a graphical user interface. The Profile Editor allows the user to add conditions (e.g. a specific time of day) to sets of user preferences.

5.2.3 Context Gathering and Interpretation

The Context Gathering and Interpretation (CGI) component processes context information. It receives low-level context information from various sources, and infers higher-level context information so that it can be more easily used by other components. As such, it follows the *Context Management Framework* described in Chapter 4, and exposes both the *Context Consumer* and *Context Provider* interfaces.

Some examples of raw user context data that the CGI interprets are the following:

- presence and mood of the user;
- location information obtained by various means such as GPS or the presence of known Bluetooth devices in the vicinity;
- movement information generated by a pedometer;
- fitness-reltated information about the user, such as heart rate;
- application and service usage information.

Subsequently, the Context Provider interface of the CGI allows the CGI to send structured data to the Recommender for recommendation and learning purposes.

The rest of this section describes the context gathering, interpretation and provision flow.

5.2.3.1 Context Gathering

The CGI component receives context information through its Context Consumer interface through several sources:

- The Modality Function provides information about devices discovered in the vicinity of the user.
- Applications provide information about their usage by the user.
- Any Context Provider.

5.2.3.2 Context Interpretation

Once raw context information is gathered, CGI can perform its main task: infer higher-level context information. This information will be made available to any other component. In particular, the inference part of the Recommender has to produce the recommendations and control actions by contextual reasoning about the applications or services, the personal profiles, and the current environmental situation and context interpretations. In order to meet application or service and user specific recommendation performance levels, the provision of robust and grounded situational context information about a user to the inference part of the Recommender is of utmost importance. The CGI applies various pre-processing, feature extraction, feature selection and classification schemes.

A simple example of the CGI in action is the inference of the user location from device discovery information obtained from the Modality Function. Whenever the Gateway Function discovers a new device in the vicinity of the user, the Modality Function reports this to the CGI. The CGI then consults the *Individual Profile Management* (IPM) component for finding appropriate interpretation schemes that can be applied to the raw data available about the newly reported device. Based on the information returned by IPM, a chain of feature extractors, classifiers, etc., that can handle the data is created. If a new Bluetooth device is discovered in the vicinity of the user, the MAC address of the device is provided to the CGI, and will be compared to the lists of addresses of known user devices stored in IPM. If, for instance, the MAC address of the newly discovered device matches that of the user audio car kit, it can first be inferred that the user is close to this device, and thus that she is currently in her car.

Whenever an entity disappears (such as when a Bluetooth device is switched off), this will also be reported to the CGI. This has the implication that any processing that is dependent on the information supplied by such an entity is no longer possible. The associated selectors, classifiers, etc., can therefore be removed.

The learning part of the Recommender has the ability to learn and update interpretation schemes accordingly and it will forward these to the IPM for storage. Context Gathering and Interpretation will be subscribed (as *Context Consumer*) to changes in all the categorisation schemes it uses. Whenever the Recommender updates one of these in IPM, this component notifies Context Gathering and Interpretation of the changes.

5.2.3.3 Context Provision

Finally, the inferred higher-level context information is made available to all components that may request it through the Context Provider interface of the CGI.

5.2.4 Individual Profile Management

The requested functionality of the Contextual Personalisation Architecture requires the ability to manage context-dependent user profiles of different applications and services. Two major requirements have to be considered. First, a means has to be provided for describing, managing, querying and providing *context-dependent user profiles and preferences* – applied in specific user situations such as a specific user location, user activity, or a specific time of day, or combinations of different context variables. Second, one has to take into account that different applications may use different semantics for describing user data. For instance, one application could use the identifier *zip_code* whereas another service uses the identifier *postal_code* to denote the same information about the user's address. The following scenario depicts the requirement of dealing with context-dependent user preferences via an example application.

> *A businessman has a mobile phone. Typically, his preferences concerning mobile phone usage depend on his situation. Whilst in a meeting, he does not want to be notified by incoming news from his news feed service. On the other hand, when at home, he wants to be notified of all incoming news immediately. When driving his car, he wants to be notified of incoming news immediately* only *if speech output is available, since he does not want to be distracted.*

The support of such scenarios requires management, inquiry and delivery of context-dependent user profiles and preferences including the creation, modification and deletion of user data (i.e. general information about the user, user behaviour and user preferences).

In the following sections, an approach to structuring a user profile that considers the above requirements and enables the inclusion of situational constraints is described. Second, a proposal for a profile management component and related interaction with clients is detailed.

5.2.4.1 Profile Structure

The overall high-level user profile structure is depicted in Figure 5.16. Each user profile can be separated into different sections that include application-dependent user information. An application could also be interpreted as a virtual application – i.e. a group of actual applications that adhere to the same semantics and hence are capable of processing the related user data. Each application-related section includes at least one so-called profile subset, the *Default Profile Subset* (*Default Subset*); more details will be described while introducing the basic elements of the profile structure: *Profiles*, *Profile Subsets*, *Qualifiers* and *User Data*. The dependencies of these basic elements are depicted in Figure 5.16.

- *Profile*. A user profile contains all persistent data about a user – user data required by all applications and services the user is subscribed to.
- *Profile Subset*. A Profile Subset contains an application-related subset of all persistent user data. Besides the actual user data, it also contains Qualifiers. A user can have several Profile Subsets, which can be distinguished into Default and Conditional Profile Subsets. However, there has to be exactly one Default Profile Subset for each supported application including

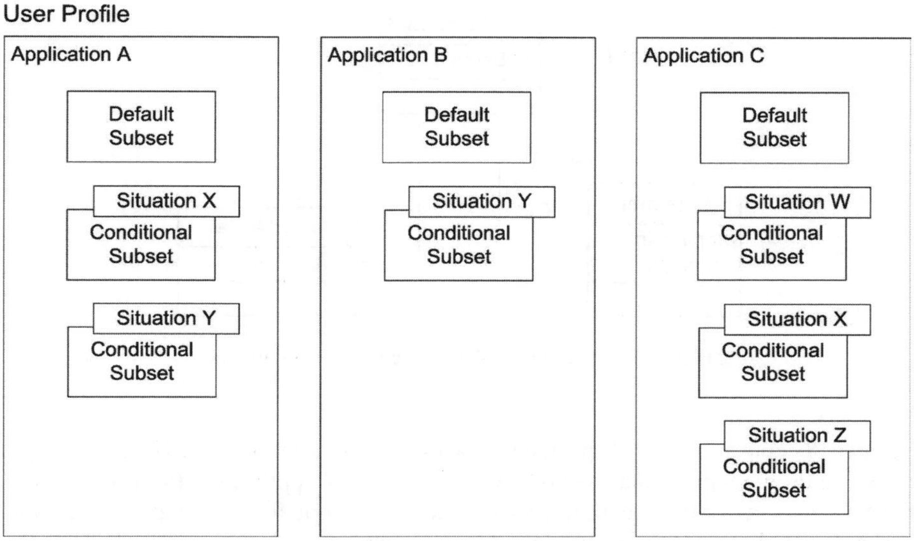

Figure 5.16 Illustration of User Profile structure.

default user information (see Figure 5.16). Furthermore, there can be several Conditional Profile Subsets related to each application. A Conditional Profile Subset includes situational constraints that describe the situation in which the user information within this Profile Subset shall be applied. For instance, the contextual constraint could express the location of the related user, a user activity or the time of day. When no user context is available or no Conditional Profile Subset matches the current user's situation, the Default Profile Subset will be applied.

- *Qualifier.* A Qualifier contains metadata about a Profile Subset. In particular, it identifies the application and the situation in which the Profile Subset shall be applied. The application identifier could, for example, be implemented as a URI that points to a unique application schema describing required user data. In this way also a group of applications adhering to this schema could share the available user data. On the other hand, the situation description could be implemented as a logical conjunction of attribute–value pairs and relation operators, where (i) the attribute is the type of context, such as location, (ii) the value represents the actual context value, such as at home, and (iii) the relation operator could, for example, be *equal*, *notEqual*, etc.
- *User Data.* The structuring of user data should be considered as application-dependent. Using RDF triples (consisting of subject, predicate and object) for representing this user data could be one way to deal with this heterogeneity. RDF triples allow more freedom to use application-dependent semantics than using attribute–value pairs, since it does not restrict an application developer in the development and in the use of vocabularies.

The application identifier of the qualifier in Figure 5.17 has always to be specified. In contrast, the context qualifier is optional. When the context qualifier is empty, the Profile Subset represents a Default Profile Subset, otherwise a Conditional Profile Subset.

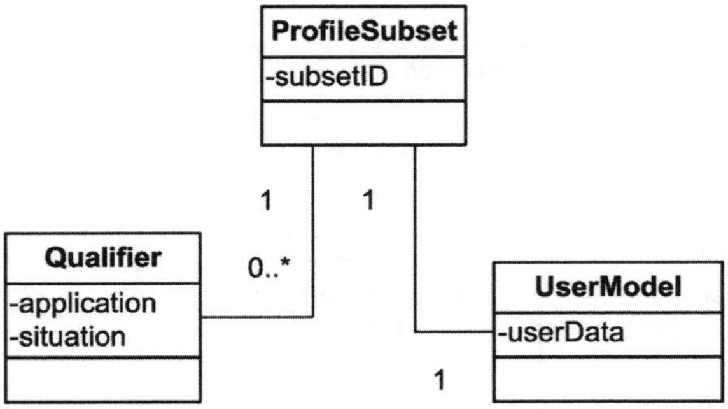

Figure 5.17 Profile Subset structure with qualifiers.

Figure 5.18 depicts an example profile following the above description. The example profile contains two Profile Subsets that are both related to the same application. Both are Conditional Profile Subsets, one for the situation 'user is at home' and one for the situation 'user is in her car'. The depicted example user profile is not complete, since the mandatory Default Profile Subset is not depicted. Both Profile Subsets contain a very small example user model in RDF notation. Prefixes and data type attributes are left out in the example user model for simplicity. For the Profile Subset (on the left-hand side), the user preference for the preferred news genre is *Traffic News* and the preferred output type is *Text*. By contrast, the preferred news genre of the Profile Subset on the right-hand side is *Sports* and the preferred output type is *Video*.

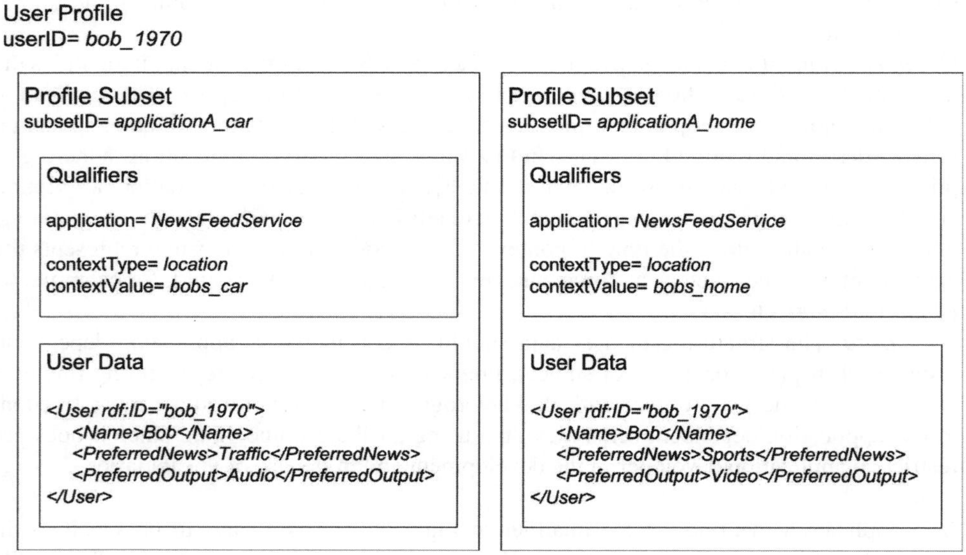

Figure 5.18 Profile excerpt example.

5.2.4.2 Component Structure

The Profile Management system can be decomposed into different subcomponents that can be mapped to the required functionality. These subcomponents are introduced below.

- *Profile Manager Core.* This subcomponent provides the external interfaces for user data management, inquiry and updates to the outside world – i.e. to the Profile Editor, service platform components such as the Recommender for learning and inference purposes, and applications. Furthermore, it manages and delegates incoming requests to additional subcomponents, such as the *Profile Subset Manager* or the *Profile Subset Search.* Besides delegating tasks, it is itself responsible for the creation, maintenance and deletion of user profiles, and communication with the database layer for profile storage and retrieval.
- *Profile Subset Manager.* This is the subcomponent for the creation, modification and deletion of Profile Subsets and related functionality. Furthermore, it handles the management of user data, since user data is directly related to Profile Subsets and not to the overall profile.
- *Profile Subset Search.* This subcomponent is responsible for the provision of the best matching Profile Subset, the user data that best matches the user's current situation. If a profile management client requests user data from the profile management component, the *Profile Manager Core* triggers the Profile Subset Search subcomponent. That subcomponent then uses algorithms to match the situation qualifier on the Profile Subset with the user's current context in order to find the best matching Profile Subset.
- *Profile Storage.* The Profile Management component should not specify any database-related requirements or constraints in order to enable easy usage and exchange of the underlying database management system. In this way, the Profile Management component could also abstract from additional functionality such as the distribution of profile data. Hence, the underlying database management system can also provide means for distributed profile data storage which is decoupled from the Profile Management development itself.
- *Profile Editor.* This should provide means for the management of a user profile, related profile subsets and context-dependent user preferences.

5.2.4.3 User Data Inquiry

In principle, two different options exist on how a user data inquiry could be processed. The main difference between these two options is related to the user context inquiry. In the first option (Figure 5.19), this is done by the requesting service platform component, such as the Recommender. In the second option (Figure 5.20), it is done by the Profile Management component itself.

The Profile Subset search process for the former option (Option 1) takes place as follows:

1. An application or service that requests user data from the Profile Manager provides the identifier of the related application or service and information on the user's current context. The application or service has therefore requested the user's context from the architecture beforehand; i.e. from a Context Provider (see Figure 5.19).
2. The Profile Management internally triggers the Profile Subset Search mechanism. This mechanism first checks whether the given application is supported; i.e. whether there is at least a Default Profile Subset containing default user data for this application.

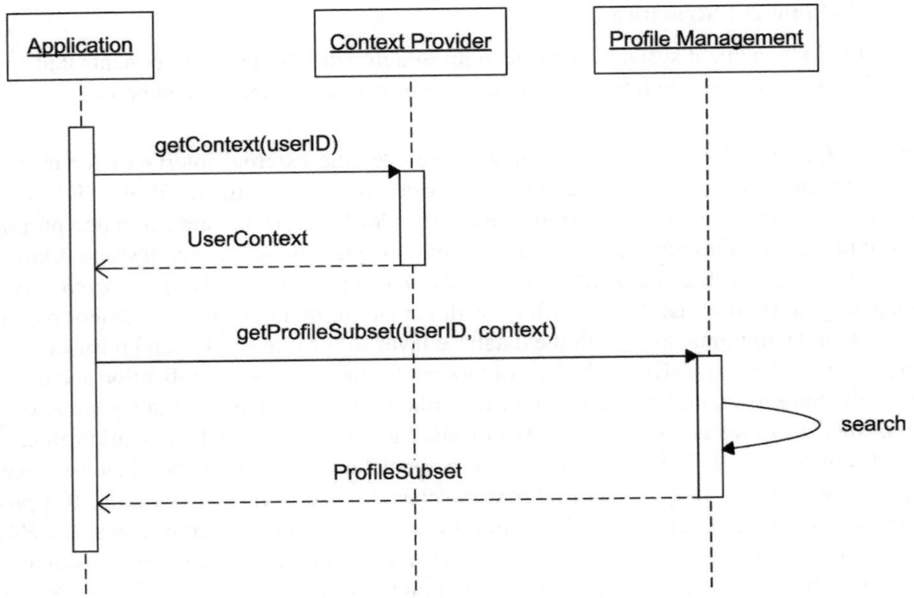

Figure 5.19 User data request (Option 1).

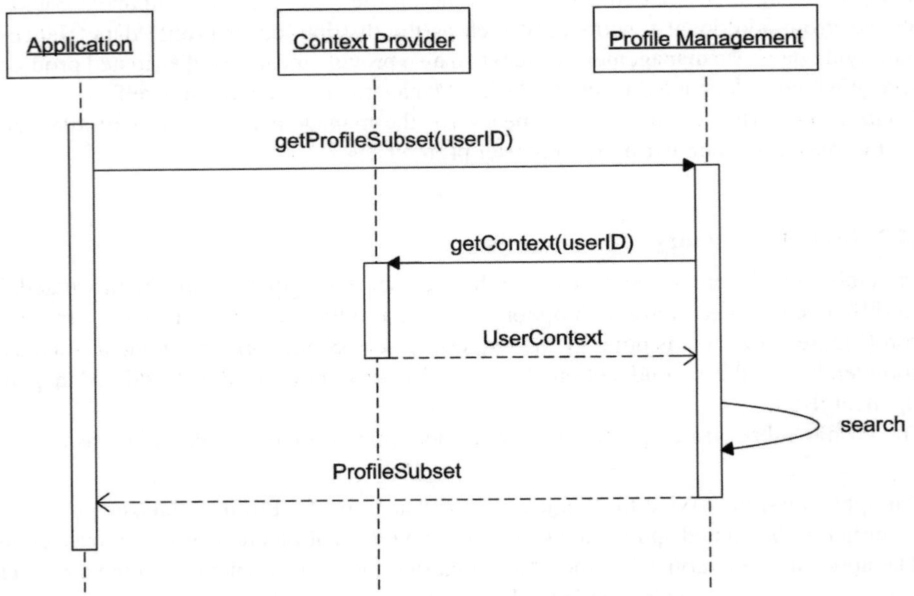

Figure 5.20 User data request (Option 2).

3. If the application is supported, the given user's current context is matched against the situation qualifier of the Conditional Profile Subsets (if any exist). If the application is not supported, the requester cannot be served successfully.
4. If the context matching step is successful, the corresponding Conditional Profile Subset is returned. If not, the default user data of the Default Profile Subset is returned.

Figure 5.20 depicts the second option (Option 2). The Profile Management itself organises the retrieval of the user's current context from a Context Provider such as *Context Gathering and Interpretation* (CGI). Since the Profile Management can look into situation qualifiers of existing Profile Subsets in order to evaluate which user context attributes are useful for the Profile Subset search process, the *getContext()* request could possibly be restricted to few context attributes and hence speed up this process a lot. For instance, when the user (or a learning mechanism) has never specified a situation qualifier that describes the appliance of a Profile Subset based on the user's location, requesting the user's location from a Context Provider such as CGI would be superfluous. The second advantage is that clients requesting user data do not necessarily have to deal with context retrieval. This is particularly interesting for applications that are not part of the Contextual Personalisation Architecture (in contrast to platform services that are part of that architecture), since complexity is taken away from application development.

5.2.5 Usage Record Provider

The purpose of the Usage Record Provider (URP) is to store user actions annoted by the user context at the time when the action was realised. The main objective is to gather data to be used by the learning part of the Recommender.

The URP stores this information as a set of elementary *Usage Records*. A Usage Record created by the URP is a data structure that stores the behaviour of an entity in a specific context. It must contain all information that will be useful for learning user preferences. The main parameters each record contains are:

- the time when the action was performed;
- the context of the user or group in which the action was performed;
- the identifier of the user, group or application or service to whom this Usage Record is relevant;
- the action performed;
- links to ontologies that allow to semantically interpret the other parameters.

Applications or services can place Usage Records in the URP. They do not have to specify the context – the URP will itself collect information from Context Providers.

Usage Records enable the learning process within the learning component of the Recommender. From the Usage Records the situations in which the user requests certain services or performed specific activities may be derived. Based on the learned behaviour of the user, the Recommender is then better able to give sensible recommendations to applications.

5.2.6 Recommendations for Personalisation

The previous sections described the various building blocks that enable *contextual personalisation*, and in particular how context and user information can be stored and managed. The output of the personalisation system is *recommendations* – pieces of information to be interpreted by the application to better adapt to the user in a given context. Recommendations can take various forms, depending on the receiving applications. For the *Modality Function* described earlier in this chapter, they are a selection of modalities and devices the user prefers in her current context; for instance audio through the car-kit in her car or video on the large screen television in her home. For the *Multimedia Infotainer* application described in Chapter 7, it may be preferred news topics.

The recommendations are determined by the Recommender component, which is described in Chapter 4.

The following is a typical example of how the complete system is used. Consider that the user has installed on her device various applications that all use the multimodality framework described earlier in this chapter. This allows these applications to use different devices and modalities to present content to the user. Initially, no information about preferred modality exists for the user. Thus, the Modality Function will be able to make only limited inferences for proposing the most suitable modality to the user. For instance, it will be able to propose high-quality and high-bandwidth video on the large screen television, or lower-quality low-bandwidth video on the user's phone while at home, but will not have information about the preferred device. However, when the user makes a manual modality selection, it will be sent to the *Context Gathering and Inference* component, which in turn will create a Usage Record. This record will store the current user context and modality selection. Progressively, the learning components of the Recommender will be able to create a model of the user regarding her modality preferences, and this model will be stored by the *Individual Profile Manager*. Then, when the user is again in a similar context, the inference components of the Recommender will be able to use this model and recommend modalities to the Modality Function. For instance, the system may have learned that at home the preferred device is the television, while in the car the preferred device is the handsfree kit and the preferred modality is audio. A similar process will take place for all applications that provide information to the CGI and make use of the Recommender API.

Note that the personalisation process is separate from the applications. Applications only have to provide information about user actions, and will get no request recommendations related to those actions. Usage Records link actions with context information that have been processed by the CGI (not raw context information), which allows an easier generalisation of the knowledge learned about the user. In turn, the raw context information is provided by any Context Provider the CGI has access to, without applications having to take care of this. This allows applications to better adapt to the user by using all available information, without requiring application developers to take care of this by themselves.

5.3 Conclusions

In this chapter, a set of components providing multimodality and personalisation services has been presented. These components realise part of the architecture described in Chapter 3, together with the context management components presented in Chapter 4, and privacy, trust and group communication components described in Chapter 6.

These components have been used in applications described in Chapter 7. Technical and user evaluations of these applications have proved the benefits of the technical approach chosen, and the interest shown by end-users. Nonetheless, many challenges remain, from both commercialisation and research points of view.

Some improvements to data format and management appear useful or even necessary for a widespread deployment of these services. Requirements for a device description rich enough for supporting multimodality have been listed, and a description language has been created, but a common language would have to be adopted by the industry. Also, the contextual personalisation of services and applications has to deal with different service and application semantics. In this regard, an open issue is the interoperability of user data between applications and services that adhere to different semantics. The re-use and sharing of user data would be a valuable support for the user, since it can certainly ease the management of user data. For instance, a change of general user data such as the user's credit card details that are required by all pay services could be done once for all applications instead of once for each single application, assuming the user has agreed to this. Also, schema matching mechanisms could be used in order to enable the re-use of user data. However, existing mechanisms do not provide fully automated results in finding mappings between different schemas and hence need the interaction of the user or a system administrator.

Many improvements could be made to the internal mechanisms of each component, and this offers further research areas. For instance, regarding modality fusion, an interesting idea could be to combine pure user modality input with the current user's context. For example, based on the current activity of the user ('in a meeting with other people', 'jogging', etc.), a more accurate interpretation of the current user intention could be given. Regarding user modelling, the component has been designed with flexibility in mind, which eases the re-use and evaluation of different machine learning methods. In the future, additional machine learning methods other than rule learning and neural networks could be implemented. Last but not least, the relevance of different machine learning methods, and hence the resulting personalisation of applications and services, could be tested in performing user acceptance studies.

5.4 Acknowledgements

The following individuals contributed to this chapter: David Bonnefoy (Motorola SAS, France), Mathieu Boussard (Alcatel-CIT, France), Nermin Brgulja (University of Kassel, Germany), Alexander Domene (Fraunhofer FOKUS, Germany), Olaf Drögehorn (University of Kassel, Germany), Giovanni Giuliani (HP Italiana, Italy), Ralf Kernchen (University of Surrey, UK), Sian Lun Lau (University of Kassel, Germany), Jean Millerat (Motorola SAS, France), Bernd Mrohs (Fraunhofer FOKUS, Germany), Petteri Nurmi (University of Helsinki, Finland), Pekka J Ollikainen (Nokia, Finland), Mateusz Radziszewski (BLStream, Poland), Christian Räck (Fraunhofer FOKUS, Germany), Marcin Salacinski (BLStream, Poland), Alfons Salden (Telematica Instituut, The Netherlands), and Michael Sutterer (University of Kassel, Germany).

References

[1] Amazon.com. Online: www.amazon.com.

[2] Caokim S. and Sedillot S.: 'Profiles Management for Personalised Services Provisioning'. 2nd European Conference on Universal Multiservice Networks, ECUMN 2002, 8–10 April 2002 pp. 315–321.

[3] Corradini A., Mehta M., Bernsen N.O., Martin J.-C. and Abrilian S.: 'Multimodal Input Fusion in Human–Computer Interaction: On the example of the NICE project'. *NATO Science Series, III: Computer and Systems Sciences*, Vol. 198, 2005, pp. 223–234.

[4] Dey A. and Abowd G.: 'The Context Toolkit: Aiding the development of context-aware applications'. Workshop on Software Engineering for Wearable and Pervasive Computing, Limerick, Ireland, June 6, 2000.

[5] Duarte C. and Carriço L.: 'A Conceptual Framework for Developing Adaptive Multimodal Applications'. 11th International Conference on Intelligent User Interfaces, Sydney, Australia, 2006.

[6] Elting C., Rapp S., Möhler G. and Strube M.: 'Architecture and Implementation of Multimodal Plug and Play'. 5th International Conference on Multimodal Interfaces ICMI03. Vancouver, BC, 2003.

[7] Johnston M.: 'Unification-based Multimodal Parsing'. 36th Annual Meeting on Association for Computational Linguistics. Montreal, Quebec, Canada, 1998, pp. 624–630.

[8] Kernchen R., Boda P.P., Moessner K., Mrohs B., Boussard M. and Giuliani G.: 'Multimodal User Interfaces for Context-aware Mobile Applications'. 16th Annual IEEE International Symposium on Personal Indoor and Mobile Radio Communications. Berlin, Germany, 2005.

[9] Kernchen R., Boussard M., Haensel R., Moessner K. and Mrohs B.: 'Device Description for Mobile Multimodal Interfaces'. 15th IST Mobile & Wireless Communication Summit, Myconos, Greece, 2006.

[10] Kernchen R., Mrohs B., Sałaciński M. and Moessner K.: 'Context-aware Multimodal Output Selection for the Device and Modality Function (DeaMon)'. 6th International Workshop on Applications and Services in Wireless Networks. Berlin, Germany, 2006.

[11] Larson J.A., Raman T.V., Raggett D., Bodell M., Johnston M., Kumar S., Potter S. and Waters K.: 'W3C Multimodal Interaction Framework'. 2003. Online: www.w3.org/TR/mmi-framework.

[12] Oviatt S.: 'Multimodal Interfaces'. In Jacko J.A. and Sears A. (eds), *The Human–Computer Interaction Handbook: Fundamentals, evolving technologies and emerging applications*. Lawrence Erlbaum Associates, 2003.

[13] Pfleger N.: 'Context Based Multimodal Fusion'. International Conference on Multimodal Interfaces, State College, Pennsylvania, USA, ACM Press, 2004, pp. 265–272.

[14] Rugelbak J. and Hamnes K.: 'Multimodal Interaction: Will users tap and speak simultaneously?'. *Telektronikk*, Vol. 2, 2003, pp. 118–124.

[15] Schomaker L., Nijtmans J., Camurri A., Lavagetto F., Morasso P., Benoît C., Guiard-Marigny T., Goff B.L., Robert-Ribes J., Adjoudani A., Defée I., Münch S., Hartung K., Blauert J.: 'A Taxonomy of Multimodal Interaction in the Human Information Processing System'. Report of the ESPRIT PROJECT 8579, February 1995.

[16] Sievers T. and Sidnell J.: 'Introduction: Multimodal interaction'. *Semiotica*, Vol. 156, 2005, pp. 1–20.

[17] W3C: 'EMMA: Extensible MultiModal Annotation markup language'. 2005. Online: www.w3.org/TR/emma/.

[18] W3C: 'Synchronized Multimedia Integration Language (SMIL 2.1)'. 2005. Online: www.w3.org/TR/SMIL2/.

[19] Wahlster W., Reithinger N. and Blocher A.: 'SmartKom: Multimodal communication with a life-like character'. EUROSPEECH-01, Vol. 3. Aalborg, Denmark, 2001, pp. 1547–1550.

6

Privacy, Trust and Group Communications

Edited by Göran Schultz (LM Ericsson, Finland), Olivier Coutand (University of Kassel, Germany), Ronald van Eijk (Telematica Instituut, The Netherlands), Johan Hjelm (Ericsson AB, Sweden), Silke Holtmanns (Nokia, Finland), Markus Miettinen (Nokia, Finland) and Rinaldo Nani (Neos, Italy)

This chapter discusses the ability of mobile communications systems and infrastructure to adapt to an individual's needs and expectations in different situations, and consequently provide personalised capabilities to users.

Earlier in this book, the environment was where individual users interact with objects and services. In doing so, one tends to regard individuals as isolated entities on technological islands, and the interaction between people and their environments is easily ignored. Consequently, an important aspect when envisioning future communication systems is that individuals act in groups when undertaking various kinds of activity. Therefore, advances in mobile applications and service infrastructures must also allow individuals to relate to, share and interact with each other and common artefacts. The same principles of ease of use apply to groups as well as to individuals.

Section 6.1 describes an implemented approach to investigate group communication for ubiquitous applications and services. First, the key concept of a group is discussed and the existing technologies dedicated to groups in the World Wide Web and mobile environments are briefly reviewed. Next, a vision of group communication via a scenario that has guided the research in this book is described, providing requirements for the communication system as well as for the modelling of groups. The last part of the group communications section

Enabling Technologies for Mobile Services: The MobiLife Book Edited by Mika Klemettinen

presents an overview of the designed architecture of the communication system as well as some descriptions of its major building blocks.

Since a mobile system including personal devices and applications often contains considerable amounts of person-specific data about a user, it is very important that the privacy protection measures in place are sufficient to ensure users' trust in the system. At the same time it must be possible for the user to share information in an efficient and straightforward way with parties the user trusts.

The basic approach adopted is therefore that privacy control is user-centric. The user must always be in control of how her data is shared, with whom and under which circumstances. This user control is realised by user-defined privacy policies. Section 6.2 describes how the system presented in this chapter is designed to enable users to trust it.

The trust architecture permits user control over data access. The use of the *Group Awareness Function* (GAF) to connect this access to group management enables the control of access to personal data at an individual Context Provider level, and using policies integrated with the group management enables the management of data in a fine-grained way.

How well does the system correspond to the requirements of the users? How does the user visualise the privacy policies? Both these aspects have been investigated in privacy research, but not in terms of a context-based system. Section 6.3 discusses how to visualise user choices regarding groups, privacy and trust. Section 6.4 is a short conclusion.

6.1 Group Communications

6.1.1 Technological Review of Groups

6.1.1.1 Group: A Social Entity

Groups play a crucial role in human affairs, and a large proportion of human behaviour relates to groups. People form groups for many purposes: to make decisions, to solve problems together, to satisfy social needs, to undertake activities, etc. In fact, interactions between individuals can potentially occur as soon as individuals share any type of social relationship. Such relationships are an essential component of humanity, and are exhibited in groups such as families, relatives, friends, colleagues and so on. More generally, any grouping of individuals sharing interests, or undertaking activities together, results in the formation of a group.

Groups have been studied for decades in the field of social psychology with the purpose of examining aspects of group behaviours not as isolated phenomena, but as interrelated processes in social interactions. In this work, the same view is shared and the term 'group' is referred to as it is defined in [18]:

> A group is two or more individuals each aware of his or her membership in the group, the others who belong to the group and their positive interdependence as they strive to achieve mutual goals.

As this definition points out, an important property of groups is that group members are aware of belonging to the group and of the other members. Consequently, this definition leads to the distinction between highly connected groups and loosely connected, hub-like structured communities. In the latter category the entities are fully managed by a single individual (e.g. 'My Friends' groups, where members are relevant to a single person but may not know each other). In such communities communication takes place primarily between the administrator

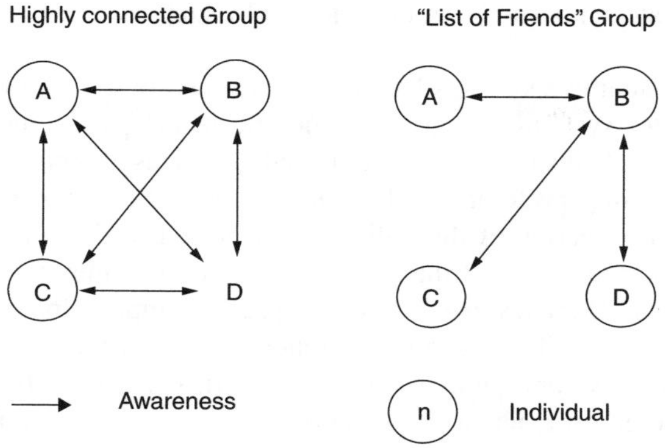

Figure 6.1 Highly connected group versus loosely connected, hub-like group.

and individual group members. Emphasis is not placed on the sharing of experiences, but rather on the categorisation of persons from the perspective of an individual. The distinction between the two categories is depicted in Figure 6.1.

The development of new communication technologies enables new types of group communications. Social relationships and the resulting groups are no longer constrained by traditional communication media. Instead, new forms of groups can appear, since everyone can communicate via a large variety of new electronic media (SMS, emails, file transfers, etc.). In a ubiquitous computing environment, groups are able to perform new types of action: they can interact on an informational basis, share and exchange personal resources and data; or interact on a physical basis, accessing objects together, for instance. The characteristics and states of any one individual can influence everyone's environment. Finally, coming together in and participating in groups can occur in a much more flexible way than was previously possible. Groups may be planned but also formed ad hoc. One user can take the initiative and invite people or people can subscribe to groups themselves. Groups may exist over longer time, or exist to perform one task or activity only.

6.1.1.2 Technologies for Groups

Groups have been studied in various fields of computer science. In particular, research directions and applications have targeted groups both in the web and in mobile environments.

Computer Supported Cooperative Work (CSCW) is a field of computer science concerned with the development of software ('groupware') to support and coordinate group activities. Groupware tools are used to perform group activities, typically involving the sharing of information artefacts (e.g. presentations, documents, applications, video streams, voice streams), allowing people to manage such shared resources (e.g. permit specific users to create such artefacts, view them, manipulate them, check their status).

In the web, services have appeared in the last few years that provide community members with capabilities to communicate via 'chat' forums (MSN groups, Yahoo! groups, Google groups) and instant messaging (IM) enhanced with presence information (e.g. 'available', 'offline', 'away'), or to share information (photos and files) and ideas. In mobile environments, social software on smart mobile devices (MoSoSo) facilitates social encounters and help people to communicate with small groups of people such as close friends, colleagues or relatives [6,23].

Lately, research in adaptive systems has been performed to develop personalised systems targeting groups. These systems use group modelling techniques – and in particular group preference learning mechanisms – to combine individual user models and model groups. In MusicFX [21], a group preference arbitration system allows the members of a fitness centre to influence the selection of the radio station played in the centre. Members specify their musical genre preferences and the group preference is computed using an arbitration algorithm. GroupCast [22] makes use of UniCast [22], an application which allows users to specify content they would like to see on peripheral displays located within their primary workspaces. UniCast runs continuously on a dedicated, peripheral display; it allows for a broad selection of content; it reacts to the location of its owner via an infrared badge system; and it is tied into and makes use of content belonging to other UniCast user profiles. The current implementation of GroupCast uses all of the items in a person's UniCast profile, with no mechanism to let people specify a subset of content they would like to be displayed automatically while entering a GroupCast public space. The lack of privacy protection is likely to promote some self-censoring.

In PolyLens [24], a collaborative filtering recommender system is introduced that recommends movies for small groups of users (two to three) based on individual tastes. It allows users that know each other to create groups and to ask for recommendations to the groups. Furthermore, INTRIGUE [2] is a tourist information server that tailors the recommendation of attractions for tourist groups. A group is modelled as a set of partitioned subgroups having similar characteristics and preferences. Attractions are evaluated for each subgroup with regard to their preferences and an average is computed for the group by combining the satisfaction scores of the subgroups in a weighted way. Finally, Masthoff [20] introduces an adaptive television system for groups, where different recommendation strategies are investigated. The work evaluates decision strategies and individual satisfaction rates for the domain of the study.

Context awareness has also been discussed and investigated in the framework of groups and communities. Erickson and Kellogg [7] assert that applications should support the fact that humans are social beings and thus are looking for other people, and are interested in sharing context, experiences and emotions with each other. Hence, personal context sharing has been proven to facilitate the undertaking of group tasks. In personal context sharing, any user can let other group members be aware of where she is, what she is currently doing, etc. Various context sharing applications, running on mobile devices, exist, such as ContextPhone [26], Reno [10] and myCampus [29].

6.1.2 Mobile Groups: Advanced Social Relationships for the Next Generation of Mobile Users

6.1.2.1 A Scenario for Advanced Group Communications

In this section, a scenario describing the different types of interactions an individual can have in various groups is presented. This scenario is examined in the following sections.

> *Alice is an active woman in her middle thirties. Her life is well-balanced between her job (she works for an insurance company) and her family (she is married and has two children). Besides the family, she also has a group of close friends. Together they like going out, shopping, or to restaurants.*

This morning Alice goes to work earlier than usual. She has planned to have her afternoon free, since she must drive her children to the sports centre where they have a football competition. Later she has also planned to go shopping in the town.

As she arrives at her office, she finishes polishing a presentation that she will give the day after. She notes that she has a couple of open issues she would like to resolve with Paul, her boss, and Maria, her colleague. Both of them will attend the meeting. Shortly afterwards, she is notified that Paul and Maria are both available for a 30-minute talk. An unscheduled meeting is arranged in the boardroom on-the-fly for the three of them. As they enter the room, the environment is adapted to the group settings: air-conditioning is set to an acceptable temperature, the projector is switched on, curtains are rolled down, resources including the presentation under discussion are uploaded on the company server and their access is restricted to Maria, Paul and Alice. The boardroom is tagged as 'not available' for a time slot of 30 minutes.

It is now noon, and it is time for Alice to have a quick lunch. She realises that she has almost an hour before she has to go back to her office. 'Plenty of time to do some socialising', she thinks. She opens an application on her mobile device and requests some suggestions for activities together with people of any of the groups she belongs to within the next hour and in the neighbourhood. The application quickly analyses the availability of her groups and their members, and offers various activities: Alice's office colleagues are finishing lunch and drinking coffee in a restaurant a few metres away. A friend of hers, Jane, is showing online photos of her last holidays in the Caribbean. The fitness group she participates in twice a week is currently inactive, and so no related activity is proposed. Lastly, two friends of hers, Tom and Tina, are currently shopping in the neighbourhood and are available for the selected duration. The application suggests to Alice either to join her friend Jane at her office, enjoying tea and cookies; or to initiate a spontaneous meeting with Tom and Tina at a nearby coffee bar that all three of them like. Alice decides to meet Tom and Tina, and sends an invitation via the TimeGems application (see Chapter 7). *Her two friends reply and accept the invitation within minutes.*

In the afternoon, Alice has to stop work unusually early, since she has to pick up her children and drive them to the sports centre where they have a football competition. On her way back home a delivery van bumps into her car. Alice is a little bit shocked but it seems that neither she nor the van driver are physically injured. Her husband, though he is busy, is informed of what happened and that Alice is unhurt.

However, she is going to arrive late at home and won't be able to pick up her children on time. Rapidly she sets up an alternative plan. Relatives in the neighbourhood are contacted and asked if they can drive the children to the sports centre. Information related to the event, as well as the itinerary, are shared with them. She also shares for a few minutes her family management rights on objects. Relatives can search in the house for objects needed by the children for the event. Sportwear is in children's wardrobe, whereas the soccer ball is in the garage.

Alice's car is a little damaged and she contacts the car insurance company. Drivers who witnessed the accident are contacted by the company. They receive a request via their mobile devices to bear witness and to join a group for reporting the accident to the insurance company. Two of them agree. An electronic form for 'responsibility agreement' is filled in by Alice and the delivery van driver and then approved by the two witnesses. Alice must drive to a repair shop. On her way Alice and her husband are notified that their children have arrived at the sports centre.

At the repair shop, Alice leaves her car for inspection and is asked to come back in two hours. Since she has now some free time she decides to go to the city centre. She will buy a little something to thank her relatives for having picked up her children. The city centre is several kilometres away. Instead of waiting for the bus, she calls up another application on her mobile device for finding people driving by and willing to share their car. Alice quickly inputs her current location, her destination, as well as other parameters such as her trust level on the car owner: a family member, acquaintances, or anyone having a car. The application indicates to her a list of possibilities. She is very happy when she realises that her friends Jane and Mary are driving together towards the city. A few minutes later her friend's car picks her up near to the repair shop and together they head to the city centre for some shopping. 'After all, the accident has a much more positive outcome than I could have expected!', Alice thinks.

6.1.2.2 Requirements for Advanced Group Communications

The scenario in Section 6.1.2.1 presents different types of interactions a user can have and illustrates how these interactions can be facilitated by the users being members of a group. From the scenario, some high-level requirements to guide the development of an infrastructure for group communications for ubiquitous applications and services (in addition to the general requirements presented in Chapter 3) can be determined.

Need to Support Group Management in the Mobile Environment

Alice interacts with different groups. She is part of ad-hoc groups formed with her colleagues, her family, her friends, etc. In fact, innumerable types of groups exist, having very diverse characteristics. Groups are very much different in terms of lifetime, their members, internal policies, etc. Therefore, a system that enables group communication has to keep the knowledge of all specific characteristics of a group, and has to adapt to each group type. Because of the variety of groups, the system has to provide group management mechanisms in order to enable creation, disposal and update of different and simultaneous existing groups.

Need for Group Awareness

In the scenario, Alice constantly needs to be aware of other people in the context of their groups. For example, she is aware that colleagues Paul and Maria are available for a talk, and later she finds out that her friends Jane and Mary are driving together towards the city and can pick her up. Drivers, who witness the accident, share information to report it to the insurance company. A system that facilitates communication and sharing of information in groups can leverage environmental information. Therefore, it has to enable group awareness. Group awareness refers also to the use of context information related to a group that enables the provisioning of ubiquitous applications and services in order to address the group's concerns and needs. Context information characterises the situation of a person or a group of persons, a place, or an object that is considered relevant to the interaction between a user (or group) and an application. Therefore, a system that enables group awareness has to gather information about all entities relevant to characterise the situation of and within a group (person, place, etc). All group-related information needs to be stored in a way accessible to applications, to produce a frequently updated interpretation of the context of a group. This group context is further used to unobtrusively select service categories and applications suitable for the group. Typical examples are: presence of another group member in the vicinity, and new information posted by a group member.

Need for Adaptation and Learning Group Information and Group Context

To facilitate communications within groups, applications and services can be adapted to best match the requirements of the group members. This is illustrated when Alice creates a group with her colleagues and meets with them in a meeting room. To facilitate the group meeting experience, the room lighting, air conditioning system, etc., can be set to levels acceptable for all group members, matching the group preferences. Thus, a system that enables group communication must manage the (sometimes implicit) rules of each group member (policies, preferences) as well as actual parameter values. Such information can be acquired from the sum of user profiles, and thus group information and group context must be gathered, aggregated, interpreted, created and published.

The system may provide a mechanism to learn the information (preferences) of a group. The information is both derived from the single preference of every member and learned from the group behaviour. Information or preferences related to a group that are stored in the group profile and used by an application to provide activities must be learned as the group is active.

The system also needs to be able to learn social networks and find good candidates for permanent groups, as well as the creation or rejection of ad-hoc groups. Some group applications suggest ad-hoc groups to the user. The user will either accept or reject these suggested groups. While not essential to the application, this information can be used to learn social networks. Furthermore, when strong components have been found in the social network, creating a permanent group can be suggested to the user. The system could for example learn that Alice's friends can pick her up every week. These permanent groups can be used in other group-aware applications.

Need for Trustworthy Communications (Trust)

In the scenario, a group of witnesses forms an ad-hoc group that needs to share private information and share it in a trusted way with the insurance company. A system that is truly ubiquitous and provides group awareness will encompass numerous parties (users, service providers) and support communications within and towards groups. To provide these kinds of interactions in a trustworthy and private manner, mechanisms need to be deployed which ensure that individual as well as group data is protected against threats, like identity theft, personalised spamming or eavesdropping. These mechanisms must be flexible enough to allow various degrees of privacy and trust within groups, since different groups may not have the same relevance for different individuals. But even this might not be sufficient, in case the service itself is not trustworthy, and therefore additional user-focused protection mechanisms are needed.

Need for User-Controlled Sharing of Context (Privacy)

Alice's colleagues decided to show to Alice that they are available for a meeting. Alice shares her location with others in order to find people who can pick her up. A system that enables sharing of real-time context (e.g. location and presence) needs mechanisms that enable fine-grained user control over users' private data that may be shared with others. In order to control how context data of a user is shared with other group members, it is required to define group policies and to enforce these policies on *Context Providers*. This enables then access control to group profiles and group context.

6.1.2.3 Modelling Groups

In the ubiquitous computing environment, groups are dynamic and mobile. Also, innumerable types of groups exist. Examples of such groups are families, classmates, work colleagues, sports groups, and support groups. All these groups and others differ in their characteristics such as lifetime, mode of creation (e.g. ad-hoc or scheduled), membership policies, membership constraints, and group lifetime. In this work, the following group characteristics have been regarded as significant:

- *Group leadership.* Leadership refers to the process of managing a group and is concerned with the control of information and the power of decision. Responsibilities of group leaders may include administration of group policies (e.g. concerning group information sharing) or the specification of constraints (e.g. regarding membership in the group). Group leadership can be carried out co-operatively by all members or can be assigned to one or more specific members in the group.
- *Group membership and group membership policies.* Groups have different policies and constraints for allowing individuals to join and leave groups. Group membership may be restricted to individuals with specific characteristics such as identity or age. Furthermore, group policies may express restrictions regarding the handling of information within the group and in connection to external entities.
- *Group name and group rationale.* In order to distinguish between the different groups they belong to, users can assign names to their various groups. Also group rationales can be indicated when creating groups to further enable reasoning about the groups. These names can be user-determined, but there needs to be a way for the system to identify groups, to ensure that Group A of user A is the same as Group A of user B. Generic identity management mechanisms can be adapted to this.
- *Group preferences and group data policy.* A group preference is any relevant information about the group which is available when the group is accessing an application. The group data policy contains the rules/permissions about who can access or modify data about the group. This policy can be changed after group creation, if the appropriate permission has been granted.
- *Group state.* Groups can be in several states (different states for different users simultaneously is theoretically possible, but has not been implemented in this work). Two types of state are considered here: active and inactive. In the active state all group operations (e.g. member addition, policy changes) can be performed. In the inactive state all group information is frozen, members cannot join or leave the group and members' personal information is not shared with others, etc.

To allow management of groups, different processes like 'group creation' and 'group disposal', as well as different states like 'active group' and 'inactive group', are distinguished. Event triggers produce transitions between the processes and states. They may be defined explicitly by members (policies) as the group is created or is in the active state. In addition, event triggers are launched by explicit notification by members, or automatically by group information provided by the group profile. The processes, states and transition types shown in Figure 6.2 are discussed below.

The *group creation process* happens when users agree on creating a group (*creation trigger*). In this phase a group template is selected that provides a guideline for the creation of the group; group templates are discussed in detail in Section 6.1.3. Users who are contacted in order to become members of the group can refer to the group template to decide what kind of constraints they wish to set.

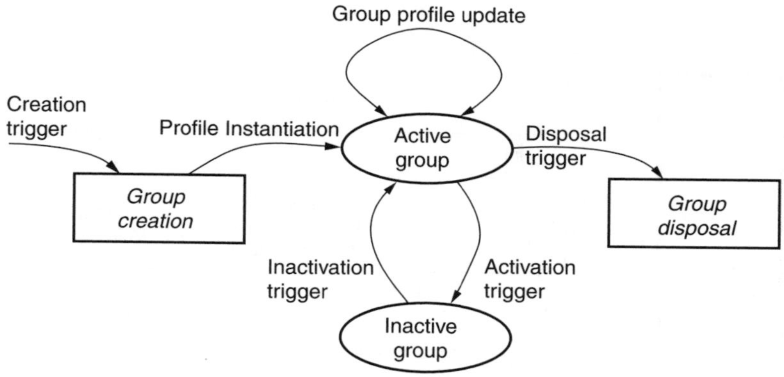

Figure 6.2 Group states.

Once the group has been created, the *group profile instantiation trigger* initiates the *active state* of the group. The group profile is instantiated using the information stored in the group template and additional information given by group members. Later it stores all information about the group and its members, such as preferences and policies that comply with the trust and privacy policies. During this phase, the group profile can be constantly updated with new information, and users can join or leave the group (permanently or not). When a group is active, the system is also responsible for interpreting group context.

In some cases (e.g. for groups that meet regularly), it may also be possible to enter a *passive state*. In this state, the group profile cannot be updated. No new information about the group and its management can be inserted. Group members must define inactivation and activation triggers for this state. By default, groups remain in the *active state*.

The second process is the *group deletion process* that is typically initiated by the group administrator, who has determined that the group is becoming obsolete. Other triggers are also possible: Groups may be dissolved as the group membership list has become empty, or as the group has not been used for a time that exceeds the predefined group lifetime. The disposal process frees the resources used by removing the group profile from the group management system. The profile is not destroyed immediately, but it is entered to a candidate pool for removal. After a pre-established time, or if the group manager system is running low on resources, the profile is removed and deleted.

6.1.3 The GAF: Group Awareness Function

In the previous section, a scenario and requirements for a group management system were described. In the Mobile Services Architecture, the group management has been implemented as the *Group Awareness Function* (GAF).

6.1.3.1 Architecture Overview

To support group communications in mobile environments, a sub-system which combines a set of core functions has been developed as part of the Mobile Services Architecture. The

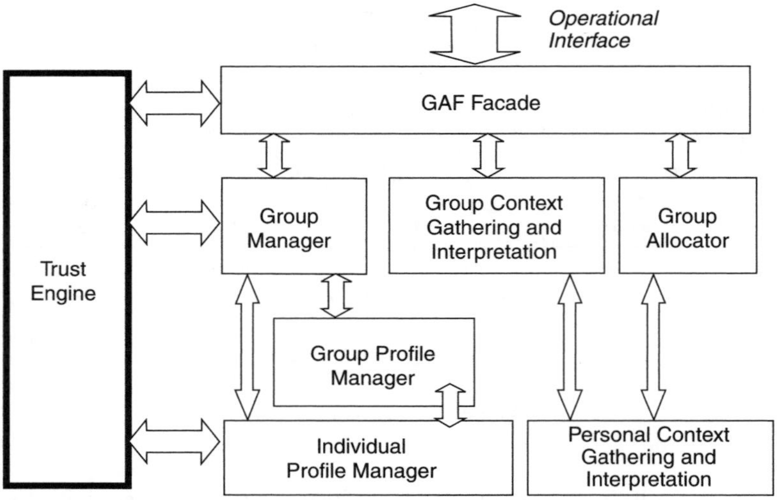

Figure 6.3 Structure of the Group Awareness Function (GAF).

challenges mentioned previously were used to define the basic functions, which were then implemented as a set of components creating the Group Awareness Function (GAF). The GAF works in collaboration with the other functions of the Mobile Services Architecture to create context-aware services. An overview of the GAF architecture is given in Figure 6.3.

The components of the GAF are:

- *GAF Façade.* The GAF Façade is the external interface of the Group Awareness Function. This is the only component, besides the *Trust Engine* (see Section 6.2.2) that the end-users have to interact with directly, if they want to make use of the group-related functionality provided by the Mobile Services Architecture components. It is also the component which anyone building group-aware applications interacts with.
- *Group Manager.* The Group Manager enables the management of operations related to the states of a group. To address the variety of groups previously described, the Group Manager enables the creation, disposal and update of existing groups and their related group profiles simultaneously.
- *Group Profile Manager* and *Individual Profile Manager.* The Group Profile Manager links and stores profile information with the related group models. Three types of information that is stored into a profile are distinguished: group characteristics, group preferences and group policies. The Group Profile Manager is associated to the Individual Profile Managers of all group members. Group preferences are aggregated from the personal preferences of the group members, being gathered from their Individual Profile Managers.
- *Group Allocator.* The Group Allocator provides support for dynamic group discovery using context information; i.e. making sure the user finds the group which is the best fit to her context.
- *Group Context Gathering and Interpretation* and *Personal Context Gathering and Interpretation.* The group context characterises the situation of the group (including the situation

of its members), and reveals the commonalities between members' personal contexts. To interpret group contexts, information about all group members must be gathered and a common context inferred. This is the reason why this component is associated to the Personal Context Gathering and Interpretation components of the group members.
- *Trust Engine*. This is a component that issues certificates, stores and enforces both user and group policies, and provides methods to modify/retrieve such policies.

The components of the GAF collaborate to support context-aware group applications. In the following, deeper descriptions of each of them are given. The Trust Engine component is described separately and in more detail in Section 6.2.2.

6.1.3.2 GAF Façade

The first component is the GAF Façade. Though it does not directly implement any of the functions of the Group Awareness Function (GAF), it is a central component since it is the external interface of the GAF.

The GAF enables group creation and assists the user in managing the group lifecycle. The GAF Façade has been designed as the central front-end for the GAF components. It is accessed by any external components which require operations on groups. The GAF Façade is the primary operational interface of the Group Manager. It provides basic control functions to create, modify and remove groups. It can easily be extended to include other group-related components as well.

The GAF Façade integrates the Group Manager, the Group Profile Manager and the Trust Engine – and via the Trust Engine, to some extent, the *Privacy Display Widget*. The latter is a graphical component, designed for end-user terminals, that interacts with the Trust Engine to present policies to the user graphically and allows the user to modify policies, resolve conflicts in policies and notifies the user if access to a data item is required when the policy states that the user must be asked every time. Therefore, it is not a component of the GAF per se. See the definition of the Privacy Display Widget in Section 6.3.2.

The GAF Façade uses the Trust Engine for access control and delegates method calls to the Group Manager after approval from the Trust Engine. The Group Manager, in turn, uses the Group Profile Manager, which is actually an instantiation of the Profile Manager. The Group Manager also uses the Trust Engine to set policies on group data; and the Trust Engine, in turn, uses the Group Manager in the case of group-related access control decisions. See further discussion in Section 6.2.6.

6.1.3.3 Group Manager

The Group Manager component is responsible for creating, managing and deleting groups. In its simplest form, this component is responsible for maintaining a list of members. However, when placed in the context of everyday life, a group becomes more than just a list of members. For example, a group may become active (engaged in a joint activity). In this case, the group as a whole would have a situational context. A group itself may act as a Context Provider, providing context information about the group. This means that the Group Manager needs to interact with components that collect, aggregate and provide group context. Since the management of group information includes the personal group member information as well, policies about the data management need to be in place. Moreover, there may be membership constraints and policies associated with joining a group, which means that the Group Manager has to interact with components that enforce access control on group data. The interactions and interworking between the group data-related policies and the individual data policies are outlined in Section 6.2.6.

The Group Manager can create, delete, activate and deactivate a group, add and remove members, and retrieve and modify data about the group. All data about a group is stored in the *Group Profile*, managed by the *Group Profile Manager*. When a group is created, several types of policy (for the group as a whole or for individual members of the group) can be defined. Some of these policies may be changed after the group has been created, others remain constant once a given condition has been fulfilled. More information about group and individual policies and how they are managed is provided later in this chapter.

Data about the Group

The Group Manager (and therefore the GAF Façade) provides the ability to store, update and retrieve the following data about a group:

- *Name*. A human-readable name of the group.
- *Rationale*. A human-readable explanation about the purpose of the group and the reason why it was formed.
- *Member list*. A list of members that are part of the group.
- *Principal*. The identifier of the principal of the group. The principal 'owns' the group and may or may not be a member.
- *Status*. The status of the group denotes whether the group is active or inactive.
- *Characteristics*. Characteristics are properties/facts about the group, consisting of a name and a value.
- *Preferences*. Contains the preferences of the group as a whole and/or the individual members. Each preference consists of a name and a value.
- *Group Data Policy*. Contains the rules/permissions about who can access or modify data about the group. This policy can be changed after group creation, if the appropriate permission has been granted. A policy describes which data about the group can be given out to what entities and under which conditions. A policy could, for example, contain a rule that 'access to the group member list is restricted to the group members', but it could also state that 'the group context can be given out to other named groups'. Note that the group data policy may conflict with the individual data policy; this case is described in more detail in Section 6.2.7.
- *Group Membership Policy*. Contains rules that the individual members have to accept before they are allowed to become a member of the group. This policy can be changed after group creation only if there are no existing or pending members in the group. This will be further elaborated in Section 6.2.6. Another membership related policy describes what context information the user has to share with the group – either for the benefit of the individual members, or because this information has to be accessible to derive the context of a group. Such a policy would, for example, state that the locations of the individual members have to be shared with all the group members.

- *Membership constraints.* This policy describes which constraints apply to members when they want to join the group. Membership constraints can, for example, state that the potential member has to be of a certain age or living in a certain neighbourhood.
- *Group Templates.* The Group Manager, and therefore the GAF Façade, uses the concept of Group Templates to simplify group creation. Basically, this allows for setting multiple values (e.g. multiple policies) for a group at the same time.

Two operations are particularly important for the Group Manager component: the creation of a group, and the joining of a new group member. These two operations are further detailed below.

Creating a Group

The Group Manager deals with the creation of new groups, which can be created by anyone. The creator of the group is called the group principal, and may or may not be a member of the group. Upon creation, the principal defines the policies that apply to the group. To assist the principal in this task, the Group Manager uses a Group Template which contains default policies that can be amended if required. As soon as a group is created, a Group Profile is instantiated. Static values from the Group Templates, such as characteristics of the group (e.g. name, rationale and description) are taken from the Group Template and stored in the Group Profile.

Becoming a Group Member

The Group Manager controls the addition of new members to a group. When a user joins a group, the user has to confirm that she is interested in becoming a member of the group, has to comply with the Group Data Policy and Group Membership Policy, and finally has to fulfil the membership constraints. These membership constraints can be verified, after the Group Membership Policy has been approved, since the group (or the Group Manager application) has then gained access to the preferences and characteristics of the potential member, which are stored in the individual user profile.

6.1.3.4 Group Profile Manager

The Group Profile Manager component implements a data record that contains information related to a given group (the *Group Profile*). The profile can contain information about the group, but also about the group's devices or the group's behaviour(s) while accessing services.

Group information consists of group characteristics, policies and constraints, and group preferences.

Group characteristics, policies and constraints are information about the group that are always applicable whatever the context is. This information is used for managing the group. Group preferences are information about a group, which are true (or relevant) in the group context. A group preference can be a piece of data (e.g. network plans), which relates to a context of use (e.g. public transportation) that is used by an application as a rule ('when the group is in the public transportation, provide network plans').

A group preference can also be a model, which is learned from the group behaviours. Group information is either passed to services and applications via the *Operational Group Context* or to be browsed by the group users.

Group Profiles are managed by the Group Profile Management component. Similarly to the Personal Profile Manager, the Group Profile Management component is composed of a single component that offers functions to create, delete and modify profile subsets and included data. It also allows querying of Group Profiles for matching profile subsets, and for managing access to the actual data. Finally, the component administrates group preferences by defining several representation layers for information: *Profile*, *Profile Subsets* and *Entries*.

The internal structure of a profile consists of Profile Subsets and Entries. A piece of information is stored as an entry in a key–value pair form. Entries related to a specific application and a context are contained in a Profile Subset. Therefore, each subset is described by qualifiers identifying the application and the context.

6.1.3.5 Group Context Interpretation

The group context both characterises the situation of the group and reveals the commonalities between members' personal contexts. The component in charge of providing a description of the group context is the Group Context Interpretation component.

Typically, the context of any entity is composed of the different elements that are essential or relevant to characterise its situation (location, time, environment, etc). Therefore, in order to enable comparisons between contexts, a context model that assesses these context elements has to be created. When it comes to groups, a group context model provides a description of the context elements that are significant for interpreting the context of any group. The context model in the GAF is built around a set of core context elements, listed in the following:

- *Activity*. The activity element gives indications about what the group is doing. An activity can imply other activities. For example, 'while going out with my friends to the theatre' (our main activity), 'we take the bus' (second activity), 'while chatting together' (third activity).
- *Device*. The device element characterises the ability of a group to communicate. It lists the devices that the group members can use and share to communicate together.
- *Location*. The location of a group describes the site occupied by the group (room, building, etc). In the current system, it is determined from the average position of all the group members and the maximum distance of one member to the centre, when considering the spatial coordinates of all the members. The group location consists of the smallest ensemble that comprises the average position of the members and the maximum distance to the centre; also other algorithms for defining the location of the group could be applied.
- *Vicinity*. This element provides information about the group's environment, including nearby and related entities (other groups or individuals) or objects. The vicinity is closely related to the group location, and as a consequence the vicinity can be characterised only when the group location is determined.

The group context also contains the identity of the group, which consists of a unique identifier; the list of group members currently active in the group; and a timestamp that specifies when the context description was constructed.

Group contexts are constructed from group members' personal contexts. When a group context is requested, the group membership list is inspected. All personal contexts of the group members are requested by the *Personal Function* components and are gathered within a timeframe (Personal Context Gathering, A in Figure 6.4). When the timeframe is exceeded

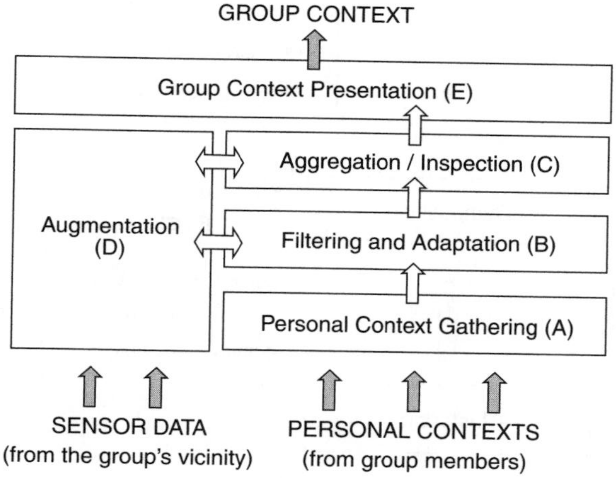

Figure 6.4 Group context interpretation mechanisms.

(for instance, due to terminals being unreachable), missing personal contexts are not taken into account.

Though the gathered personal contexts are composed of the same context elements as the personal context model, they still can display inconsistencies (e.g. differences in unit of measure) or incomplete information (e.g. parameters missing). The filtering and adaptation process (B in Figure 6.4) allows the aligning of elements of personal contexts.

Further, group context elements are aggregated or inspected. Whether aggregation or inspection is done depends on the type of context group that is requested. When a group context is needed that characterises the group's situation, group members' context elements are aggregated (C in Figure 6.4). When commonalities between group members are revealed, some of the personal context elements are inspected. Both processes are carried out by applying a set of predefined rules. When data is missing to apply some of the specified rules, the data can be gathered from the group vicinity (D in Figure 6.4).

If data is missing, it may result from group members' privacy concerns. Therefore, no additional request is made to group members to complement their personal contexts (since it would be refused a second time as well). Missing data that are relevant to characterise, for example, the vicinity of the group are queried from the group environment instead. This implies that missing data does not hinder determining the group location. The group context is assembled according to the type of requested group context, characterising the group situation or depicting the commonalities between group members (E in Figure 6.4).

What follows is a more precise description of the aggregation and inspection mechanism. This is made visible outside the GAF by using RuleML [27] as the rule representation language.

For each context element, a set of conditional rules are defined to infer a value from the personal context elements. A conditional rule establishes a relationship between a set of formulas, called the premise, and an assertion, called the conclusion.

In the premise of a rule, the number of predicates is fixed. This can cause a problem as the number of group members varies from group to group. For instance, let's consider the

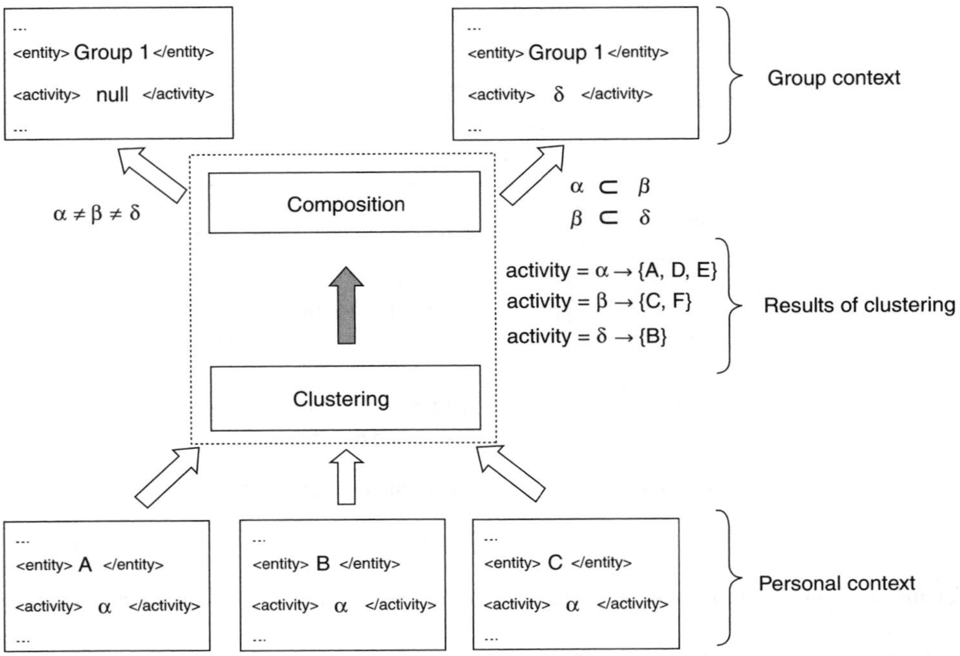

Figure 6.5 Aggregation and reasoning scheme.

following simple rule.

$$\begin{array}{l}\text{IF}(\text{act}(A) = \text{act}(B) = \text{act}(C)) \\ \text{THEN}(\text{act}(\text{GROUP}) = \text{act}(A))\end{array} \tag{1}$$

The rule states that if in a three-member group all members (A, B, C) have the same activity (act; e.g. going to the theatre), then this is the activity of the group. However, if the next group is composed of four members, this rule cannot be applied any more and a new one is needed to infer value of the group context activity element.

In this work, a method that clusters group members and computes the context element value from the clustered values for each context element to infer is adopted. This process is depicted in Figure 6.5. Personal contexts are gathered that describe the activity of the group members. In the clustering phase, rules are applied that group individuals in clusters of common activity. In the example, when clustering ends, three clusters (activity α, activity β, and activity δ) are defined.

In the second phase (composition), clusters are inspected and rules are used to determine whether activities can be composed (i.e. how they are related to each other). When no commonalities between the clusters can be found, the group activity element is set to null. An example of group context obtained after aggregation is given in Figure 6.6a.

In the inspection, a member requests a list of group members that have a similar context. For instance, a group member wants to know who else is in the same building. This process

```
<group_context>
    <identify> 12ZTU </identify>
    <group_members>
        <member> John </member>
        <member> Alice </member>
        <member> ... </member>
    </group_members>
    <timestamp> 2005-01-16T15:03:16.235+01:00</timestamp>
    <location>
        <room>
            <value> R1234 </value>
            <type> boarding room </type>
            <timestamp>2005-01-16T15:02:45.145+01:00</timestamp>
        </room>
        <building> ... </building>
        <address> ... </address>
    </location>
    <device> ... </device>
    <activty> ... </activty>
    <state> ... </state>
    <vicinity> ... </vicinity>
</group_context>
```

(a)

```
<group_context>
    <identify> 12ZTU </identify>
    <group_members>
        <member> John </member>
        <member> Alice </member>
        <member> Mark </member>
    </group_members>
    <timestamp> 2005-01-16T15:03:16.235+01:00</timestamp>
    <location>
        <building>
            <value> Alpha building </value>
            <type> private building </type>
            <timestamp>2005-01-16T15:02:45.145+01:00</timestamp>
        <building>
        <address> ... </address>
    </location>
</group_context>
```

(b)

Figure 6.6 Examples of group context descriptions.

is similar to aggregation. Clusters of group members are constructed for the requested context element (e.g. activity). Later, the context of the requester is used to select the cluster(s).

When inspection is carried out with several context elements (this corresponds to a request like 'find people of my group of friends who are in town AND are not sleeping'), the composition phase is also responsible for merging clusters (e.g. the cluster of individuals currently in town and the cluster of individuals that are not sleeping). An example of a group context obtained after inspection is given in Figure 6.6b.

6.1.3.6 Group Allocator

The Group Allocator is the GAF component that provides support for smart dynamic group discovery using context information. It encapsulates the complexity associated to group discovery behind a simple interface. When a group is requested for any application, the Group Allocator infers the optimal groups given the relationships between the individuals, and returns a ranked list of groups.

The Group Allocator component consists of three subcomponents: *Relevance Assessment*, *User Clustering* and *Feedback Gathering*. Each component is discussed in more detail in the following.

Relevance Assessment

The Individual Profile Manager component in the platform stores models for group allocation for each user and application.

When a user or an application requests a group, the appropriate model is retrieved from her profile and handed to a recommender system. This model contains the variables that are considered relevant for that kind of group, as customised for that user. The inference itself is done by a modular plug-in, which currently considers rule-based systems and Bayesian Networks.

As an example, the car-sharing application that Alice uses in the scenario presented in Section 6.1.2.1 creates groups to share a car ride to a common destination. The group discovery is based on a model which, when stable, contains restrictions on the pick-up and drop-off locations of the users, their preferences (e.g. smoking tolerant or kind of music), traffic condition at their pick-up point, etc.

As for most reasoning systems, these restrictions need to be discretised. Furthermore, absolute values are seldom of interest, since groups often require relative comparisons between the members. For this reason, parameters are defined as Boolean operands over discrete thresholds, such as:

$$[\text{User.position} - \text{Candidate.position} \langle \text{ 500m}]$$
$$[\text{User.MusicPrefence} == \text{Candidate.preference}]$$

The result of the inference is a relevancy score, which indicates how suitable a given candidate is when considered in a group with the requester. By repeatedly performing this operation on the pool of available users, a relevance matrix is obtained where each position indicates the relevance between any two users. Note that such relevance is not necessarily symmetrical.

It is clear from this step that limiting the pool of candidates to consider is of capital importance, in order to avoid overloading the discovery component. For that reason, the pool is pre-filtered according to user-defined hard policies, such as trust requirements.

User Clustering

The Relevance Assessment determines how each user relates to other users, but this is not sufficient to group users into optimal groups.

Consider, for instance, the case of users A and B, with very low relevance for each other but 'joined' by C, who is very close to both of them. Without C, there would be no reason to link A and B, unless the group type demanded a minimum number of members, in which case A–B might make as much sense as any other random link. Still, if one is able to detect the joining influence of C, the resulting group might be much more convenient at least for C.

Likewise, if the model for a user is not mature enough (or not evolved enough from its stereotyped starting point), relevant relationships might be overlooked. In that case, it is likely that a transitive relation exists such that 'what joins B and C is a still unknown parameter which also binds A and B together'.

In order to visualise these effects, a Force Directed Placement algorithm, as presented in [8], has been implemented.

Figure 6.7 shows the result of placing users in the car-sharing scenario. The nodes are initially placed randomly, and the system iterates towards a stable solution where all nodes repel each others equally, but also attract themselves according to the relevancy of the relationship.

Given this weighted graph, it is now relatively easy to identify the nuclei of closely related users, and the problem is reduced to a clustering scenario, where numerous approaches already exist.

The clustering process can be greatly simplified if restricting conditions apply. For instance, groups for car sharing need to include at least one car and have at most as many members as the car's capacity.

Feedback Gathering

Given the nature of the relevance assessment step, learning algorithms that can evolve the initial model defined by the user's stereotype are considered. To this end, one needs to collect feedback on the user's satisfaction, which will then be used for model training.

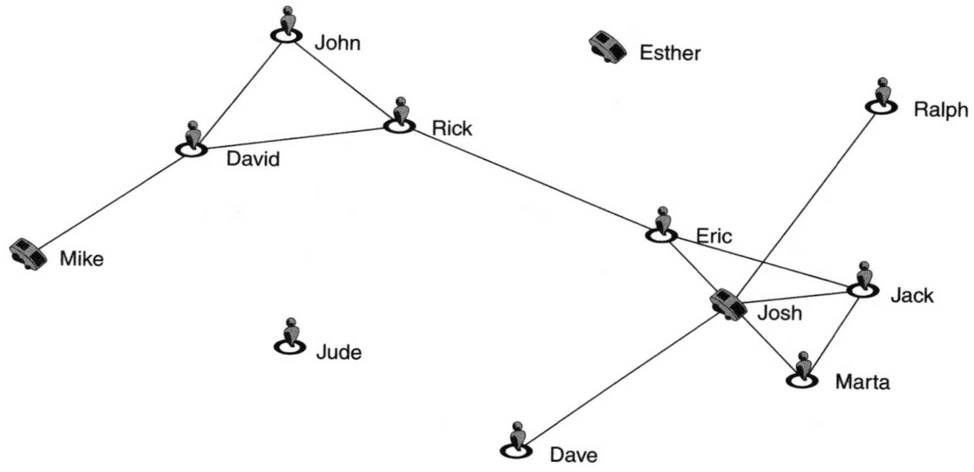

Figure 6.7 Relative positioning using the Force Directed Placement algorithm.

6.2 Privacy and Trust

6.2.1 Basic Trust Architecture

From a privacy and trust management point of view, the Mobile World can be seen as consisting of systems and Context Providers that hold personal information about the users. Our *mobile system* is defined as a user's personal device (or devices) and applications run by the user.

At the core of the mobile privacy and trust architecture of the Mobile Services Architecture are the so-called *Trust Engines*, which are special trusted components that take care of handling the definition and enforcement of the user's trust settings, and so access to the user's data. The Trust Engines are used to set up and enforce the user's privacy settings in all data repositories; i.e. Context Providers covered by the mobile framework. The privacy policies can be specified on the level of individual users, or for established groups of users. The Trust Engines enforce privacy settings for groups in co-operation with the *Group Awareness Function*, which is responsible for aggregating the privacy settings for mobile groups based on the privacy settings of individual users.

6.2.2 The Trust Engine Concept

The Trust Engines are the fundamental components that are used to implement the privacy and trust management in the Mobile Services Architecture based system. Each user has one or several personal Trust Engines, each of which corresponds to a different role of the user that she may take while interacting in the system. The Trust Engines form a network of trusted components that control on a system-wide level how the user's information is shared among the mobile users and applications.

When a user sets up or modifies her trust preferences, she does so with the help of her personal Trust Engine, the so-called *User Trust Engine*. This manages the user's privacy policies and takes care of propagating the policies and any updates to relevant policy repositories in the

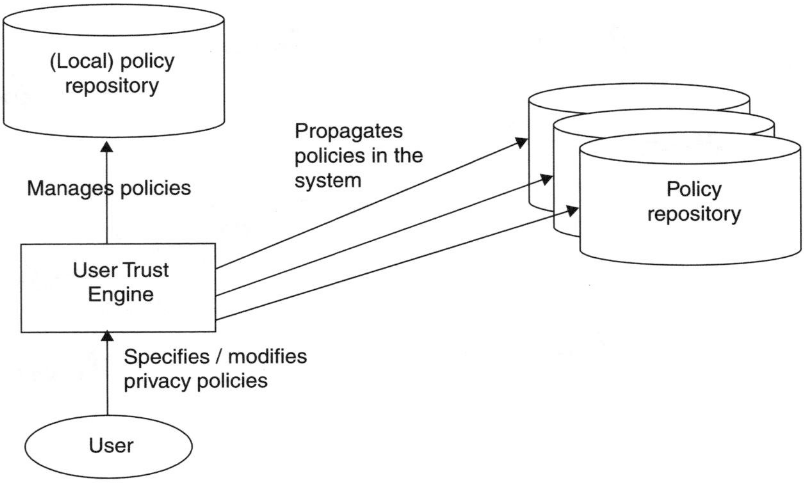

Figure 6.8 Role of the Trust Engine in setting up privacy policies.

system. These policy repositories can then be used by other components hosting user data to determine whether or not these data may be released upon specific data requests. The role of the Trust Engine in setting up and modifying privacy policies is depicted in Figure 6.8.

User data in the mobile system is hosted by different data retrieval and storage components, so-called Context Providers (CPs; see Chapter 4). As much of the information handled by CPs is related to individual users, privacy policies have to be used to control how the data may be released by the CP. To accomplish this, each CP is equipped with a dedicated Trust Engine, which has the task of enforcing privacy policies set by the users whose data is hosted by the CP, and make the access control decisions needed to determine if a CP may or may not release some data in response to specific data requests, as depicted in Figure 6.9.

The Trust Engines are trusted entities in the system, which have a twofold role in the trust and privacy management framework. On one hand, the Trust Engines act as *Policy Decision Points* (PDPs), which can be used by the user to specify her trust preferences by defining privacy policies that control the release of any of the user's private data. On the other hand, the Trust Engines act also as *Policy Enforcement Points* (PEPs), which have the responsibility of making access control decisions based on the privacy policies defined by the user. How the initial trust is enforced will be discussed later in this chapter.

There are two kinds of Trust Engine: *User Trust Engines* and *CP Trust Engines*. In addition, the *GAF Trust Engine* is a special kind of CP Trust Engine for handling group context and profile data. Below, all of these Trust Engines are described in more detail.

User Trust Engine

This acts as a Policy Decision Point with which a user can create and specify and modify privacy policies controlling the access to her data. The policy relates to one user, but may contain rules for different pieces of data, retrieved from different Context Providers. If a user has different roles in the system, she may have dedicated User Trust Engines associated with her different roles, one for every role (mother, employee, etc.).

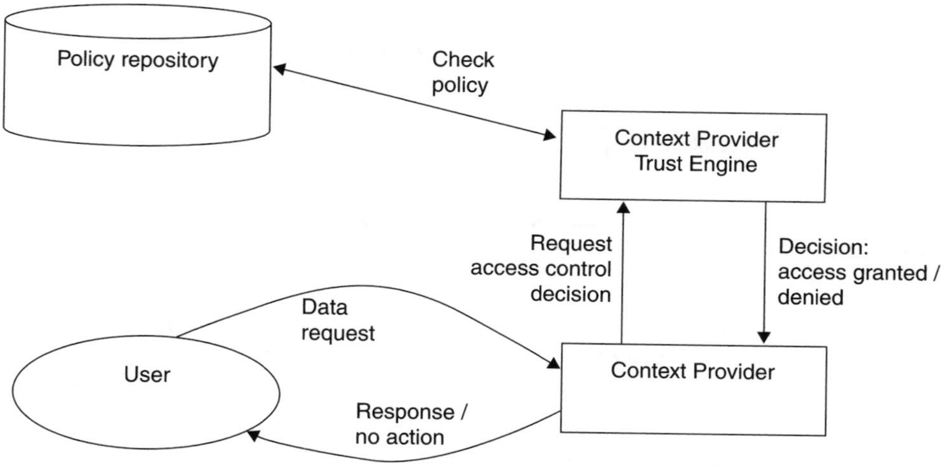

Figure 6.9 Role of the Trust Engine in enforcing privacy policies.

The User Trust Engine takes care of uploading the relevant parts of the user's policies to a policy repository or directly to Policy Enforcement Points. These PEPs are in practice CP Trust Engines having custody of the user's data items that are handled by the respective CPs. The Trust Engine is also responsible for propagating subsequent changes to privacy policies to relevant targets holding policies that are affected by the changes.

Context Provider Trust Engine

This acts as the Policy Enforcement Point for individual context and profile data held by the respective CP. When a CP receives a data request, it is redirected to the CP Trust Engine. This then looks up the policies related to this request from the policy database, and determines based on them if the data request can be allowed.

If a policy related to a data request states that access to a data item can be granted only if the requesting user is a member of a specific group, the evaluation of the request requires interaction with the GAF (and therefore the GAF Trust Engine). This is necessary, since the CP Trust Engine does not know the group memberships of the requesting user.

If a user changes her privacy policies, the User Trust Engine notifies the CP Trust Engine of the changes that affect those data items that are under the latter's custody.

GAF Trust Engine

A special type of CP Trust Engine is the GAF Trust Engine. It is also a Policy Enforcement Point, but in this case for group context and group profile data. The behaviour of the GAF Trust Engine is very similar to the behaviour of the CP Trust Engine, but this one has special privileges that enable it to decide whether a group-related request can be allowed. In general, only the GAF Trust Engine has access to the group membership lists.

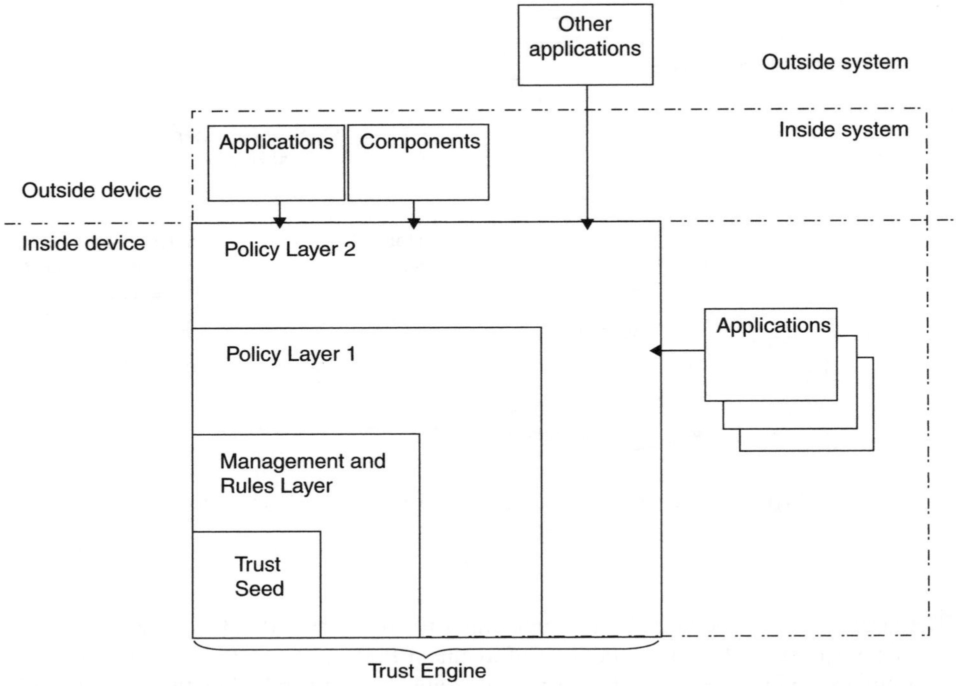

Figure 6.10 The layered structure of the Trust Engine.

6.2.3 The Trust Engine Structure

The internal design of the Trust Engine is a layered structure, which implies that there is a strict hierarchy between entities residing in different layers. It also means that components by default have different levels of trust to each other. The layers are the *Trust Seed* (TS), the *Management and Rules Layer* (MRL) and the *Policy Layers*, as depicted in Figure 6.10. At system start-up, the user sets up and approves the fundamental core policies, which are placed inside the MRL. This layer is locked before further actions are possible (e.g. the loading of policies to the Policy Layers).

The Trust Seed

At the core of the Trust Engine architecture is the so-called Trust Seed (TS). It is a cryptographic identifier which is unique to each user role and cannot be altered. All trust inside the system is inherited from the Trust Seed.

The TS is used to validate the content of the Management and Rules Layer, and it works as a fundamental parameter for generating keys used to create certificates and establishing encrypted trust relations between components of the Mobile Services Architecture based system.

Management and Rules Layer

The MRL contains system-specific policies related to the device, a rule engine to enforce them, and a policy management component. The Trust Seed and the MRL are locked before the device can access the outside network.

Policy Layers

These layers contain most of the policies. These include policies concerning privacy, components and applications. Policies residing in these layers are strictly controlled by underlying core policies. This means that modifications to the policies are possible only within the constraints set by the core policies on the inner layers of the Trust Engine.

6.2.4 Secure Initialisation of the Trust Engine

When a system is started for the very first time on a device, the user must have full control over all running applications. Each new application will be launched under the full control of the Trust Engine through a dedicated API layer. The user interface is also a special application under control of the Trust Engine. All applications running on behalf of the user have an identity which can be traced back to the user's Trust Seed that acts as the foundation of all of the user's trust associations in the system. Both users and applications are identified with a unique ID, which is used as the key to these associations (and which is associated with a cryptographic signature).

All the applications, installed on the mobile device or running outside the mobile device, will interact with the mobile framework through the use of Application Programming Interface (API) layers depending on the granted access rights and privacy policies defined by the user.

6.2.5 Privacy and Trust Policies

6.2.5.1 Policy Language

The policy language used by the Trust Engines has to have the capability of specifying the necessary properties of privacy policies related to users and groups. These include the following:

- *What to share* – Identification of the data item that the policy concerns.
- *With whom* – Specification of the target of the policy. This can be one or a combination of the following:
  - *anybody:* the data item can be shared without restrictions;
  - *users or applications:* the data item can be shared with one or more named users or applications;
  - *groups:* the data item can be shared with specified groups of users.
- *When* – Specification of the conditions under which the sharing of data is possible. This can be one of the following:
  - always;
  - never;
  - after approval (ask user every time).

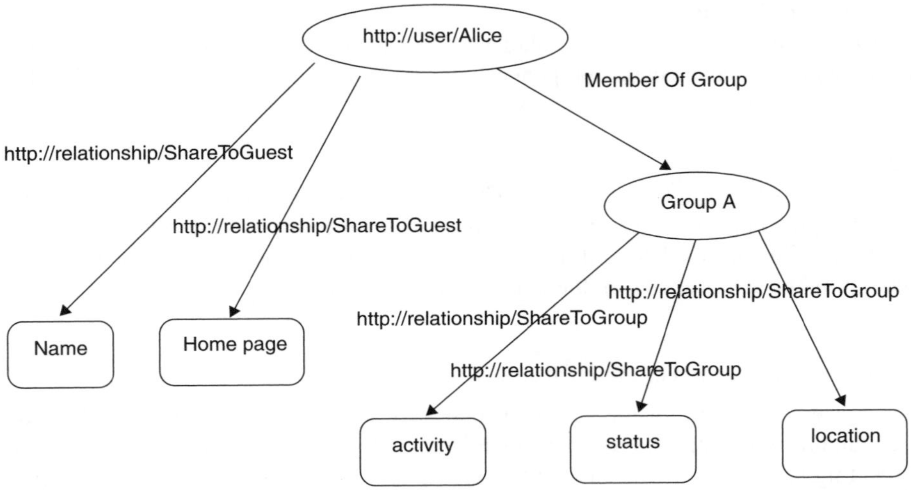

Figure 6.11 Example of a possible policy definition in RDF.

Since the rules language is based on the system requirements, the evaluation of rules provides only two possible outcome options: *success*, which results in the requested data being delivered; or *failure*, which results in non-delivery.

There are several possibilities for creating such a language; RuleML, mentioned in Section 6.1.3 about the GAF, is one such possibility. In this work, RDF (Resource Description Language) is used. In RDF, the rules are modelled as triples specifying the entity requesting the data, the data item to be delivered, and the entity that owns the data. When a request comes in, the Trust Engine evaluates the request, and acts as specified in the policy.

One example of a policy for a user Alice who wants to share both her name and homepage address to guests, and her current activity, status and location information only to the Group A that she is a member of, could look similar to the one depicted in Figure 6.11.

6.2.5.2 Policy Enforcement

Applications request data from Context Providers in a controlled manner. The CP Trust Engines, which act as Policy Enforcement Points, prevent any access to personal data which is not explicitly allowed by the privacy policies governing the requested data items. Figure 6.12 presents the message sequence chart of a user application requesting access to data from a CP.

Whenever a user application attempts to obtain access to data that resides inside the Mobile Services Architecture-based system, the Policy Enforcement Point decides whether the request can be allowed. The evaluation of the access request involves the following steps:

1. The application requests data from the CP by sending an authenticated data request to the CP interface. The target of the request can be either the user's own or another user's data item.
2. The CP forwards the request to the CP Trust Engine, which acts as the PEP to check if the requesting user is authorised to receive the requested data.
3. The PEP checks the *Policy Repository* for the data owner's privacy policies.

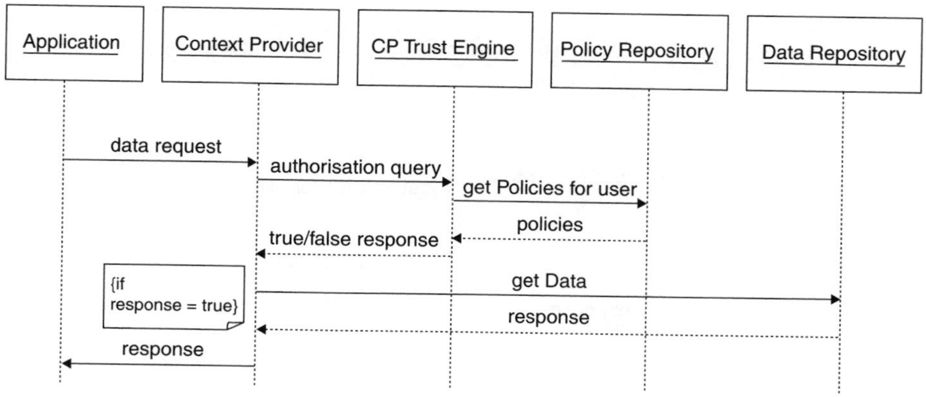

Figure 6.12 Policy enforcement.

4. If the policies permit the user to have access to the requested data, a positive response is sent to the CP, otherwise the request is denied by sending a 'false' notification.
5. The CP checks the response given by the PEP and reacts accordingly. If the response grants access, the CP gets the data from the *Data Repository* and sends it to the requester, else not.

6.2.5.3 Policy Management

Privacy and trust policies are initialised and updated by the user on her local device. The specified policies are then distributed to remote Context Providers that provide access to personal data of the user. The policies can be distributed either by storing them on a policy server, or directly uploading them to the policy repositories of the CP Trust Engines. The remotely stored policy sets are always uploaded from the local set, and user intervention is called for when a discrepancy between user and group policies is detected.

The latter modification of privacy policies is under the strict control of the initial Management and Rules Layer policies. Changes to existing policies are possible only if the MRL policies allow the policies to be modified.

The message sequence chart in Figure 6.13 shows the interactions that take place at start-up when the Trust Engine is used for setting the policies for the user. When the user has chosen her preferences, the GUI sends the data to the *Policy Decision Point* (User Trust Engine) that stores them in the *Policy Repository* for the userID in question.

Therefore, every userID has an own set of policies, which is uploaded appropriately to the relevant servers. These servers contain the policy repositories that the PEPs (the CP Trust Engines) check when evaluating access requests. The set of relevant servers covers the user's mobile environment; it includes the locations where user data are stored and the groups that the user is a member of.

6.2.6 Group Policies and Policy Enforcement

In addition to the individual policies of the user, two types of privacy policies are used for managing the privacy settings for the mobile groups: *Group Data Policies* and *Group Membership Policies*.

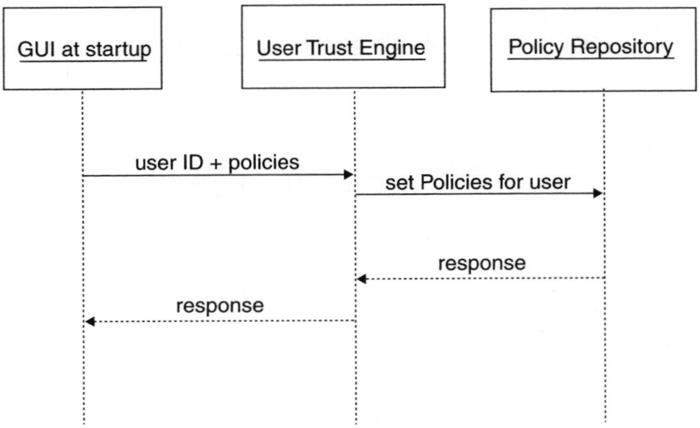

Figure 6.13 Setting user privacy policies.

Group Data Policies

These contain the group data access control rules about who can access or modify data about the group or data jointly owned by the group (e.g. data in a joint group directory). A requestor that wants to access data of a group, or tries to modify data about a group, needs to have the appropriate permission – otherwise access is denied. The default behaviour, if no rules are defined, is to deny access to group data.

Group Membership Policy

This consists of the rules of the group. A new group member has to comply with those rules before joining the group as outlined in Section 6.1.3. Such a policy typically states which data items the group members have to share with the group, in order for the group to be able to fulfil the purpose of the group and, in GAF terminology, to derive a common group context.

A Group Membership Policy can be defined only before any members are added to the group. If a group already has members, or has pending memberships, the policy in the system cannot be changed. Any change would then require consent from the group members and would make the system very complex (e.g. a group might no longer function because one member is on vacation and has not given his consent to the change).

The Group Membership Policy is presented to the user wishing to join the group via the Trust Engine and the Privacy Display Widget (PDW). The user can then accept or reject the presented rules. Only if this request is accepted, the process of adding a member to the group can continue. If no Group Membership Policy is defined, then a member can be added to the group immediately. By default, the Group Membership Policy is specified by the group creator.

In situations in which the entities requesting access to a data item are members of groups, both personal and group policies have to be considered. There is a defined policy enforcement precedence order which the Trust Engine follows when evaluating access requests.

The Trust Engine first considers global policies that apply to the special user 'anybody'. This special user applies to all mobile users in the system. If a global policy exists for the requested data item, this policy is enforced.

If no matching global privacy policy is found, entity-specific policies are considered next. Entity-specific policies are such privacy policies that are specified for explicitly named entities. If there is a policy for the requesting entity ID, the policy is enforced, otherwise group-specific policies are considered.

Group-specific policies are such policies that are specified for specific mobile groups and target all member entities of the group. Group policies are matched against all groups of which the entity requesting the data is a member. If a matching group-specific policy is found, that policy is enforced. Otherwise, the Trust Engine denies the access request.

6.2.7 Policy Resolution in Conflict Situations

In the setting described above, it is possible that some of the matching policies for a data item are in conflict. This situation may occur, for example, if the requesting entity is a member of more than one group that have a group-specific policy for the requested data item. The group policies are regarded of equal status; i.e. there is no policy hierarchy. Which one of the matching group policies gets enforced? To solve this question, a *restrictive* or *permissive* approach can be taken.

In the *permissive* approach, access is granted if any of the matching group-specific policies would allow the access. This approach has the advantage that not all policies and their conditions need to be evaluated for each request. It is sufficient that the policies are evaluated in a sequential manner. If a policy, which grants access, is encountered, the remaining policies need not be evaluated any more. The permissive approach is, however, somewhat questionable, since the user has to explicitly specify that she does not want her data to be shared with the specified group.

The *restrictive* approach would be that access is granted only if all matching group-specific policies allow the access. This approach would provide more protection for data privacy, but it would also require all matching group-specific policies to be evaluated before making the access decision. This approach might be preferable if a requesting entity is in the intersection of two groups, where for one group access is granted and for the other denied. Otherwise, it would be possible for the data item to 'leak' through this entity to the rest of the denied group.

An alternative approach could also be that in case of policy conflicts, the user is explicitly requested to resolve the conflict through the Privacy Display Widget. The user could then review the situation and determine whether the request should be allowed. A combination of the restrictive approach together with prompting of the user might seem the most appropriate approach in this case. However, one must also consider that when user prompting is involved, the time required to respond to a request might in the worst case be comparatively long, and the usability of the whole application might be reduced if the user is prompted too often. The Privacy Display Widget based approach is described in Section 6.3.

Another conflict may arise if a global policy conflicts with any of the other policy types. If a matching global policy is in conflict with entity-specific or group-specific policies, the global policy takes precedence over any other policies. This means that in every case the global policy will be enforced, regardless of other policies. The global policy would then reside in a potential policy hierarchy on a higher level than the entity-specific or group-specific policy.

If an entity-specific policy and a group-specific policy are in conflict, the resolution of the conflict is very straightforward. Since the entity-specific policies have precedence over the group-specific policies, the entity-specific policy will be enforced in any case. This is a very natural solution, since if a user has expressed her trust or distrust to a specific entity by specifying an entity-specific policy, this trust or distrust of the user should be honoured in any case.

6.2.8 Trust Management Systems

The term *trust management* is used for many very different purposes. It can refer to key distribution protocols, reputation systems, privacy policy management systems or contractual relationships. It can also refer to modification or establishment of a binary trust relationship (i.e. a cryptographic key used has a correct certificate chain) or to fine-grained trust systems that take into account many aspects. In this work, the focus is on 'electronic' trust; thus, the legal aspects of trust or with trust influencing factors that cannot be captured by an electronic system are not dealt with. The following definition of *electronic trust management* contains all these aspects from Gradison and Sloman [9]:

> Trust management is the activity of collecting, codifying, analysing, and presenting evidence relating to competence, honesty, security or dependability with the purpose of making assessments and decisions regarding trust relationships for Internet applications.

In the following, different trust management purposes are covered.

There is a large range of *key distribution protocols* available in the Internet and in the mobile environment. The list below is by no means exhaustive and can only give a brief glimpse of the whole field:

- Internet Security Association and Key Management Protocol (ISAKMP) [15] RFC 2408 gives security concepts necessary for establishing Security Associations (SAs) and cryptographic keys in an Internet environment. Internet Key Exchange (IKE) [17] RFC 2409 offers a protocol to negotiate and provide authenticated keying material for creation of a security association. IKE is used for IP Security (IPSec). Both ISAKMP and IKE were replaced and merged into IKEv2 [14] RFC 4306.
- Multimedia Internet Keying (MIKEY) [16] RFC 3830 has a slightly different focus. It offers a key management protocol that supports key management for secure real-time transport that is used together with multimedia applications like streaming. There are several extension for MIKEY that recently became RFCs.
- Key Management for groups is outlined in Group Secure Association Key Management Protocol (GSAKMP) RFC 4535 [13] or in the Group Key Management Protocol (GKMP) RFC 2094 [11] and RFC 2093 [12].
- Generic Bootstrapping Architecture (GBA) [1] Technical Specification 33.220 provides a common application-specific shared secret to an application server and a mobile terminal based on the cellular authentication of the user.

Reputation systems are a popular method for trust establishment in peer-to-peer systems. The basic concept is that knowledge about the user's behaviour on previous transactions is processed and summarised to give some indication on the trustworthiness of the transaction peer. The reputation providing node can be distributed or central node (e.g. eBay, PeerTrust and TrustMe). The processed experience factors and weighting factors may vary depending on the actual purpose of the reputation system itself. Some of those reputation systems offer user anonymity or pseudonymity that reveal only the relevant ‘trust data’ to the requestor (e.g. [30]). Reputation systems have several major problems:

- partners who build up a good reputation and then perform some major fraud;
- cross-recommendations, where two or more accounts give each other a good recommendation to build up or improve a reputation;
- account switching or the blank account problem, where a newcomer to a reputation system has no history.

These problems are well-known and often addressed by procedural means of establishing a valid account. Another approach is to use social network analysis [28] in the reputation systems.

The most popular approach for privacy preferences management is the Platform for Privacy Preferences (P3P) that was developed by W3C [32]. With P3P, web sites can express their privacy practices in a standard format that can be retrieved and processed automatically. The need for manual user interaction is decreased and still informs the user of site practices. P3P offers machine and human-readable formats to express the privacy preferences. The XML description allows automating the decision-making. However, P3P does not provide the actual APIs or the product, but there are LDAP-based extensions for databases that offer some support on this.

One main problem of trust management systems is that they need some form of foundation for the trust. Trust has to start somewhere and most trust management systems do not offer a reliable trust foundation to build upon. With the existence of a trust seed, a layered and granular trust model (as in the Mobile Services Architecture) can be established, trust can be managed and new trust relationships can be derived.

6.2.9 Protection and Auditing of the Trust System

The basic trust and privacy framework of the Mobile Services Architecture provides the needed mechanisms for controlling the access to data and applications in the mobile system. However, these mechanisms work properly only if the application code of the applications and the components involved can really be trusted.

Faulty implementations and bad software engineering (e.g. enabling buffer overflows) or viruses that sneaked in during code writing cannot be completely prevented. Today, security patches are often released to remedy these problems, but this common practise is neither secure nor efficient. In order to protect the system from these kinds of failure, a second line of defence is required. The task is to detect if the primary protection methods are failing, and initiate appropriate countermeasures in order to avoid unauthorised data access that would be possible due to the failure. In the following, intrusion detection and run-time security issues are discussed.

6.2.9.1 Intrusion Detection

A faulty application that is performing a malicious action often behaves in a way that is not typical for the application. Such behaviour towards a Trust Engine or a Data Repository can be detected by an anomaly detection system and can be used to implement intrusion detection

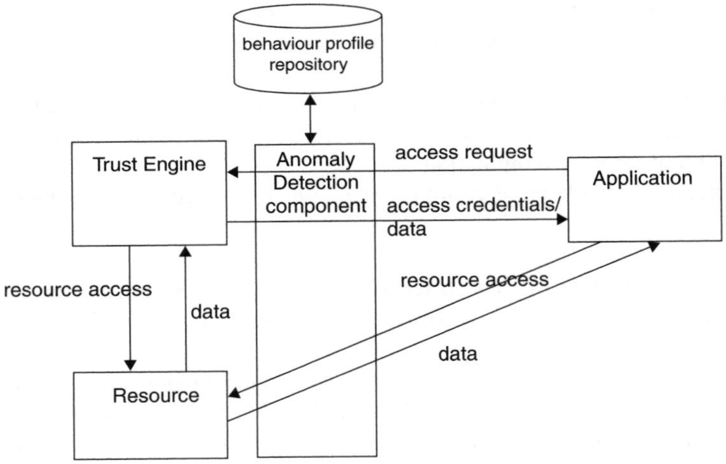

Figure 6.14 The anomaly detection component monitoring the application behaviour.

functionality. This functionality could be used to protect the device from malicious software and intrusions.

The intrusion detection functionality can be implemented by introducing an *Anomaly Detection* (AD) component, which co-operates with the Trust Engine of the device in question. The AD component monitors all access requests and resource accesses issued by the applications. Based on the observations, it builds behaviour profiles that describe how the applications request access to and use the resources. The basic set-up of a Trust Engine enhanced with an anomaly detection component is depicted in Figure 6.14.

After a sufficient training period, the profiles are used for detecting cases in which an application acts maliciously towards the device. Incoming access requests and resource accesses are compared with the profiles; if significant deviations between the incoming observations and the stored profiles are detected, an alarm is raised (e.g. by notifying the user of the device).

The rationale behind this concept is that usually applications show a fairly static behaviour pattern; the access requests and resource accesses that the applications issue are targeted at a limited set of resources. If, however, a malicious application or an attacker succeed in misusing a faulty application or in utilising vulnerability in the application code to take over control of the application, it is very likely that the behaviour of the compromised application changes radically. The compromised application may, for example, attempt to obtain access to resources that it would never do during normal operation, or the frequency and type of accesses may change, so that suddenly the compromised application starts to request more write access, when it earlier issued only read requests.

Due to this change in the behaviour of a compromised application, it is possible for the anomaly detection component to detect that the application has been compromised. As a countermeasure, the AD component can notify the Trust Engine. It can, in turn, initiate countermeasures by, for example, denying (temporarily) the access to the resources for the compromised application, restrict the access, or ask explicit confirmation for the action from a system administrator or a user.

To not only observe anomalous behaviour of resources requested, the applications themselves are restricted to a *Sandbox*, as described in the next section.

6.2.9.2 Run-time Security of the Trust Engine

The Trust Engine creates a *Trusted Computing Base* on which all applications run. The container in which all mobile applications are executed is the so-called *Sandbox*. A Sandbox controls all access outside its perimeter and enforces privacy policies for every access. Each application has its own 'personal' storage area. To allow more flexibility, several parts of an application, *bundles*, may be loaded into the same Sandbox if they have the same trust level (e.g. applications from the same vendor). From a storage management point of view, this bundling does not provide substantial advantages.

The Sandbox term comes from Java [31] and 'bundle' comes from OSGi extensions for Java [25]. The framework presented here is not, however, limited to Java applications. The Sandbox model can be supported by secure hardware that stores the most trusted data, and a secure operating system that provides the control of the Sandbox and prevents an application requesting outside of its authorisation. Figure 6.15 illustrates the Sandbox high-level stack view.

In the current literature, several approaches to a secure computing environment are presented.

- Trusted Computing Platforms contain a trusted area of hardware, controlled software and trusted or protected areas of memory and storage.
- In the Java approach, each application runs in a separate guarded Sandbox environment, creating problems for more involved forms of interaction. The OSGi approach allows multiple applications to live in a common Sandbox, potentially sharing data beyond the user's control. The method is based on renaming applications and giving them controlled rights within the Sandbox.

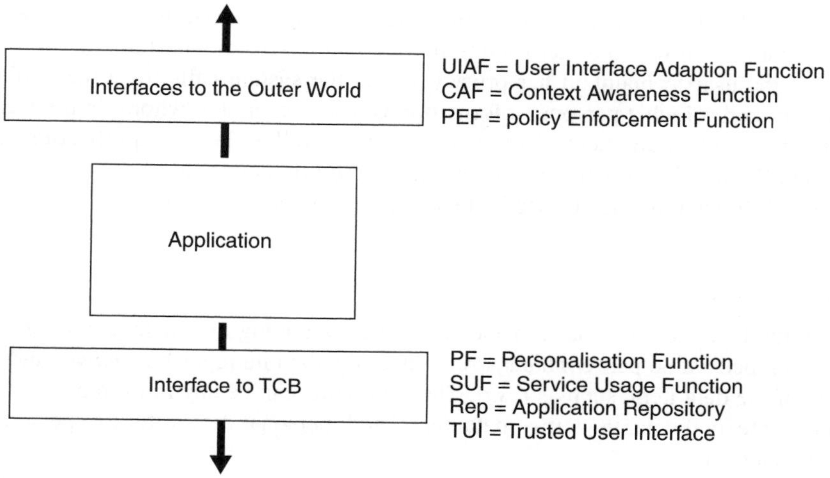

Figure 6.15 Sandbox control of applications.

- In the Symbian Platform security approach, each application that requires more than the basic functionalities needs so called 'capabilities'. These capabilities, basically access rights to resources or functions, need to be signed. Only if the signature validation is successful, the platform allows an application to exercise its rights (i.e. execute). Each application has its own dedicated storage area. Very sensitive user data can be protected by a special capability (*ReadUserData*).

The Sandbox approach presented in this book is closely related to the OSGi and the Symbian approaches. If the user's mobile system is split between different locations, the Sandbox can contain encrypted channels with endpoints following common user-set policies for external interaction. A new application in an existing mobile system must always be started up within a Sandbox in order to prevent applications gaining unauthorised access to personal user data.

6.3 Visualisation of User Choices Regarding Groups, Privacy and Trust

Typically, users are only willing to reveal or receive personal, real-time context data, if certain conditions are met:

- The people with whom the context information is shared have some kind of relationship with the sharing person (i.e. they are not complete strangers).
- The information shared helps in feeling connected, and has true added value.
- The shared information is not provided in too much detail.
- The context information to be shared is visualised in a compact and comprehensive way.
- It must be possible to disable the sharing or receiving of such data at any time.

There are situations in which the user wants to share some of this data with other users – or groups of users – within the system in order to gain increased value from the applications. The increased value for the user could, for example, be receiving information on events nearby her location or finding out which of her friends are free to go to the movies tonight.

Allowing the user to maintain her privacy within the system calls for a mechanism that helps the user control what personal information is disclosed and to whom. In the following, the assumptions and requirements for (Section 6.3.1) as well as the concept (Section 6.3.2) of the so-called *Privacy Display Widget* are presented. Additionally, some remaining challenges related to usability and privacy tradeoffs are discussed in Section 6.3.3.

6.3.1 Setting the Scene

Let us assume that each user has a personal profile containing user data as reference–value pairs. A reference–value pair contains a reference (the data item, such as *name*) and a value (the value of the data item, such as *Lisa*). The data items can be anything from static data to media and contextual data; the terms 'data items' and 'personal data' refer to a person's static and contextual data.

A user can share her data by setting up a number of rules, stating which data items to share with whom. The total set of rules a user has created are contained in a privacy policy. Each user has an individual policy that contains all the rules she has set up for her information sharing in the system. In addition, there are also different types of group policies that indicate how data is shared within a group and outside the group, as was discussed earlier in this chapter in the Section 6.1.3.3.

The most basic approach to setting privacy policies is that information is either shared or it is not shared. This approach with a clear on/off situation can work for some types of information, but it is far from applicable to all situations. Thus, the rules in the user policy contain the following information:

- with whom the data is being shared;
- which specific data items are being shared;
- to what extent they are being shared.

The extent of which the data items are being shared can be set by the user with *accept*, *deny* or *ask*: 'accept' indicates that the data item is always shared, 'deny' that is it never shared, and 'ask' that another user who wants to access the data item will initiate a request to the owner of the policy.

As a default, no data items are being shared until the user sets a rule in her policy stating that it should be shared, or accepts the rules of sharing for a group she wishes to join. This is feasible, since some people might not bother to take the time to learn and modify their privacy settings [5]. When accepting the rules for a group, the user will be alerted if she already has rules set up for one or several of the group members that are conflicting with the access modes set by the group policy.

> *For instance, perhaps Alice doesn't want Trudy to find out that she and Bob are in the same location, so she sets up a rule stating that Trudy should be denied access to her location. If, at a later time she decides to join a car pool group, where members have to share their location in order to be picked up and Trudy is part of that group, she will be alerted that there is a policy conflict. When a policy conflict occurs, the user will be aided in solving the conflict by having the conflict and different options of action presented to her.*

In order to protect the user's privacy, from the very instant she starts her mobile device and logs on to the system for the first time, the default policy should be not to share any data. The user should opt-in to share her data, either by setting up rules in her policy stating what to share with whom, or by accepting policy changes pushed from the system. In order to avoid extensive user settings, it is also possible to inherit profiles. However, this has not been implemented in the current system.

To make the user aware of the potential information flow when opting in to pushed policy changes (for instance, when joining a group), she should be informed on why the information sharing in question is required and be asked to accept or deny requests for information she is not already sharing through her existing policy. To keep the user in control of what rules she has already accepted and/or denied, a log of all previous disclosures should be available.

Given a request for sensitive information, even in the same context, two different users are very unlikely to react in the same way. Since context is an important factor, pre-engaged privacy settings for situations that have not yet arisen should not be forced upon the user. This is because it is hard for the user to know how she will react to the situation before it occurs. Rules for information sharing should therefore be set as a part of system usage. An example of this is how new rules are proposed when wishing to join a new group which has a group policy. The user will then have to accept adding the rules stated in the group policy to her personal policy in order to become part of the group.

The approach of pushing rules to the user, such as applying a set of rules when joining a group, could possibly also be used for applications that require certain user data. This could be useful when the user first starts using the application, since nothing will be shared as a default, and setting up a lot of rules would be time-consuming. In those cases, accepting the pushed rules from 'safe' entities such as applications within the system makes usage simpler. A possibility would also be for the service provider to push a set of suggested rules to the user at first startup of the device. In general, however, rules should be set and pushed by the user. This is partly due to the fact that users often intend to click the 'OK' button to accept certificates, for example, when using a web browser, without actually reading what they are accepting, just to get on with the task they are trying to perform. Since privacy is important, information overflow in the form of pop-ups should be limited so that the user would not share information by accident.

Based on several user interviews [3,19], the user needs to know what information is currently being disclosed. Displaying to the user who is allowed to access what information should make the user able to control her current privacy situation as well as to plan changes to it when required. There are two ways to look at shared data; either by looking at a user who can access a specific data item or by looking at what data a specific user has access to. Which approach is most useful for the user depends on the situation as well as on the visualisation concepts. The system should therefore support both approaches to privacy management in order to show the current information flow always in the most efficient way according to the current situation.

However, users can be reluctant when it comes to making settings and configuration unless they have to. This can be dangerous when those settings regard security or privacy. Another issue regarding privacy settings is that they are based on context, and for a user it can be difficult to set her privacy preferences for a situation she is not in right now. Thus, the privacy function has to be easily accessible, visible, and also offer settings which can be made during the everyday use of the system. In Section 6.3.2, a visual control solution – *Privacy Display Widget* – is introduced and remaining challenges are discussed in Section 6.3.3.

6.3.2 Privacy Display Widget

As described above, privacy in the system (as part of the Mobile Services Architecture) is managed by user-controlled policies. In order to fully comprehend her privacy situation, the user needs to be aware of what rules she has in her policy and what they mean in terms of disclosing her information. To be able to understand and to control her privacy situation, she needs a way to manage her policy and edit or create the rules needed to maintain the level of privacy she desires. Based on these needs, a visual component for managing privacy, the Privacy Display Widget (PDW), was designed. The PDW works as a graphical user interface (GUI) client communicating with the other privacy-related components of the system.

Since privacy plays a central role, it is important that this component is easy to use and doesn't interfere more than necessary with the user's main tasks in the system. *What* private information is provided to *whom*, *when* and/or in *what context* is information that is easy to access in the PDW, in order to give the user a good overview of her current information-sharing status. This is also important related to persuading the user that she can trust the system and

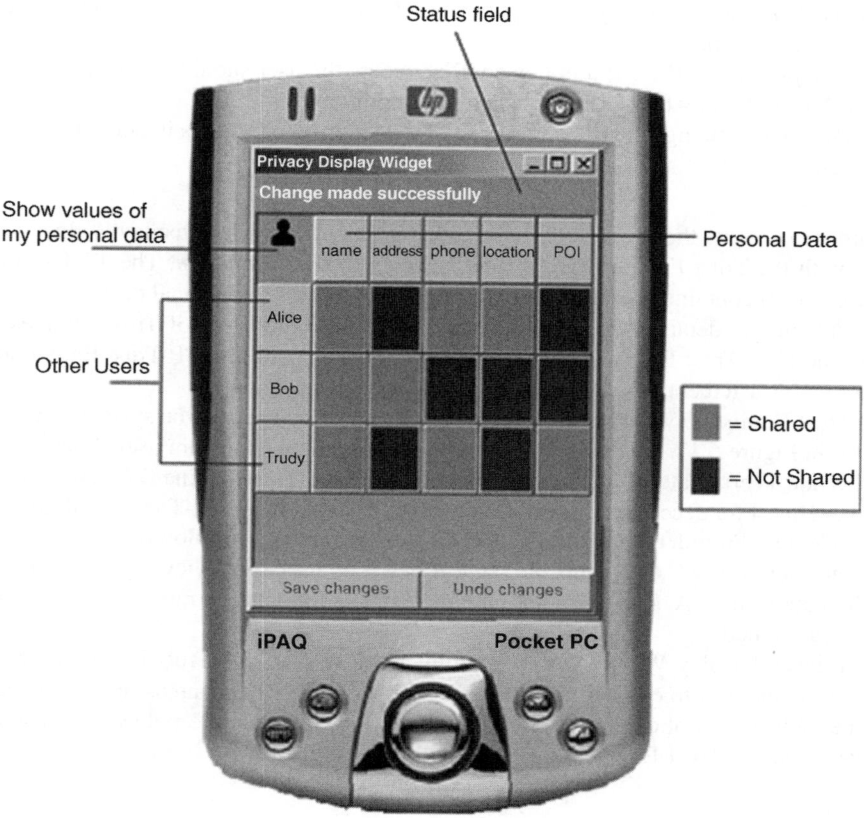

Figure 6.16 Privacy Display Widget (PDW) reference implementation GUI.

feel to be in control. When being able to view and understand her current privacy situation, the user also needs to be able to change her rules to suit changes in her ongoing system usage. Therefore, the PDW allows the user to alter her privacy settings in an intuitive manner.

The following main user requirements influenced the design of the PDW (see the reference implementation GUI in Figure 6.16):

- *View policy.* The user should be able to see who has what access (allow/deny/ask) to her data items. She should also be able to get an overview of her total amount of data sharing.
- *Modify policy:*
  - *Change access mode on existing rules*: Changing the access mode to the data items the user already has rules for.
  - *Create new rules:* Adding rules (i.e. setting an access mode for another user) for the data items that do not yet have rules.
  - *Accept requests to policy changes:* When joining a group, the user will have to accept the group policy (i.e. rules regarding what data the members of the group will share with each

other). This can potentially cause policy conflicts if rules have already been set up for a user in the group.

– *Handle policy conflicts:* If a policy conflict occurs the user should be made aware of the possible conflict by being shown which rules conflict.

- *View data items.* Being able to view the list of data items the user has in her personal profile and their values.

In order to execute the set of rules in the user's policy, the Privacy Display Widget communicates with the Trust Engine (TE) of the *Privacy and Trust Function*. The TE has a policy database (PDB) containing user policies for regulating access to personal data.

As explained in detail in Section 6.2.3, there are two main types of Trust Engines in the system: the User Trust Engine located in the user's device, and the CP Trust Engine located together with trusted entities (i.e. Context Providers) in the network.

The Trust Engine plays a role in all data requests made by users in the system. An example is shown in Figure 6.17. When user A wants to access a data item from user B, user A has to first generate a certificate (using a unique key) on her User Trust Engine (*TE_User*) and use it to send a request for access to user B's data to the CP having that data. The CP will then check with its Context Provider Trust Engine (*TE_CP*) to see if user A is allowed to access the data item from user B. If *TE_CP* grants the request based on user B's policy, the CP will respond with the data to user A. If not granted, the response will contain a message stating that the request was denied.

The Privacy Display Widget works as an interface towards the Trust Engine in the sense that it allows the user to edit her policy stored in the User Trust Engine using a graphical user interface. When the policy is updated in the *User Trust Engine*, the policy is dispatched to the concerned CP Trust Engines; this is shown in Figure 6.18. As in the previous example,

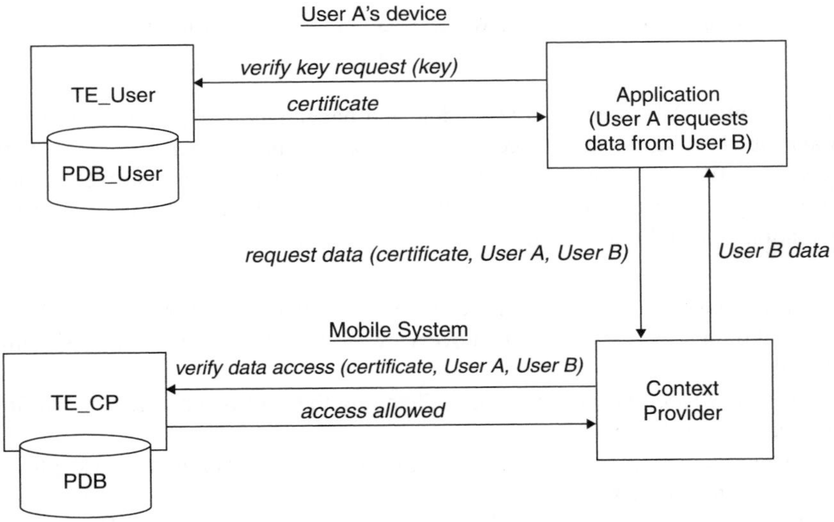

Figure 6.17 Trust Engine controlling user data access.

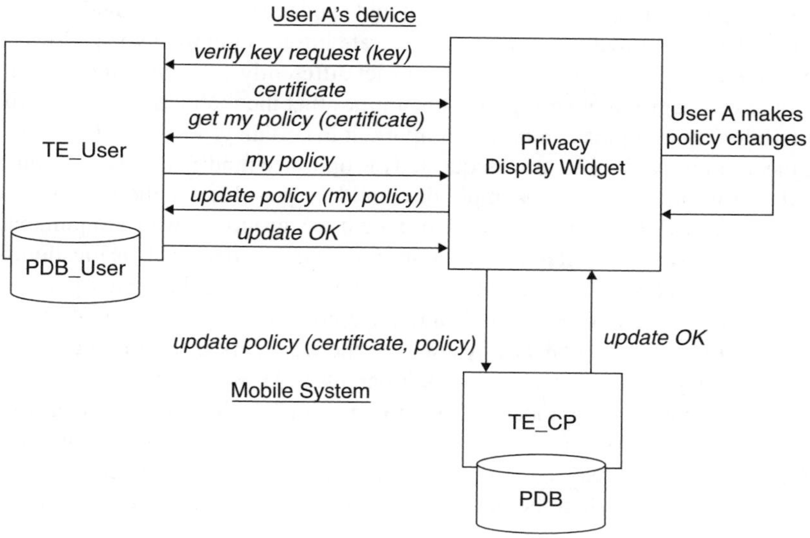

Figure 6.18 Policy update on Trust Engines using the PDW.

a certificate needs to be generated to identify the user when communicating with the Trust Engines.

6.3.3 Remaining Challenges

Requirements affecting the design of applications and services involving privacy and trust considerations come from two sources: the *requirements on the system*, which are derived from the mechanisms used, and the *users*, which might not explicitly express requirements but nevertheless have a set of requirements. These sources of the requirements are naturally the same for most of the other areas described in this book as well, but in the areas of privacy and trust it tends to be more challenging in finding out what the users actually want, what kinds of tradeoffs the users are willing to accept in privacy and usability, and how to develop privacy and trust mechanisms – including the user interaction and interface solutions – matching the complex requirements.

During the design of the Privacy Display Widget, a number of requirements referred to above emerged. Parts of them were taken into account in the initial Privacy Display Widget design and reference implementation, but many requirements included open issues that need further consideration.

Based on several interviews and user tests carried out [3,19], the users wish to be in control of the privacy and trust settings, but not to configure them – something that can partially be resolved using persistent policies as described in this book. Users also want to be involved when new privacy rules are set and new trust relations are established, but they do not want interactions with the system to be disruptive to any other task they may be engaged in. This also goes for the management of group relations: Users typically want to see a little more about other users than they want to give out themselves. Furthermore, users are not homogenous in their preferences. Some feel they have nothing to hide and are willing to provide quite accurate information; others are unwilling to provide accurate information to anyone. However, users tend to care about the groups they belong to, and act differently when interacting with a group of friends than with a group of strangers. This implies that the interaction between the Group Manager and the Trust Engine can be very important in certain types of application.

During the user interviews [3], users requested the option to modify the content and accuracy of the disclosed information. For example, it should be possible for the user to change her primary email address when being required to share it to a group she wants to join, but doesn't trust enough to provide her regular address. For more dynamic data such as location, it should be possible either to choose how specific the location should be (i.e. country, city or street name), to display no information or to display false information. The issue of showing their location was brought up as a concern by some of the teenagers that participated in the user interviews, since they did not always want their parents to know their whereabouts.

Also in general related to the location, major concern regarding location data was that the users wanted to be in control of who has access to their location and what granularity of the location (e.g. 'downtown area' or 'at this restaurant') other users have access to. The users would also like to control when the data is being accessed, in a specific context such as 'at home' or 'at work'.

How to best offer the user control over the granularity of information? For example, the user might want to have an easy way to turn off the location or some other sensitive information for all information receivers without having to change the individual policies. A possible solution for this could be to have a set of different profiles of rule sets that can be used for the information receivers. Another solution could be to have some kind of overriding rule set that would automatically override the rule sets in the user's policy when it is activated. This solution would let the user switch to, for example, a more private mode very quickly. It might, however, interfere with the user's group memberships, since these might call for sharing of certain data items. This could, of course, be altered using the information accuracy (e.g. giving the location as 'Stockholm' instead of 'Main Street 5').

During interviews, users expressed concerns about the system sending automated messages. They did not like the idea that the system could make automated decisions without their consent; negative feelings about automated system actions have also been a common concern in other studies [4]. This behaviour reduced the sense of control and did not support natural social interaction, such as calling to apologise for being late rather than sending an automated message. It was recommended that the system would suggest actions to accept or deny, and preferably explain why they should be taken. One solution to restrain the automation of information distribution in the system is to let the user opt in (select who gets information) or opt out (select who doesn't get information).

In the final evaluation [19], the users were testing the *FamilyMap* application, a map-based social application, where one can place different kinds of note and make them visible for selected other people. This application included integrated privacy protection (a simplified Privacy Display Widget) for the users. The biggest concern of the users related to their privacy was that they would reveal something private like their address or phone number while placing a new note on the map. However, pointing some note to certain known users or user groups was noted as a good thing, even though it reveals this personal information.

When sharing information to groups, the users requested that they would like to know who is in the group they are sharing information to. The users also wished to know in general what is going on and therefore wanted an easy access to their privacy information to see what data and data values they are sharing to which other users.

Some of the issues discussed in this section are rather well-known challenges, but there are still no feasible and generally applicable solutions to be used – especially in the easily complex and context-dependent mobile environment. The Privacy Display Widget and other similar visualisation solutions are paving the way for more user-friendly approaches to control privacy and data, but tradeoff between usability and privacy poses further challenges for future research and resulting systems.

6.4 Conclusions

Privacy, trust and group communications discussed in this chapter are elements containing both technical and user aspects. Social relationships are reflected in trust models and privacy requirements that have to be respected, taking into account that the Mobile Services Architecture described is distributed and the users of it are mobile. Visualisation of currently applied values resulting from user choices and a mobile dynamic architecture is challenging already without the limitations of a small screen. The *Group Awareness Function* shows an implementation of generic functionality in order to support a group lifecycle. When a group is active, it means that it is involved in an activity and thus it has a situational context that can be shared. Furthermore, policies need to be in place and enforced automatically to keep the user in control of this context sharing. The user has to make statements in her privacy policy about how and to whom to share her context within groups to be created, and how her individual context may be used to infer group context. Also, the user must be made aware who is seeing her context. Therefore, she must be aware of the other members of the group, even when a group is active for just a short time.

Representing policies, choices and group awareness to the mobile user is challenging. The *Privacy* and *Group Display Widgets* show example solutions on how to do this on a small screen. The infrastructure needed for setting and enforcing privacy of user-controlled data available through content or *Context Providers* is controlled by the *Trust Engine*, but even more importantly through a general view of how mobile devices and their service elements are started up and extended with new applications without compromising security – the necessity for trust and privacy.

Once the full palette of multimodality is exploited, the privacy aspects will re-emerge in new forms, and trust based on learning and reputation have more possibilities and challenges. Giving the user the feeling of control will remain difficult, as new groups of users enter the mobile world with very disparate backgrounds and technical understanding.

As examples of how group formation and handling works under these premises, the Chapter 7 describes two applications, *MobiCar* and *TimeGems*; both are prototypes of classes of applications. *MobiCar* deals with the multifaceted problem of forming context-dependent groups, while *TimeGems* illustrates how a user in a family setting might use groups in her daily life. Additionally, *FamilyMaps* brings out other aspects as discussed in Section 6.3.3.

6.5 Acknowledgements

The following individuals contributed to this chapter: Stefano Campadello (Nokia, Finland), Olivier Coutand (University of Kassel, Germany), Peter Ebben (Telematica Instituut, The

Netherlands), Ronald van Eijk (Telematica Instituut, The Netherlands), Johan Hjelm (Ericsson AB, Sweden), Silke Holtmanns (Nokia, Finland), Theo Kanter (Ericsson AB, Sweden), Sian Lun Lau (University of Kassel, Germany), Miquel Martin (NEC, Germany), Björn Melén (LM Ericsson, Finland), Markus Miettinen (Nokia, Finland), Rinaldo Nani (Neos, Italy), Petteri Nurmi (University of Helsinki, Finland), Mateusz Radziszewski (BLStream, Poland), Marcin Salacinski (BLStream, Poland), Göran Schultz (LM Ericsson, Finland), and Esa Turtiainen (LM Ericsson, Finland).

References

[1] 3GPP: 'Generic Authentication Architecture (GAA); Generic Bootstrapping Architecture'. Technical Specification 33.220. Online: www.3gpp.org/ftp/Specs/html-info/33220.htm.

[2] Ardissono L., Goy A., Petrone G., Segnan M. and Torasso P.: 'INTRIGUE: personalized recommendation of tourist attractions for desktop and handset devices'. *Applied Artificial Intelligence, Special Issue on Artificial Intelligence for Cultural Heritage and Digital Libraries*, Vol. 17, No. 8/9, pp. 687–714. Taylor & Francis, 2003.

[3] Conaty G. (ed.): 'Initial Scenarios, Requirements and Guidelines: User-centred approach for the design of future mobile services and applications'. IST-MobiLife Project Deliverable D6b (D1.1b), February 2006. Online: www.ist-mobilife.org.

[4] Cranor L.F., Reagle J. and Ackerman M.S.: 'Beyond Concern: Understanding net users' attitudes about online privacy'. Technical Report TR 99.4.3, AT&T Labs-Research, 1999. Online: www.research.att.com/library/trs/TRs/99/99.4/.

[5] Cranor, L.F. and Reagle J.: 'Designing a Social Protocol: Lessons learned from the platform for privacy preferences'. In MacKie-Mason J.K. and Waterman D. (eds), *Telephony, the Internet, and the Media*. Mahwah, NJ: Lawrence Erlbaum Associates, 1998. Online: www.research.att.com/~lorrie/pubs/dsp/.

[6] Dodgeball: Online: www.dodgeball.com.

[7] Erickson T. and Kellogg W.A.: 'Social Translucence: an approach to designing systems that support social processes'. *ACM Transactions on Computer–Human Interaction*, Vol. 7, No. 1, 2000, pp. 59–83.

[8] Fruchterman T.M.J. and Reingold E.M.: 'Graph Drawing by Force-directed Placement'. *Software Practice and Experience*, Vol. 21, No. 11, 1991, pp. 1129–1164.

[9] Grandison T. and Sloman M.: 'A Survey of Trust in Internet Applications'. *IEEE Communications and Survey*, Vol. 3, No. 4, 2000, pp. 2–16.

[10] Iachello G., Smith I., Consolvo S., Chen M. and Abowd G.: 'Developing Privacy Guidelines for Social Location Disclosure Applications and Services'. ACM International Conference Proceeding Series, Vol. 93. Proceedings of the 2005 symposium on Usable privacy and security (SOUPS).

[11] Internet Engineering Task Force (IETF): 'Group Key Management Protocol (GKMP) Architecture'. RFC 2094. Online: www.faqs.org/rfcs/rfc2094.html.

[12] Internet Engineering Task Force (IETF): 'Group Key Management Protocol (GKMP) Specification'. RFC 2093. Online: www.faqs.org/rfcs/rfc2093.html.

[13] Internet Engineering Task Force (IETF): 'GSAKMP: Group Secure Association Key Management Protocol'. RFC 4535. Online: www.faqs.org/rfcs/rfc4535.html.

[14] Internet Engineering Task Force (IETF): 'Internet Key Exchange (IKEv2) Protocol'. RFC4306. Online: www.faqs.org/rfcs/rfc4306.html.

[15] Internet Engineering Task Force (IETF): 'Internet Security Association and Key Management Protocol (ISAKMP)'. RFC2408. Online: www.faqs.org/rfcs/rfc2408.html.

[16] Internet Engineering Task Force (IETF): 'Multimedia Internet Keying (MIKEY)'. RFC3830. Online: www.faqs.org/rfcs/rfc3830.html.

[17] Internet Engineering Task Force (IETF): 'The Internet Key Exchange (IKE)'. RFC2409. Online: www.faqs.org/rfcs/rfc2409.html.

[18] Johnson D.W. and Johnson F.P.: '*Joining Together: Group theory and group skills*, 3rd edn. Englewood Cliffs, NJ: Prentice-Hall, 1987.

[19] Kurvinen E. (ed.): 'Results of Service and Application Evaluation'. IST-MobiLife Project Deliverable D14 (D1.9), December 2006. Online: www.ist-mobilife.org.

[20] Masthoff J.: 'Group Modeling: Selecting a sequence of television items to suit a group of viewers'. *User Modeling and User-Adapted Interaction*, Vol. 14, 2004, pp. 37–85.

[21] McCarthy J. and Anagnost T.: 'MusicFX: an arbiter of group preferences for computer supported collaborative workouts'. ACM 98 Conference on CSCW, Seattle WA, 1998, pp. 362–372.

[22] McCarthy J.F., Costa T.J. and Liongosari E.S.: 'UniCast, OutCast & GroupCast: Three steps toward ubiquitous peripheral displays'. International Conference on Ubiquitous Computing (UbiComp 2001), 30 September – 2 October 2001 Atlanta.

[23] Nokia Sensor: Online: www.nokia.com/sensor.

[24] O'Connor M., Cosley D., Konstan J.A., and Riedl J.: 'PolyLens: a recommender system for groups of users'. Proceedings of the ECSCW 2001, Bonn, Germany, 2001, pp. 199–218.

[25] Open Service Gateway Initiative (OSGi) Alliance: Online: www.osgi.org/.

[26] Raento M., Oulasvirta A., Petit R. and Toivonen H.: 'ContextPhone: a prototyping platform for context-aware mobile applications'. *IEEE Pervasive Computing*, Vol. 4, No. 2, 2005, pp. 51–59.

[27] RuleML: Online: www.ruleml.org.

[28] Sabater J. and Sierra C.: 'Reputation and Social Network Analysis in Multi-Agent Systems'. Proceedings of the First International Joint Conference on Autonomous Agents and Multiagent Systems, 2002.

[29] Sadeh N., Gandon F. and Kwon O.B.: 'Ambient Intelligence: the My-Campus experience'. School of Computer Science, Carnegie Mellon University, Technical Report CMU-ISRI-05-123, July 2005.

[30] Singh A. and Liu L.: 'TrustMe: Anonymous management of trust relationships in decentralized P2P systems'. IEEE International Conference on Peer-to-Peer Computing, 2003, pp. 142–149.

[31] SUN–Java: Online: java.sun.com.

[32] World Wide Web Consortium: Privacy Preference Platform (P3P). Online: www.w3.org/P3P/.

7

Reference Applications

Edited by Dario Melpignano (Neos, Italy)

This chapter presents examples of reference implementations of services and applications – researched and developed to demonstrate the value and practicability of the new application and service concepts (see Chapter 2) and the related technical building blocks (see an overview in Chapter 3) in this book.

The application design and implementation work around the Mobile Services Architecture is based on a set of end-user stories (*scenarios*; see the Appendix), highlighting novel mobile services using different technology layers and components. Insprired by the scenarios, the following reference applications were developed:

- *Context awareness and management* (described in Chapter 4). Context-aware applications to the areas of social interaction, healthcare and ubiquitous services:
  - ContextWatcher (Section 7.2);
  - Personal Context Monitor (Section 7.3);
  - Proactive Service Portal (Section 7.4).
- *Multimodality and personalisation* (described in Chapter 5). Applications designed to focus on distributed personal devices, personal area wireless connectivity, multimodal interfaces, gathering of personal preferences and adaptation of the application based on user preferences:
  - Multimodal Infotainer (Section 7.5);
  - Wellness-Aware Multimodal Gaming System (Section 7.6);
  - FamilyMap (Section 7.7).

Enabling Technologies for Mobile Services: The MobiLife Book Edited by Mika Klemettinen

- *Privacy, trust and group communications* (described in Chapter 6). Applications with a strong focus on group communications aspects, and services and concepts enabling users to best participate in and make use of their changing social contexts in everyday life:
  - TimeGems (Section 7.8);
  - MobiCar (Section 7.9).

These applications and services have been evaluated in practical service trials and studies (see Chapter 8).

7.1 Introduction

The *user-driven application-oriented approach* followed in this book focuses on helping people in managing communication, relationships and information in their private as well as professional lives.

Following the visions expressed in the scenarios introduced in Appendix, people will be connected to their family and friends as well as feel safer and more secure from their childhood to their later years with new kinds of ubiquitous services. In the near future there will be devices and services which become part of people's lifestyles. Young persons can use trendy, good-looking devices, which may act as wristwatches, locators, route finders, telephones, messengers and health sensors – all this at the same time. Traditional limits between different device classes become fuzzy: A wristwatch can save one's life by being capable of noticing, for example, a heart attack. In the same way, a significant change in application boundaries has to be expected: going back and forth from private activities to professional tasks is already symptomatic for people's everyday schedules.

The challenges around lifestyle-supporting mobile services demand easy-to-use devices people can and like to use. The usability and design of both the devices and the services become critical especially for elderly people. The role-based trust and associated privacy constraints, set by the user, must be strictly adhered to, while at the same time situations like a sudden heart attack and legal intervention results in a fall-back scheme of the security model. Combining newest location-sensitive technologies to advanced sensors results in mobile services, which allow people to keep their privacy, let them be selectively available in their work and leisure time, and feel safe.

But the importance of future services is not only in handling of emergency situations. They can help people loose weight, eat in a healthier way, track their fitness improvements, alcohol consumption, etc. All this can help in serious and costly problems, in addition to improving the quality of lifes of people in all age groups and cultures.

In recent years, a range of impediments have hindered the emergence of successful mobile applications and services. Examples of such impediments have been the low bandwidth of the connections and difficulties in the usability of the mobile applications on limited devices. The deployment of the new heterogeneous networks (3G and beyond) offering higher data rates and versatile local- and wide-area access methods, together with the new function-rich terminals, provide huge potential for novel mobile services and applications.

The following sections show a variety of utility and lifestyle applications that take advantage of the new possibilities in connectivity, terminal capabilities and technological components presented in this book. Primarily, the applications are included neither to show breakthrough concept innovations nor user interfaces polished to the level of commercial solutions, but to show implementations that are actually following the architectural and technological frameworks introduced in this book.

7.2 ContextWatcher

7.2.1 Description

Most mobile phone calls start with the sentence 'Where are you?' The increased mobility of conversation partners in phone calls created the necessity of asking where someone is before starting the conversation. In the near future, a new generation of mobile applications will allow users to get insight into the whereabouts of friends, family and colleagues (in short, buddies) prior to making the actual phone call, or even at any desired moment, of course with mutual consent.

The ContextWatcher is one of these applications that make it easy and unobtrusive for people to share information about their whereabouts, or any other piece of context information they might want to share, ranging from body data to pictures, local weather information, and even subjective data such as moods and experiences. The goal of the application is to offer people new ways to keep (and stay) in touch without the need for direct interaction. One example is to browse a friend's on-line photo book every week, and the contextual descriptions with the photos give a good impression of her activities, providing food for conversation next time one meets her.

ContextWatcher supports two different ways of context sharing:

- *Via an augmented buddy list*. This is near real-time sharing of context information between two persons. Before sharing, two persons have to agree to become buddies and to share specific types of context information. At any moment in time, buddies can see, for example, where their friends are, which books they read, what the weather is in their friend's location, how their friends are feeling, and who they are with. This information is presented in the form of a buddy list that is common for most Instant Messaging applications, but in ContextWatcher it also provides information about the whereabouts and more detailed information in the context menus (Figure 7.1). Additionally, buddies can be grouped in an ad-hoc fashion based on context information, so that, for example, buddies who are at home can be easily separated from those who are in the office, and the user can send the last group an SMS with an invitation for a drink after office hours. But buddies can also be grouped based on other criteria such as the type of relation, the current weather for the buddy's location, etc.
- *Via a personal blog*. In a blog the focus shifts from providing context information in a timely manner to providing context overviews that are human-readable and informative. In other words, where in the buddy list the task is to show the last observed location of a buddy, in the blog the location time series needs to be summarised in a single sentence; e.g. 'Today I travelled from Enschede (home) to Amsterdam and back'. What is in the blog can be fully configured by the user, and overnight the blog entries will be automatically generated based on the observed context information: travels during the day, the visited places, the local

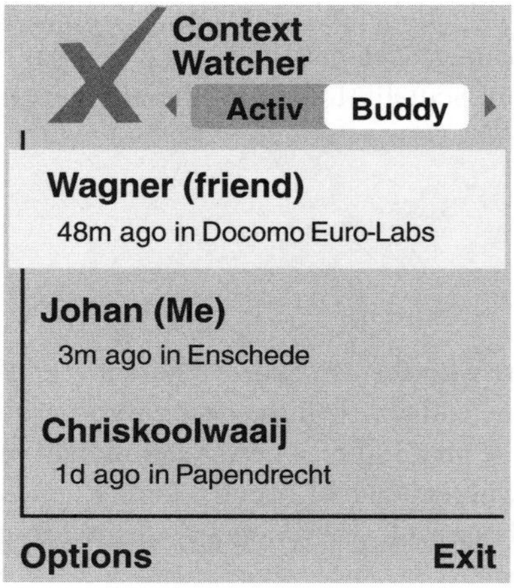

Figure 7.1 Example of a context-augmented buddy list.

weather, the people encountered, and the books read. Such a blog might be very interesting to browse for relatives who live at a distance, to become up-to-date with your activities in a very unobtrusive manner. An example blog entry is provided in Figure 7.2.

Both ways of context sharing show how easy it is to keep in touch with friends, family and colleagues, using innovative mobile applications. The only requirement for the user is to have a ContextWatcher-enabled mobile phone with her, running 24 hours a day, 7 days per week.

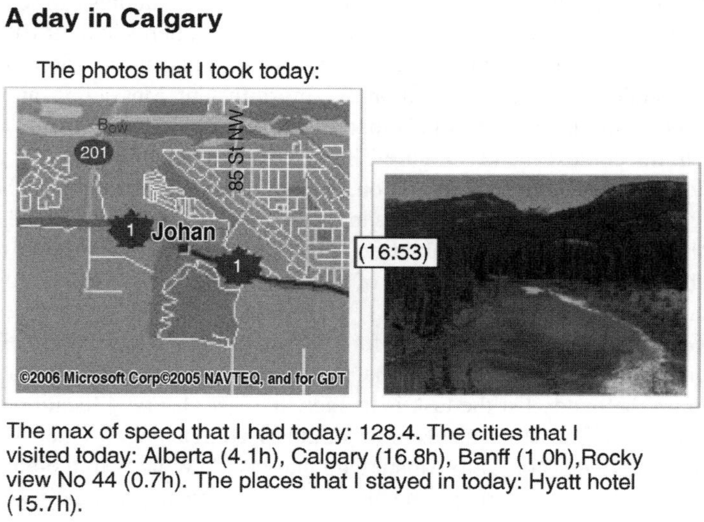

Figure 7.2 A sample blog entry for a day in Canada.

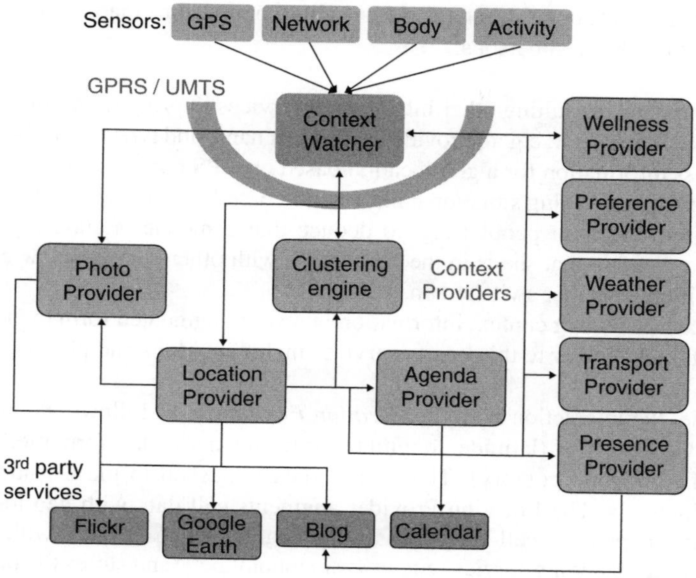

Figure 7.3 Architecture impression for the ContextWatcher application.

Buddy lists and blogs are updated automatically; the only effort for the user is to configure them according to her desires.

7.2.2 ContextWatcher Setup

The ContextWatcher is a thin client application, written in Python and running on Series 60 mobile phones, including many models from Nokia, Samsung and Lenovo. The application handles the user interaction, and integrates locally with attached context sensors, such as GPS or heart-rate sensors, and remotely with a palette of different *Context Providers* and *Context Reasoners*. An architecture impression is shown in Figure 7.3. The ContextWatcher application is modular in set-up, being able to combine different components and tabs from different developers, resulting in a dynamically configurable application set-up and automatic updates on component level. This way multiple versions of the same application can be offered (e.g. light, target group specific, and full), the application is easy to extend by new developers, and it is easy to maintain for existing developers. Due to the modular set-up, the ContextWatcher is in fact one application that can be accommodated to multiple application scenarios in which context gathering, sharing and reasoning is a key element. The ContextWatcher was originally targeted to support family members to stay in touch in an unobtrusive and entertaining manner, but has grown to an application that can easily be configured to support other scenarios, from a bird spotter community sharing their latest and most favourite bird spottings, pictures and sounds in a real-time fashion, to business people having their one-click business report with all their travels and business experiences.

Interaction with the distributed network of Context Providers (CPs) is via the mobile GPRS or UMTS networks. Data read from the local sensors will be pushed to the remote CPs. These

CPs can augment context information, reason with context information, and share context information with authorised persons.

- *Augmenting* means consulting other information services to provide additional information about a context parameter; e.g. to provide the product name and type for an EAN-13 barcode, or the address information for a geo-location based on GPS coordinates.
- *Reasoning* means providing situational information based on different context data streams, optionally from different people; e.g. to deduce that someone is currently in a business meeting from the fact that she is in the same room with other people, a few colleagues and one boss, with the beamer switched on.
- *Sharing* means providing context information in raw or augmented form to other authorised persons (e.g. buddies), or to third-party services including blogs and photo albums.

For example, the interaction with the *Location Provider* is as follows. Source data comes from either a GPS receiver (latitude, longitude) or from a network information service in the phone (cell-ID, network, country). This source data is pushed to the Location Provider at regular time intervals. The Location Provider augments cell data with geo-location using a large, operator-independent cell-ID database, and augments geo-location with address information using a geo-decoding service (e.g. www.mappoint.net), and stores the information in a database. This information is then used by other components including the clustering engine, the *Photo Provider* and the *Environment Provider*.

The clustering engine analyses the location time series per person and tries to find the frequently visited places of a person, in order to be able to translate raw location information (geo-location or street) to concepts that are more meaningful to the user and her buddies. In conversations, people usually do not talk about absolute locations, rather about relative locations: my office, the home of Marko, or close to the church. These relative locations are easier to interpret, and hence better to use, when communicating about location in a buddy list or in a blog. The clustering engine analyses location streams every night, and it can extend existing clusters or find new ones, which are presented to the user by the ContextWatcher for naming. The user can also relate a cluster to a location ontology that covers most concepts, including home, office, hotel, sport, etc. Figure 7.4 shows an example of a detected office presence together with a meeting with a colleague.

The Photo Provider stores pictures made with the camera phone, and augments them with tags, titles and descriptions automatically, based on the situation in which the photo was taken. The situation information is obtained from other Context Providers that are queried for information around the time the picture was taken. This way each picture can be tagged with automatically recorded street and city information, the geo-position, the speed and direction of movement, the name of the location cluster (is this a home or an office picture?), the local weather, and the people who are nearby.

The Photo Provider has been integrated with Flickr [5], one of the largest public image servers of this moment, where the context information is submitted as tags, and the descriptive text is automatically generated; e.g. ‘I was on a [business trip] together with [Henk] and [Bernd] in [Oulu] and I took this picture of the [Aleksanterinkatu] while travelling [with public transport] to the [summer school]’, where all the information between brackets is auto-generated. This means that one action on the mobile phone is enough to send a richly described picture to a remote image server, enabling others to easily find pictures of their liking, for

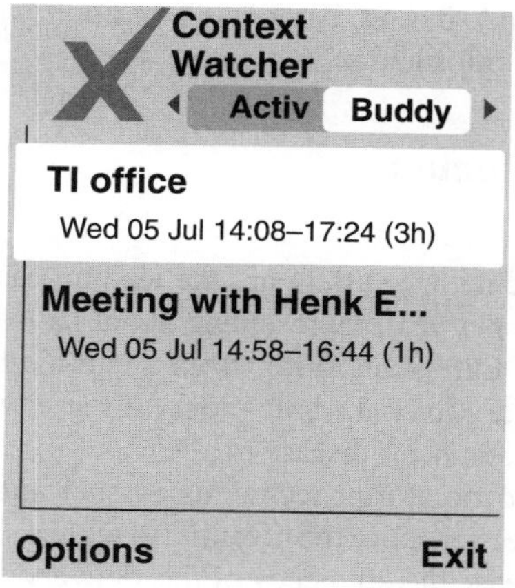

Figure 7.4 Example cluster and meeting, with the time spans.

instance by browsing the context tags to separate the home pictures from the office pictures. Rich context queries become feasible: 'Provide me with all pictures together with Bernd in Oulu while it was snowing'.

The Environment Provider provides both current weather conditions and forecasts with granularity on the city level. Since the city information can be obtained from the Location Provider, there is no need for the user to specify the city any more: One click is enough to update the weather information for any current city. The weather information is then presented in the ContextWatcher and stored for later reference, so that, for example, the blog generation engine can query the Environment Provider for the conditions along the location track of a certain user for the previous day (Figure 7.5).

Many features other than the ones described in this section are available in the reference implementation, including management of social relationships, sport scenarios with real-time

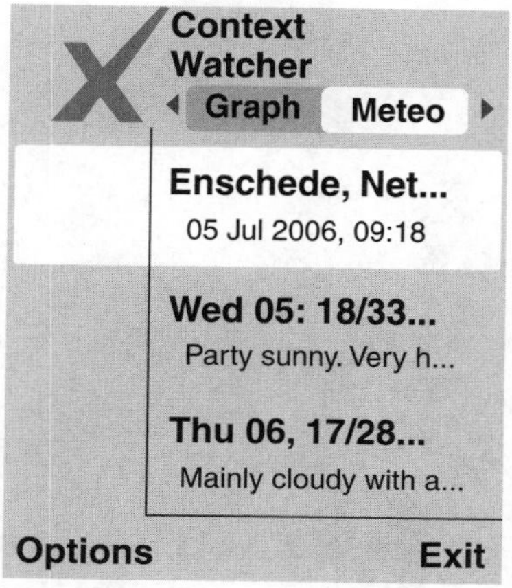

Figure 7.5 The current local weather in the ContextWatcher.

body data, sound recording and sharing, barcode recognition with the camera phone, usage logging, etc.; see [2] for more information.

7.3 Personal Context Monitor

7.3.1 Background

Diseases of the heart and circulatory system are the leading cause of death in Europe, with more than 1.9 million people per year, representing about half of all the deaths in European countries. More than 20% of Europeans suffer from a chronic heart and circulatory disease in many forms, such as coronary (or ischaemic) artery disease, valvular heart disease, stroke, high blood pressure or rheumatic heart disease [4].

Innovation in preventive and monitoring technologies, such as the Personal Context Monitor application presented in this section, have the potential to reduce expenses, to deliver healthcare services remotely and to increase the efficiency of this delivery.

Effective integration of healthcare applications and support services can stimulate self-care and self-responsibility to improve the quality of life of the citizens by enabling safer independent living and increased social relations.

7.3.2 The Application

The Personal Context Monitor, shown in Figure 7.6, has been designed as a solution to monitor and process in real time essential physiological parameters such as electrocardiogram (ECG) and heartbeat. These parameters are complemented with data that can be interpreted to detect

Figure 7.6 Personal Context Monitor equipment.

physical activity, body position, falls and location. Further modular extensions are designed to consider also body temperature and blood pressure.

Built to take advantage of the approach and infrastructure of the back-end components described so far, the Personal Context Monitor has been implemented in the first stage in Symbian (Figure 7.7), to support wireless real-time transmission of the data stream via Bluetooth to personal computers, smart phones or PDAs for storage, and provides at a glance both real-time and batch analysis of the data stream that can be easily interpreted by a doctor.

The application relies upon a hardware prototype built around state-of-the-art components and a streamlined technical design of limited costs, and to be compatible with evolving healthcare standards (e.g. Bluetooth SIG for Medical Devices 1); see Figures 7.8 and 7.9.

In the trials that have been performed, the Personal Context Monitor supports up to 60 hours of continuous wireless transmission, and parallel internal storage of data on SD (Secure Digital) memory card storage for up to 192 days, with the capability of encryption and protection of the collected, recorded and transferred information.

Small, mobile and versatile, the device has shown the capability of being worn comfortably during normal daily activities that require remote and wireless monitoring, for the purpose of cardiac rehabilitation, cardiovascular screening, cardiovascular monitoring, and also specific fitness and health management where off-the-shelf cardio frequency devices do not provide sufficient information.

7.3.3 Data Format

The application supports data structures in the European Data Format (EDF) [3] and has been tested also by implementing the Open eXchange Data Format (OpenXDF) [8]. The advantage of relying on open standards for the digital storage of time-series containing physiological signals and annotations is evident. The Personal Context Monitor supports EDF as the current de-facto standard for exchanging polygraphic, physiological data recordings. The already tested next step is the support of OpenXDF, which is based on XML as the most widely accepted standard for the digital storage of data. OpenXDF adoption has the purpose to foster cross-device information exchange and compatibility, and it allows unlimited expansion possibilities with further Context Providers/data sources, supporting as well the following additional features.

- Both European standard and proprietary information can be stored in the file without affecting compatibility.
- There are no limitations on annotations or scoring information.
- There are no limitations on patient information.
- There are no limitations on string lengths, and only loose limitations on sample sizes (1, 2, 4 and 8 bytes) and frequencies.
- Text encoding and language support is not limited by XML.
- There is robust, uniform, Y2K-compliant date/time representation based on ISO8601 [6].
- Data encryption is supported.

7.4 Proactive Service Portal

The Proactive Service Portal application picks up the idea of *Proactive Service Provisioning* and integrates a number of concepts of the Mobile Services Architecture presented in this

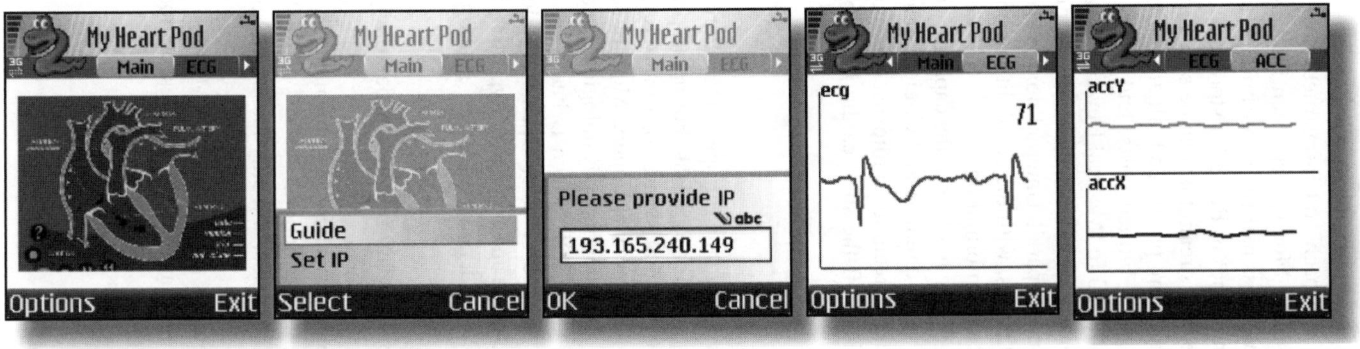

Figure 7.7 Screenshots from the implementation of the Personal Context Monitor in the Symbian OS.

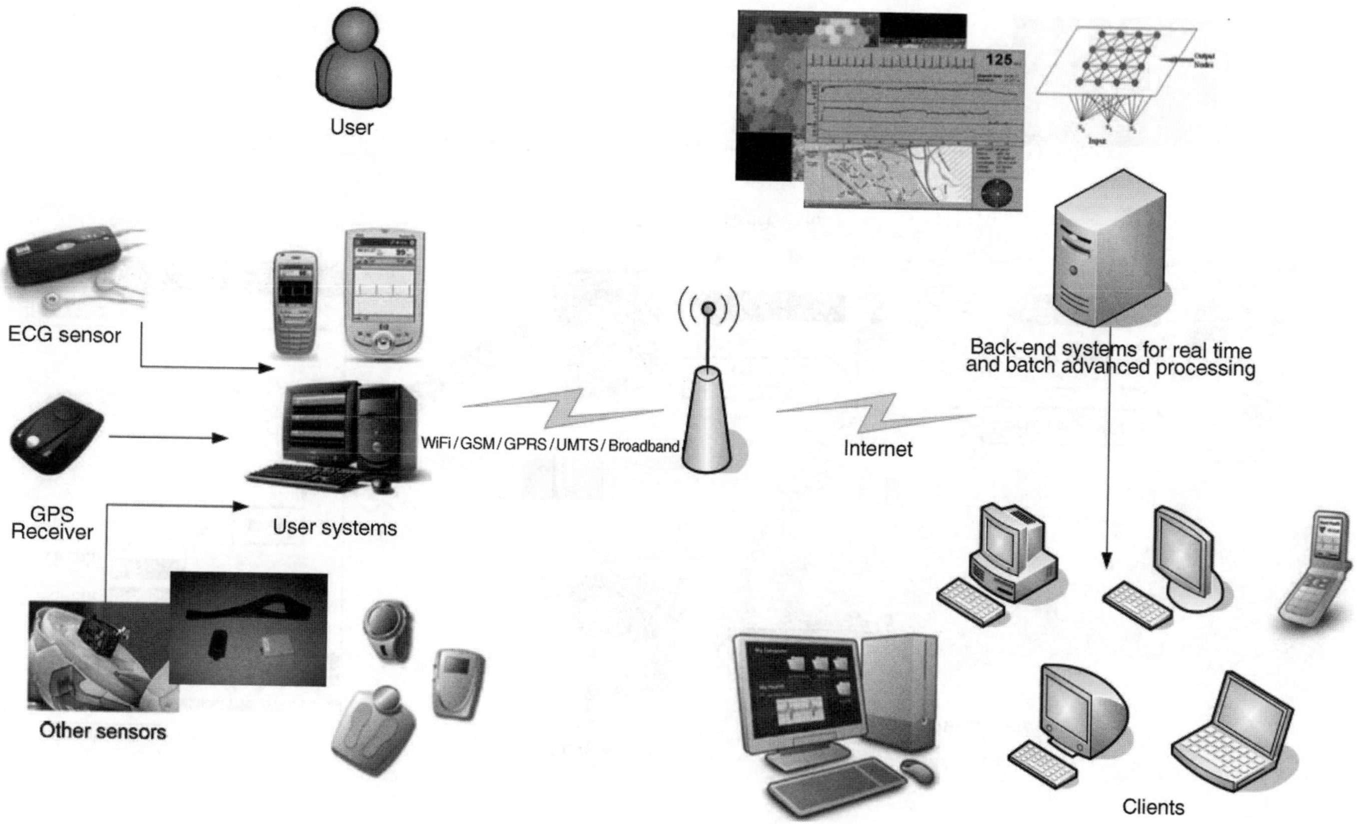

Figure 7.8 Personal Context Monitor setup.

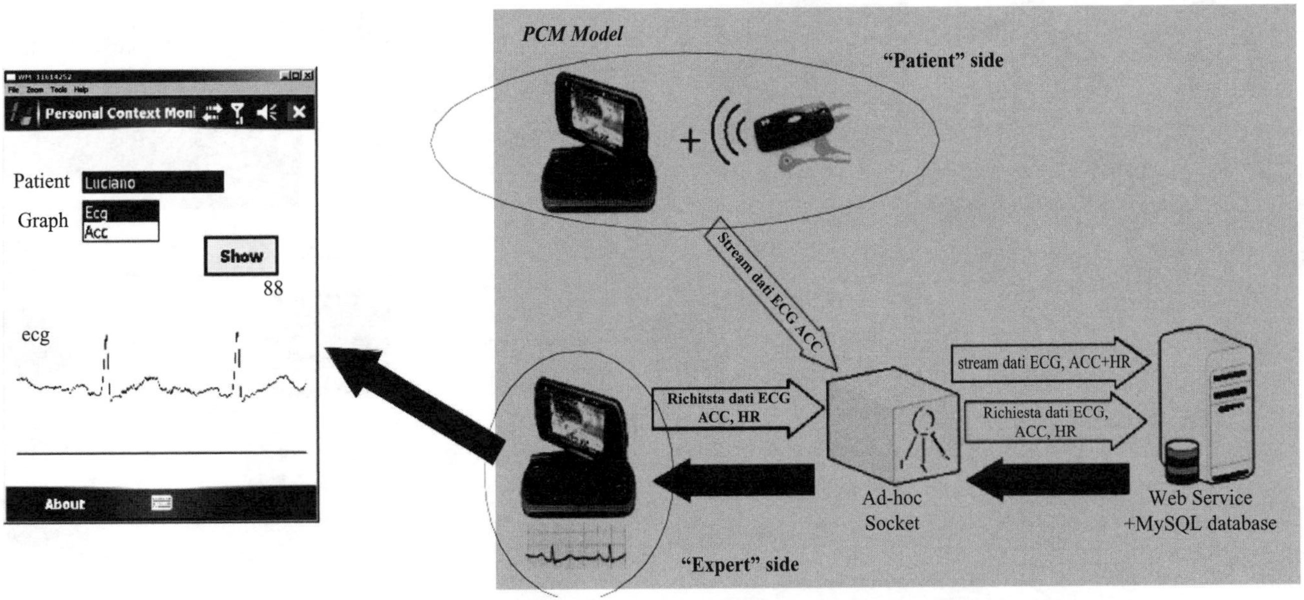

Figure 7.9 Personal Context Monitor architecture and implementation in Windows Mobile OS.

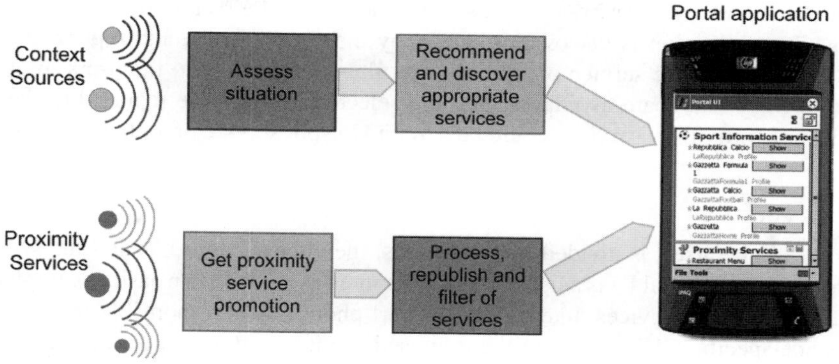

Figure 7.10 Service discovery paradigms.

book. The basic idea of the Portal is to relieve the user from browsing through endless index lists to finally find a desired service; or to avoid having the user having to type in lengthy URLs with the restricted keyboard capabilities of a mobile device. From the user's point of view it appears as a simple, but personalised and dynamic service portal.

To realise this concept, the Proactive Service Portal makes use of two different service discovery paradigms (Figure 7.10):

- the situation-based service recommendation and identification;
- the proximity-based self-promotion of services.

Beyond the service discovery mechanisms, the application implements a portal management and control function, which is responsible for collecting the services considered useful from the discovery plane and preparing the application user interface look and feel, adapted to the users' device capabilities.

7.4.1 Situation-based Recommendation and Identification

The situation-based service recommendation gathers available user's context information and tries to infer her current situation. For this situation, a number of service types are identified which are considered useful by the *Recommender*. The recommendation can be based on typical usage patterns, on information about the past behaviour of the specific user, on the users' preference, as well as on context information in general. For example, a person who is standing at the bus stop would obtain information on public transport services, whereas a person close to a theatre could get recommendations on available entertainment services.

The available services of the recommended service types are identified through standard service discovery mechanisms. A best-matching algorithm identifies the most suitable ones, which are then forwarded to the portal.

7.4.2 Proximity-based Self-promotion of Services

The proximity-based self-promoting service discovery picks up service advertisements, which are provided in the vicinity of the user. These services are actively promoted via local radio networks (Bluetooth, NFC, etc.) or via passive promoters like RFID tags or (two-dimensional) barcodes. The promotion is discovered locally by the user's device and republished to the Portal Service. There the service promotions are filtered according to the users' preferences to ensure privacy and spam avoidance, and the selected services are reported to the Portal. Typical sample proximity services could be a WiFi Hotspot service or a photo print service.

7.4.3 Portal Service

The Portal Service itself is divided into two parts, the client (Portal UI) and the server side (Portal Server). The Portal UI is a light-weight user interface, which can easily be implemented on any typical mobile devices, like PDAs or smart phones. It is responsible for the creation of the device-specific GUI to present the proposed services and to enable the user to choose an appropriate service for execution. The services are divided into categories based on their characteristics. Portal UI also provides the notification and user feedback mechanism about the services used. This feedback is used for a learning process that allows recommending as relevant services as possible in the future. The Portal UI communicates only with the Portal Server, so it is easy to switch the Portal UI to another implementation of the Portal Server.

The Portal Server is designed as a network service, and it controls the *PSP Service Discovery* and processes the identified list of useful services according to requirements from the device, from the device-specific application and from the user's preferences. It is also responsible for the feedback of service usage information. The other goal of the Portal Server is to avoid overhead in the Portal UI that should be a thin client. The Portal Server also minimises the network traffic from the Portal UI. The Portal Server communicates with other underlying CMF components.

Figure 7.11 shows sample screenshots for two possible situations:

- The first screenshot shows a situation where the user is at home in the morning and about to leave her house. The system assumes that she might be interested in public transportation services, since it detected a traffic jam on her way to the office. Furthermore, it discovered, among others, the TV remote control service locally offered by her kitchen TV via Bluetooth.
- The second screenshot actually shows the situation when she is in the office. Obviously, the transportation services have disappeared, but the most relevant office services are offered, such as the discovered printer, the coffee purchase service and a direct beamer access.

7.5 Multimedia Infotainer

The Multimedia Infotainer is an advanced multi-functional client–server system for use on mobile devices or personal computers. It is able to discover surrounding environment parameters and enables the following functions:

- Provide *user profiled information* from configurable resources through manageable content feeds (news services, blogs and other sources of data).
- Play *multimedia content,* through multiple devices or modalities, from remote streams or from local files utilising gathered environment configuration data with the built-in discovery engine.

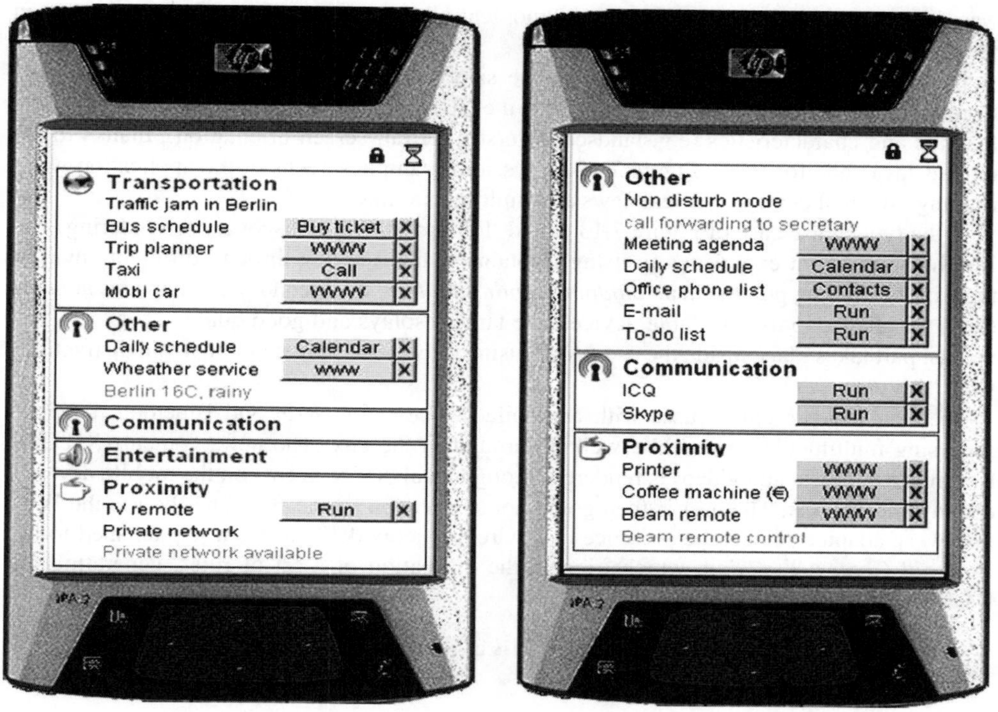

Figure 7.11 In the morning at home (left) and in the office (right).

The implemented version of the system brings a new look in the access to multimedia devices and their usage. It has been integrated with the context-aware multimodal component platform exploiting functions and functionalities introduced in Chapters 4 and 5, allowing the application to perform complementary presentation on more than one device at the same time, utilising all available multimedia means in the vicinity. Reference implementation functionality covers the main multimodal features accompanied with content delivery and presentation.

The client side has been built on top of the core components, such as the *Gateway Function* and the *Modality Function*, and provides an easy-to-use user interface. Application supports discovery of devices as well as information and selection for content presentation. User preferences can be stored, modified and further used as a base for content retrieving. The application uses current environmental situation and available devices to present content and to notify users through proper available modalities.

In order to provide an effective application functionality and demonstration of the multimodal features, a server accompanying the client application has been implemented. The server is used to retrieve content from all the content channels defined by the user, including Instant Messages. The content is then delivered to the client application for further presentation and processing.

The client application user interface is based on Java with CDC AWT libraries to present content and to provide user interaction elements. Choice of the particular interface style used was driven by the need to provide at any moment all the functions presented on the screen with standard controls.

The application interface looks almost the same on both device types used – iPaq and Communicator – with just small differences in the GUI part depending on the device's technical features and characteristics (e.g. landscape versus portrait screen orientation), thanks to the unified Java code for CDC. Figure 7.12 shows some sample screens of the application when running on a real device with real news and multimedia files.

In the typical use case of Figure 7.13 called 'Full Loop' (from discovery to rendering), the user has a different experience in using Infotainer while moving through different environments: The starting point is in an *outdoor mobile situation*, the second phase happens at *home* (with the chance to use available devices like large displays and good quality speakers), and the last part takes place inside the user's *car* (using only audio interaction to avoid distractions while driving).

The scenario presents a user with a mobile device (with WiFi and Bluetooth support) accessing multimedia news while walking around in the city. The news content (XHTML text, images, music and video) is rendered through local device agents on the mobile device's display and speaker. The user is then going home. As soon as she enters her house, the WiFi discovery adapter detects the presence of a wireless network. This event is published to the *Personal Context Function*, which triggers the evaluation of a set of rules, generating the following inference:

Detected WiFi network X + WiFi network X is in my house

→ I am at home

Inside the profile, associated to the state 'at home', the user has stored the preference to 'propose the usage of the large display and the audio system of the media center'. The system prompts these options to the user, who can select whether to optimally render the multimedia news or to use the personal device, for instance for private messages.

Later, the user leaves home and enters her car. The Bluetooth discovery adapter detects the presence of a BT device – the hands-free audio system of the car. The publishing of this event triggers the evaluation of another set of rules, ending up firing the following inference:

Detected BT address Y + BT address Y is in my car

→ I am in my car

The user's profile inside the Personal Context Function associates the state 'in my car' to the preference 'always use audio instead of text messages'. Moreover, the car computer offers services for speech and gesture recognition. Therefore, when a new piece of information is received, the user can say 'open' and the new textual message is rendered as audio. In addition to voice commands, simple gestures on the car console touchpad permit browsing through the news.

As referred to above, the main flow of the application is based on the core context-aware multimodal components and their functionalities, which can be represented in a diagram with a loop presented in Figure 7.13 (hence the 'Full Loop' name), where each component provides its specific contribution. The flow presents how the application behaves with changing contexts

Figure 7.12 Infotainer GUI on iPaq – sample screenshots.

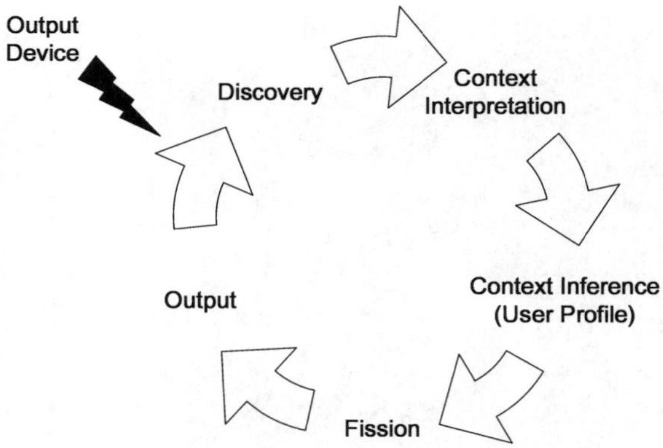

Figure 7.13 'Full Loop' use case for Multimedia Infotainer.

and devices in range to use multimodal functions. The cycle is performed according to the following flow:

1. The user starts the application and her low-level context information (available devices, GPS coordinates, reachable networks, etc.) is continuously sent to the components responsible for the context interpretation.
2. These components transform the low-level context information to high-level context (e.g. 'at home', 'in car') and forward that to context inference components.
3. Context inference processes the received information with the support of user profile data and the reasoning engine.
4. The result (context and preferences related facts, e.g. 'use only audio in car') is used for presenting the content to the user as audio or as visual data.

Typically applications will request the context-aware multimodal components for suggestions ('what modality to use?') and get responses (e.g. 'audio only') consistent with the current context and the user preferences.

7.6 Wellness-aware Multimodal Gaming System

Under the name 'Wellness-aware Multimodal Gaming System' (WAMGS), a set of different applications related to monitoring and exchanging user fitness information has been created and evaluated.

The applications were designed to study the acceptance and social aspects of the use of sensors, for evaluating technical implications of sensor usage, and for showing the benefits of the contextual personalisation approach. Thereto, WAMGS utilises user interface adaptation as well as context awareness and personalisation functionalities provided by the Mobile Services Architecture framework.

An initial version of this application featured the sharing of wellness information between football players and their family members. The system was used in football events, where football players, coaches and family members gathered together. Players wore heart-rate sensors from which wellness data was collected. Spectators standing beside the field had a mobile device on which they were able to view the data captured by the sensors (heart-rate value) and the interpretation of the wellness conditions of the players represented as 'emoticons' (emotional icons).

A later version of the Wellness-aware Multimodal Gaming System included off-line wellness analysis and improved real-time monitoring features, such as the possibility to make multimedia annotations while the spectators were watching a game. The application linked those annotation inputs to sensor readings, and presented those as bookmarks in an 'after training analysis' graph.

The final version of the WAMGS is a personalised wellness monitoring system that allows monitoring of the user's fitness situation and that creates personal training plans according to the current wellness situation of the user. The recommended training plans take into account the user's training habits derived from past trainings, the user's preferences (e.g. the preferred training devices), general user data that might influence the training (such as user's age and weight), as well as expert knowledge from professional gym coaches. Besides supporting personalised training at the fitness centre, the WAMGS also supports basic freestyle training, where users can monitor their training performance (real-time sensor data such as heart rate and running speed) on the monitor of the portal device while jogging.

WAMGS presumes that a user is wearing a heart-rate belt for measuring the heart rate and a foot pod for speed measurement. These sensors need to be accessible via Bluetooth. In the reference implementation, Suunto sensors are used and the data is transferred via a Suunto proprietary Bluetooth box to the user's portal device.

Personalisation functionalities focus on three main aspects: modality recommendation, recommendation of the next training exercises, and recommendation of motivational messages:

- *Modality recommendations* (e.g. video or audio output on the local or remote device) are based on the requested content, resource availability and user preferences.
- *Training exercises recommendations* is a list of recommendations derived from expert knowledge, user's training history and user preferences. In fact, two lists of recommendations are computed. One list is based on expert knowledge and one list is based on user's training history and user preferences.
- If the two recommendation lists above strongly differ from each other, a *motivational message recommendation* is requested by the application. This message contains information about why the user should follow the exercises recommended by the expert. For instance, the message could explain that the exercising has higher effectiveness for reaching a specific training goal on a bike instead of on a treadmill.

Figure 7.14 shows three example screenshots from the WAMGS client application running on a portal device. All three screenshots are related to a personalised training session in the gym studio. As shown in the first screenshot, the user can select a desired training duration, which is then taken into account for the generation of the personalised training exercises list. An example training exercises list is shown in the second screenshot. Each exercise contains a specific training duration and the supposed training effectiveness for the related exercise. The third screenshot shows the device and the modality selection pop-up containing resources available for displaying the training data. This pop-up window is shown to the user after she selected the next exercise.

Figure 7.15 shows three screenshots associated with a motivational message. The first one is displayed in case the expert recommendation list differs from the one generated based on

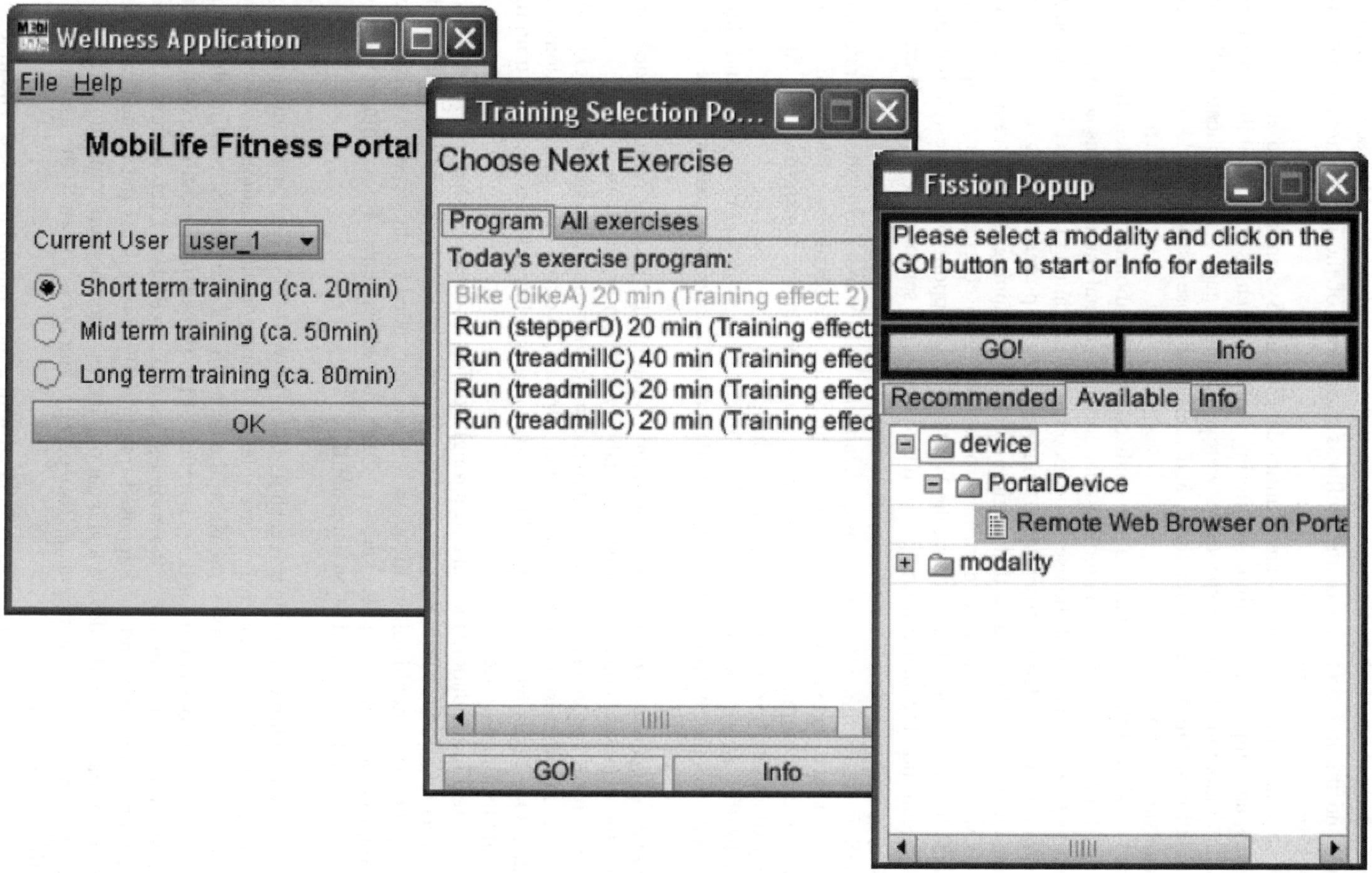

Figure 7.14 WAMGS client on portal device.

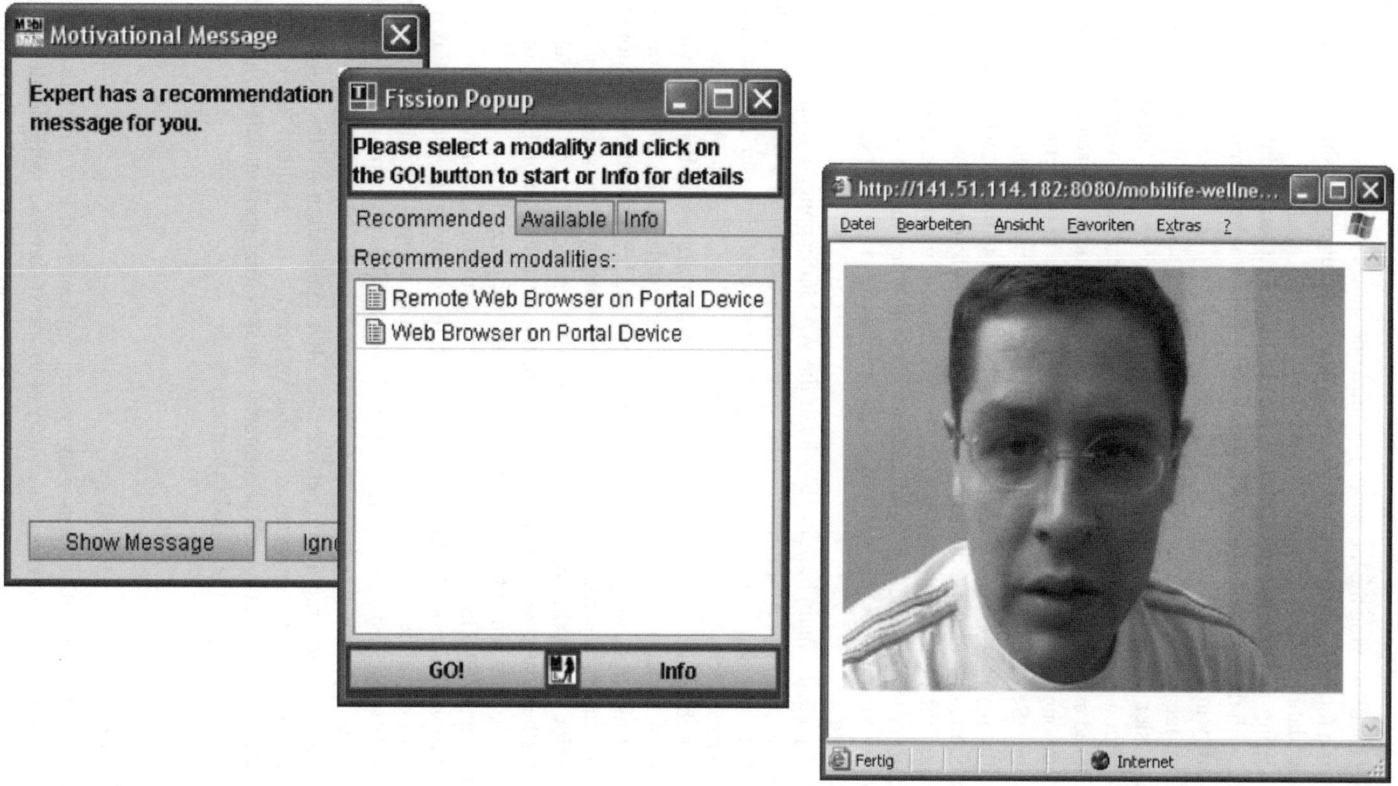

Figure 7.15 A motivational message.

the user preferences and past user's training history. The second one lets the user choose the output modality for displaying the message. The third one shows a snapshot of the motivational message (showing the professional coach) that explains to the user why she should follow the expert recommendations about training exercises.

While performing an exercise, the user can see the real-time sensor data displayed on the selected device with the selected modality. After having finished the training, the user can see her training history in a web browser.

Expert knowledge related to the WAMGS includes rules such as 'the user should first warm-up with easier exercises before performing harder exercises' or 'inexperienced users should favour biking exercises over running exercises'. These rules constitute a knowledge model that is stored as part of a user profile and then retrieved and processed by the rule-based reasoner subcomponent of the Recommender, which is located in the personalisation-related part of the Mobile Services Architecture. Learned user preferences about exercises (device, duration and difficulty level) are also used by the Recommender to predict what the user would do in a given context. Expert recommendations and user preferences-based predictions are then compared by the client application. The result of these comparisons is fed back into the Recommender in order to request recommendations about which motivational or educational message to propose to the user in case she does not want to follow the expert recommendation in the current context.

Figure 7.16 depicts the main building blocks of the WAMGS application and its dependencies in the Mobile Services Architecture. The core component of the WAMGS application is the WAMGS Client that is running on the portal device of the user and interacts with the Mobile Services Architecture. The *Context Awareness Function* and the *Personalisation Function* of the architecture are used by the WAMGS Client for a number of services. First, the Personalisation Function is used for storing the user profile of the application user. All user-specific data like age, weight and the current user's wellness situation are stored there. Second, the same function is requested for personalised training plans when the user starts a new training session. Third, the application sends logs of the selected user's exercises, exercise durations and further exercise-related information to the Personalisation Function, which in turn evaluates the logs to refine the personalisation of the training plans.

Furthermore, the WAMGS Client also utilises the functionalities provided by the *User Interface Adaptation Function* (UIAF) of the architecture. In particular, the UIAF is used for the visualisation of real-time training data on various output devices. Finally, the *Sensor Data Agent* in Figure 7.16, located on the portal device, is responsible for communication with the personal sensors worn by the user.

As a result of the depicted interaction between WAMGS application and the Mobile Services Architecture components, the application is able to provide personalised services utilising context awareness, personalisation and user interface adaptation functionalities without having to deal with the complexity of these tasks themselves.

7.7 FamilyMap

FamilyMap is a location-sensitive messaging system implemented into a map user interface. The main focus of the FamilyMap application is to assist families with small children in their everyday life, especially when on the move. Dealing with location-based, logistical and

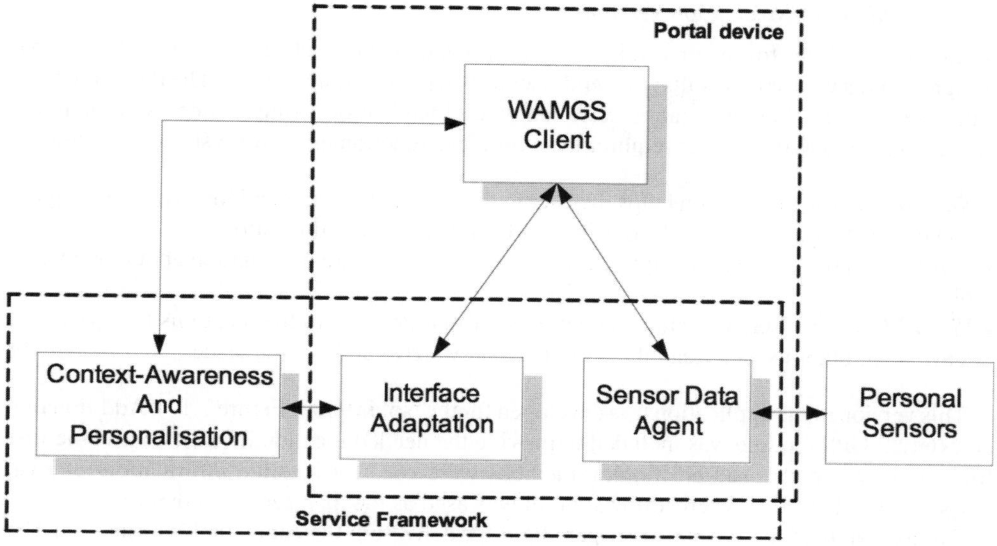

Figure 7.16 WAMGS functional components.

baby-care related issues came from the realisation of difficulties present when one tries to find necessary information from sporadically available information sources while in the city. Thus, the general objective was to solve practical problems and dynamically find locations within the city area that will be useful to the parents in their everyday life.

The implemented versions of the FamilyMap, however, were made to be useful and entertaining for any group of people needing to share information and to coordinate their activities with one dedicated mobile application and device.

7.7.1 FamilyMap Initial Version

The initial version of the FamilyMap application was implemented using the following devices:

- Nokia 9500 Communicator with wireless GPS module and Wayfinder's navigation software;
- iPaq PDA to show system messages and contextual information, as well as to give audible guidance;
- various sensors from Suunto to detect speed, distance, temperature, air pressure and altitude;
- a baby carriage with a vibrating engine attached to the bar for tactile feedback as system warning.

Field testing and analysis of the UI components of the initial version provided ideas for user-based content creation models to be implemented in the next version of the application (Section 7.7.2) and also first sketches for the multimodal version of the application (Section 7.7.3).

7.7.2 FamilyMap: Nokia 9500 Version

Main functionalities for the first fully working application were selected based on the original design ideas as well as on results from end-user tests with the initial version. The three functionalities selected for implementation were map-based navigation, point-of-interest system and messaging. Essential end-user requirements for these functionalities were stated as follows:

- *Navigation.* The application helps the user to discover unknown locations, while leaving the choice for the user to select the destination from the suggested locations.
- *Point-of-interest system.* The application allows users to share information about the visited places.
- *Virtual Post-its.* Users can attach comments, virtual Post-its, to the locations they visit and leave it for other users to read them out once they arrive at the same place.

This version of the application was developed for the Nokia 9500 (Figure 7.17). Additionally, an external GPS module was included to provide the needed location information. In the first trials, however, cell-based positioning methods were used for location information retrieval purposes. Furthermore, Microsoft MapPoint was used as the map service provider.

FamilyMap uses the client–server paradigm for delivering the contextual information. The main use of the FamilyMap is to view information about the surrounding environment. The client is in the mobile device and it communicates with the server on the network side to fetch the context information. The architecture of the client side and the general components of the Mobile Services Architecture used are depicted in Figure 7.18.

The database at the server side is used to store the points of interest that will be constantly synchronised with the client application. The server also communicates with the Microsoft MapPoint map service, processes the map data and takes care of the data delivery to the client application.

This reference implementation uses two separate sets of points of interest. One set is in the server: This set is complete (meaning that every point of interest is stored in this set) and it includes a description of the point in question. The client has a subset of these, a personalised set of points that are of specific interest to this particular user, stored in the device.

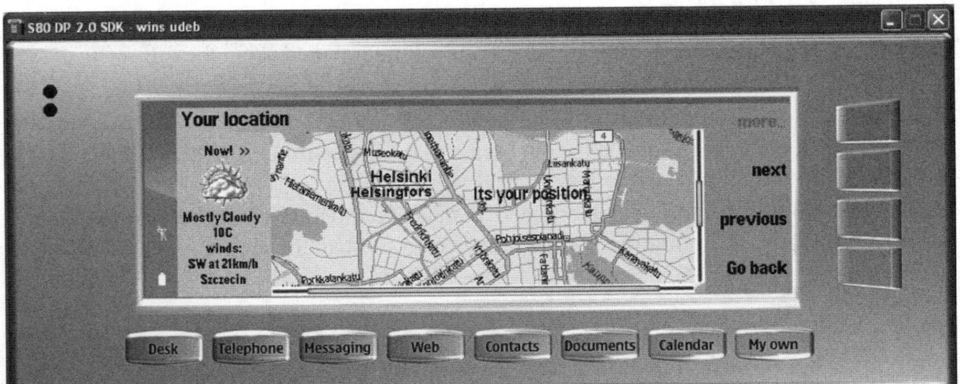

Figure 7.17 FamilyMap UI view on the Nokia 9500. (Microsoft product screen shot reprinted with permission from Microsoft Corporation).

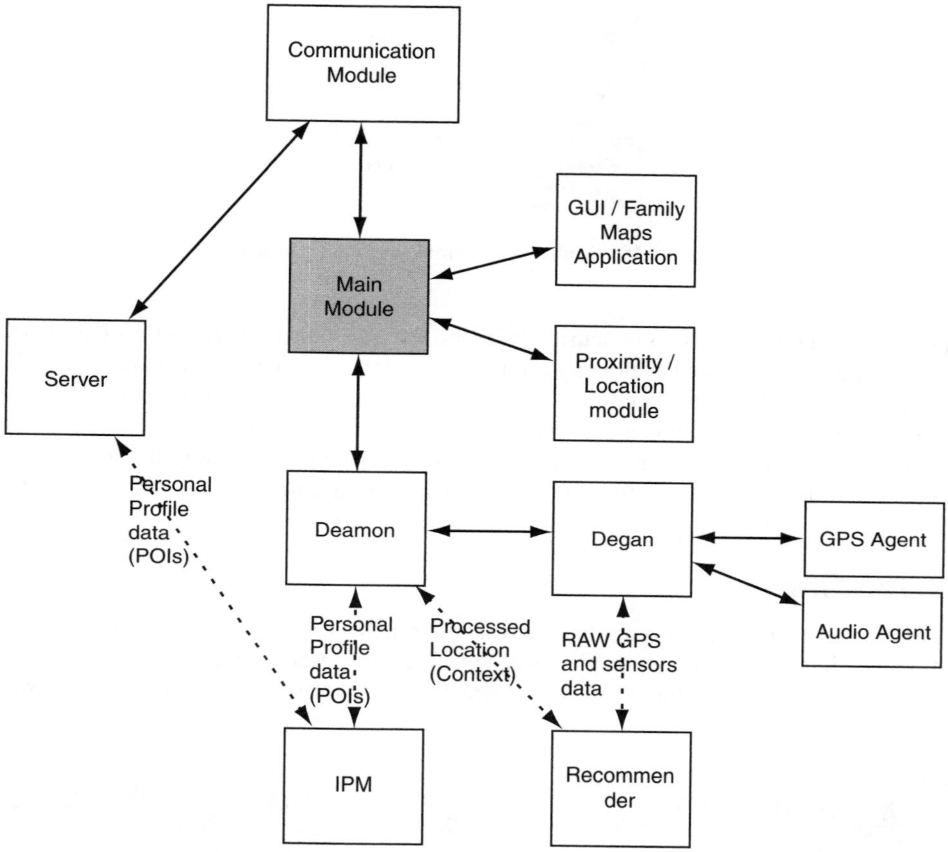

Figure 7.18 The architecture of the FamilyMap client.

7.7.3 FamilyMap: Multimodal UI Version

The multimodal FamilyMap version was developed in synchronisation with the Nokia 9500 version in order to test and evaluate the multimodality functionalities sketched based on the initial version of the application. These additional multimodal features were implemented on the Nokia 770 platform supporting speech, pen and tactile feedback. The implementation is based on the same client–server approach described in Section 7.7.2, but with the Nokia 770 Internet Tablet used as the client device due to the modality requirements. In the reference implementation, the server performs the demanding speech recognition and text-to-speech synthesis. The tactile feedback is provided by a Bluetooth-enabled wireless vibrating device attached to the user's wrist. The device configuration is shown in Figure 7.19.

The client connects to the server via WLAN or 3G broadband network, and to the GPS and the vibrating device via Bluetooth short-range network. The broadband data connection is required in order to perform the real-time audio streaming between the client and the server.

The multimodal FamilyMap version was tested with a group of experts. In the multimodal version, the main emphasis was put on enabling coordinated multimodal user interaction by

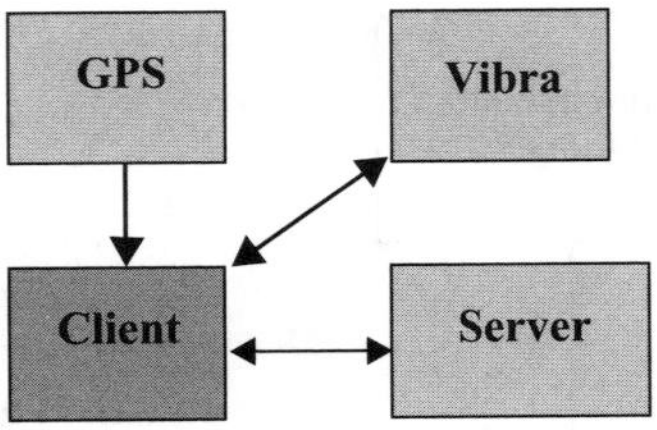

Figure 7.19 Multimodal FamilyMap configuration.

supporting different modalities concurrently. In the tests, original scenarios of the FamilyMap from previous phases of the development were reconstructed with emphasis on the multimodal interactions enabled by the Nokia 770 (Figure 7.20). In these evaluation discussions, practical challenges related to multimodal user interface usage in real-life situations were pointed out. Sometimes even the multiple modalities do not help, especially regarding the input related interaction, if, for example, the focus is completely on the children one has to take care of.

7.8 TimeGems

TimeGems is a group-centric application that highlights groups as social interaction and communication spaces. The application is designed to allow users to enrich their daily life and work experiences by providing them with extended interaction possibilities within the various

Figure 7.20 Multimodal FamilyMap UI (Microsoft product screen shot reprinted with permission from Microsoft Corporation).

groups they belong to (work, family and friends). The application primarily aims to facilitate group activity planning within the users' unused occasional and free-time slots for which a spontaneous organisation of group activities is usually very difficult.

To this purpose, TimeGems provides users with possible activities including scheduling and premise information that can be performed jointly within the available groups. The application also offers group management functions, thereby allowing users to, for instance, broaden their relationships by effortlessly creating new activity groups or managing group memberships.

Groups are an integral part of people's daily lives. Each day virtually everybody is bound to communicate and interact within a multitude of different groups such as family, friends and colleagues. Groups are a social phenomenon bringing together people with common interests, experiences and objectives for purposes such as exchanging information, sharing emotions and undertaking joint activities.

Due to the increasingly mobile nature of our lives, interacting physically within the numerous groups is becoming a progressively difficult challenge. Once group members are spatially dispersed, initiation of group meetings or group-related activities often requires intense communication and artful coordination for establishing the availability and location of group members, as well as for scheduling common time-slots and meeting places. Long-term scheduling may be seen as a possible solution for alleviating this growing challenge. Yet, this often proves inflexible and inappropriate, in particular concerning spontaneous activities and ad-hoc time slots.

7.8.1 Enabling Scenario

To better understand the functions of the TimeGems application, imagine the following scenario (related to the scenario presented in Section 6.1.2.1); see Figure 7.21.

> *Alice uses TimeGems in her daily life. It helps her to enjoy her time by discovering and organising social activities within her recreational time-slots. Today Alice had a rapid lunch. She realises that she has almost half an hour before going back to her office. She opens TimeGems to receive suggestions about possible activities to undertake with friends of any of the groups she belongs to. TimeGems enquires her preferred time-slot and analyses the availability of her groups and their members. Matching the available groups' and Alice's preference, TimeGems quickly suggests various activities: Alice's office colleagues are finishing lunch and drinking coffee in a restaurant a few metres away. Jane, a friend of hers, is showing online photos of her last holidays. Alice's fitness group she participates in weekly is inactive, so no related activity is proposed. Last, two friends of hers, Tom and Tina, are currently shopping in the neighbourhood, being available for the selected duration. TimeGems suggests Alice to either join Jane at her office, enjoying a tea and cookies, or to initiate a spontaneous meeting with Tom and Tina at a nearby cafe preferred by all three of them. Alice decides to meet Tom and Tina. She sends an invitation via TimeGems. Her two friends reply and accept the invitation within minutes.*

7.8.2 The Application

TimeGems offers to the user two primary services, depicted in Figure 7.22 as the *discovery of suitable group activities using inspection of groups' availability* and the *management of groups and group memberships*.

Figure 7.21 TimeGems.

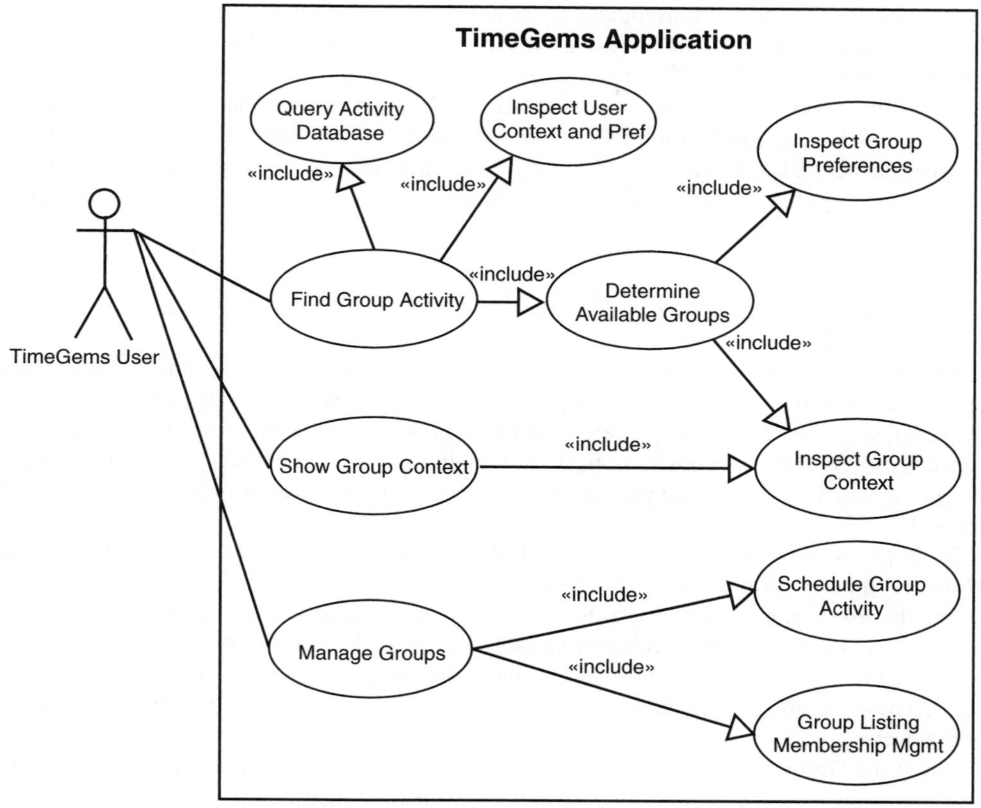

Figure 7.22 TimeGems and the GAF.

Discovery of group activities determines available user-related groups and related activities corresponding to the denoted span of time. Group's availability inspection visualises the context information of groups. Last, group management provides functions concerned with the lifecycle of groups and the scheduling of activities. Group management and context related functions are provided by the GAF (*Group Awareness Function*; see Section 6.1.3).

TimeGems' primary services are:

- discovery of available activities within the user's social groups;
- visualisation of groups' contexts;
- group membership and lifecycle management.

Discovery of group activities, enjoyable to the querying user and the available groups, is the application's core functionality. To find such social activities, TimeGems initially asks the user for her available time frame, her preferred activity domain and the group she would like to undertake the activity with. In the next step, the application resolves the user's groups and inspects the group's availability. For group members in active groups within walking distance of the user, availability and current activity might be determined. Finally, TimeGems presents this list of discovered activities to the user, sorted according her preferences. The users can subsequently inspect other offered activities with other groups.

TimeGems leverages the GAF for consuming and providing group-related functionality. The *Recommender* component formulates recommendations on the basis of learned associations between contextual information and entity behaviour (e.g. activity recommendations corresponding to a group's context). Finally, the *Activity Repository* contains information about potential user and group activities.

7.9 MobiCar

MobiCar is a group and context-aware car-sharing application. Through it, a user is able to set up a group with the objective of sharing a ride in a car that belongs to one of the members. However, even if here the example reference implementation is related to car sharing, the same concept can be utilised in other application domains as well. The novelty in the application lies in the way that the service parameters are set, and how the user group is discovered and formed.

By implementing components of the Mobile Services Architecture and notably the *Profile Manager*, the system is capable of determining the key parameters (and their values), which define the user's expectations towards the service. In doing so, the user is presented with a preconfigured set of the most relevant variables, such as destination, pick-up point, timing, preferred music, driver's experience, traffic conditions and air conditioning preferences (Figure 7.23). These variables, once the system has been trained, will match the current expectations of the user.

Using the Group Awareness Function and the Trust Engine, the application discovers the best matching groups that can fulfill the service needs, not just in terms of convenience, but also in terms of feasibility and trust.

Furthermore, in order to provide user-friendly interaction, MobiCar proposes a group discovery and pre-configuration mechanism, maintained up-to-date with the user's preferences, capable of providing an easy-to-use service with a minimum input from the user.

7.9.1 Enabling Scenario

The following step-by-step scenario illustrates the envisioned use of MobiCar in a family environment, adapted by the Trust Engine to provide the service in a trust-critical environment. In the envisioned scenario, Alice, a teenager, needs to ride a car with a group of other people, which must be deemed as trustworthy by an external authority, in this case, the parents.

Alice wants to go and see her sister's baseball match. This is at a very inconvenient time, when both her parents are working, and she therefore needs transportation. She calls up MobiCar to try and share a ride.

The system first asks for the pool of people that should be considered. Alice chooses among a set of predefined pools: 'Group of friends', 'Trusted neighbours', 'Family members' and 'Anyone awaiting a car'. Due to the restrictions set on the system by her mother, Alice gets an error when trying to include into the search pool either of the untrusted (by her mother) groups: 'Group of friends' and 'Anyone awaiting a car'. The pool is thus set to all the rest.

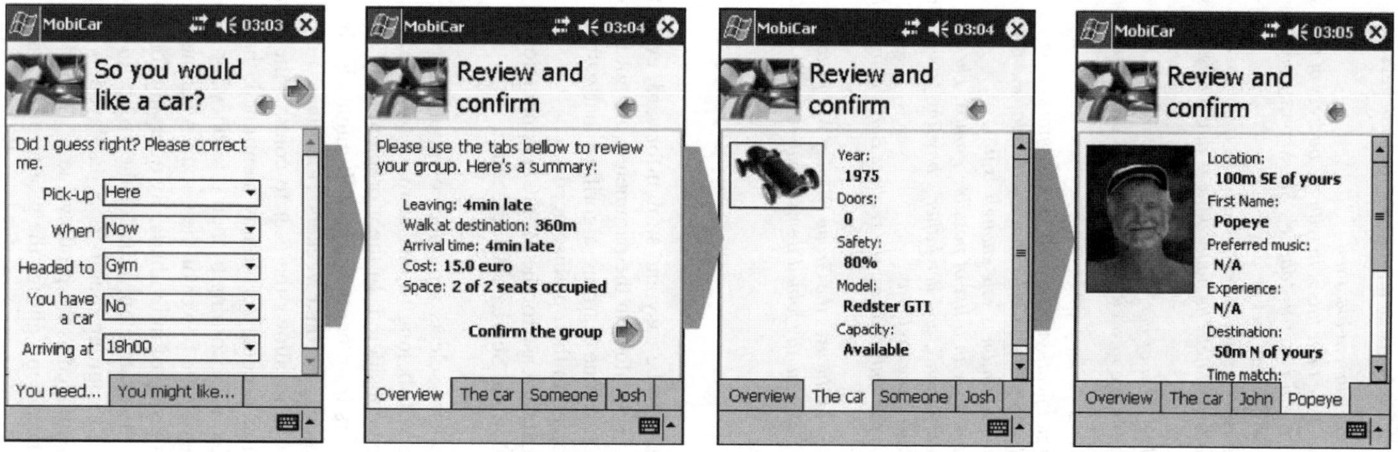

Figure 7.23 Basic usage of MobiCar.

The system presents Alice with the list of parameters that it believes are relevant for Alice:

- *The pick-up point is set to the current location.*
- *The destination is unknown (she later fills it in with the address of the baseball stadium).*
- *The type of music on the radio is set to 'dance'. She knows from experience that this is seldom satisfied, but maybe she gets lucky today.*
- *The traffic situation of the other people is set to fluid. She is never satisfied when people are stuck in traffic. They are always late! The funny thing is that the system seems to know how much she dislikes it . . .*
- *The other riders are within 2 km of her. She did not ask for this before, and once had to pay for a guy that came from the next city! The system is already adapting to the problem though.*

She proceeds and gets a notification that some of her information has to be shared with the MobiCar service in order to find a match. That is all right, she is not allowing the name though, and she considers the situation is safe enough as it is.

The system comes back to her with a list of possible groups. There is actually her friend Ben's parents going to a nearby mall, and her father's bowling friend, Bob. 'Whoa, I didn't know dad trusted Bob so much!' she exclaims.

After reviewing the details for both choices, she realises Ben's going to be with his parents, and decides to join that car.

The group is now created, and she uses the Group Display Widget to discuss the final details with the rest of them, since Bob actually joined Ben's group as well! A ride for three – that is convenient!

7.9.2 The Application

The usage of MobiCar is meant to be easy and straightforward, even if the system behind it is somewhat complex. The preset values in the parameters are not factory defaults; the group formation is inferred from a machine learning algorithm that gathers experiences collected from individual users and adapted to their precise needs.

Figure 7.24 displays a typical use case for the MobiCar application with four main functions using the components of the Mobile Services Architecture:

- *Choose pool of users.* The user selects a number of pools. The users that are eventually suggested to share the ride will belong to one or more of them. These pools are selected to customise the users to be considered. In the aforementioned scenario, this is used as a barrier to keep untrusted users from being selected. It might also be used to choose among friends by discriminating colleagues on a weekend ride, etc.
- *Choose service parameters.* This allows the user to confirm the proposed parameters and their settings, or alter them as needed. As already mentioned, these are set from the user's context and profile, and selected according to their relevancy for the user.
- *Visualise group.* This harnesses the power of the Group Awareness Function to summarise the information related to a group of individuals. By querying for the variables for a group, the system is able to show the abstracted value for the group. In the case of 'position', for instance, the average for all the group members is displayed, while in the case of 'music type' the majority vote is displayed as the group value.
- *Confirm group.* This sets the green light for the group formation in the platform to begin. Once the group is formed, the user can benefit from all the functionalities that come with the integrated group management platform.

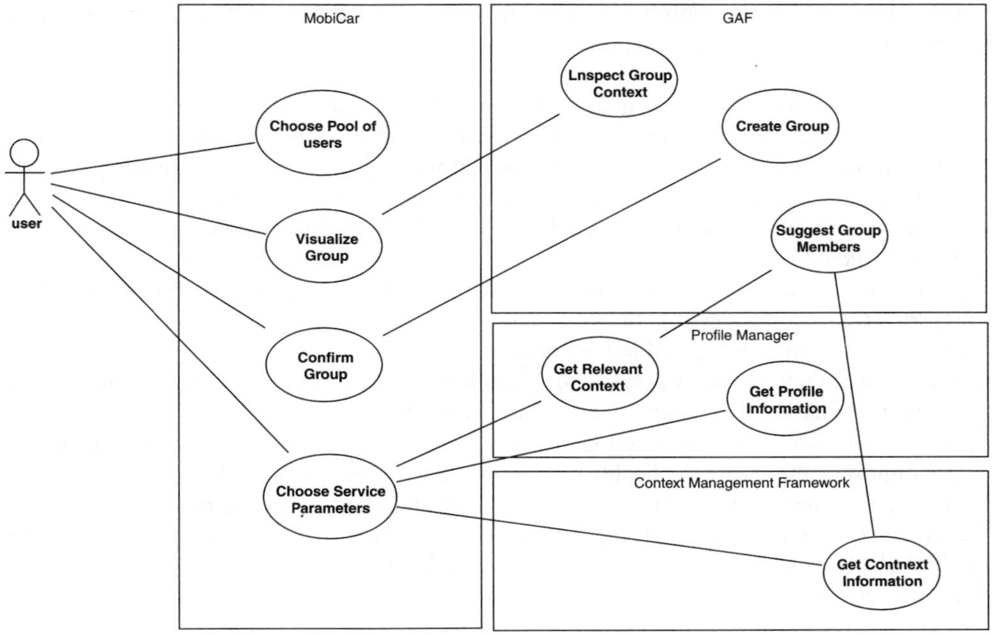

Figure 7.24 Use case diagram for the MobiCar application.

7.10 Conclusions

The applications and their reference implementations introduced here are presented in order to show what can be enabled by the technologies and architectural components introduced in this book. As a summary:

- *ContextWatcher* makes it easy for an end-user to automatically record, store and use context information.
- In *Personal Context Monitor*, ECG (electrocardiogram), heartbeat and acceleration are used to monitor the physical status of the user.
- *Proactive Service Portal* uses contextual reasoning to proactively offer services to the mobile user.
- *Multimedia Infotainer* displays news to the user in a context-aware and multimodal way.
- *Wellness-aware Multimodal Gaming System* takes advantage of the personalisation improvements to provide tailored training plans.
- *FamilyMap* application provides help to parents on the road with their children.
- *TimeGems* suggests to users potential activities within related groups, taking into account the availability of group members and their preferences regarding activities.
- *MobiCar* is a group and context-aware car-sharing application.

These reference implementations have been presented mainly to show the usefulness of the different components of the Mobile Services Architecture, but how about the general exploitability of these application concepts? First of all, most of the concepts can be as such easily adapted to various applications domains. As a simple example, the Personal Context Monitor concept has multiple potential fields for use:

- *Sports medicine* is the first potential field; here the most relevant issues are cardiovascular screening, continuous and comparable monitoring of athletic performance (team training), as well as data collection and processing also through artificial intelligent techniques (e.g. to identify the most effective personal training pattern for a specific athlete).
- Another possibility is the screening, diagnosis and treatment of *chronic cardiovascular diseases*, including atrial fibrillation, congestive heart failure and sleep apnoea, as well as *general health and fitness applications*, with an eye on monitoring and rehabilitation, where non-invasive remote support to the treatment of patients (e.g. at home, on the move, at work site) could become widespread in the future.
- *Prevention* through automated reasoning is another area for further development. Monitoring of health status based on non-invasive and open standards-compliant wearable sensors provides an added value through decision support capabilities. Information is gathered by monitoring the wearer's 'health behaviour'. This disease knowledge can be enriched over time as the system learns the patient's behaviour, for example, by monitoring and 'remembering' the heartbeat during daily activities. Biometric signals, primarily ECG, are collected in various situations and can be displayed on a central station accessed by the patient's doctor.

The Personal Context Monitor is just an early example of what can be expected. In the future, as physical inactivity and unhealthy eating patterns are responsible for a worldwide epidemic of preventable illnesses and deaths, there could be a key role for ubiquitous computing applications in promoting physical activity.

Introducing advanced context-based technologies included in most of the application concepts in this book is sometimes a challenge, especially if the functionality of the end-user application relies on automation and reasoning, since it can be difficult or in some cases even impossible to model and acquire all the relevant information needed for correct results and recommendations – or the state-of-the-art reasoning methods are simply not intelligent enough yet. One of the most important findings from the ContextWatcher user tests [7] was that – although people generally enjoyed the concept of it – the users had difficulties with the sometimes unpredictable results of the reasoning. If one wants to target commercial markets and to gain users' confidence in a system like this (or the reasoning and recommendations in, for example, Multimedia Infotainer and FamilyMap), improving the quality of reasoning remains one of the key technical challenges. Furthermore, even if the reasoning results is exactly what the past behaviour of the user and the available situational environment would assume them to be, the user might still not prefer exactly that suggestion in the current situation.

Some of the application concepts might raise potential legal and ethical concerns that should be carefully considered before exploiting the ideas commercially. In the user tests of TimeGems [7], for example, continuous location identification of group members generated criticism: people felt uncomfortable with the idea that someone or others in general would be able to track their location at all times. Yet, TimeGems crucially depends on this feature for selecting groups relevant to the user's current context. A similar and even more complex situation is related to ContextWatcher, where other information is gathered and shared. A possible trade-off for resolving this type of conflict may involve allowing users to switch off the location identification (or any context gathering and sharing) at any time, yet only permitting usage of the application's service when the location identification feature is switched on. Other concerns might be related to the health-related applications, such as the Personal Context Monitor, where sensitive biometrical information is collected. See Chapter 10 for further discussion from the legal and regulatory points of view.

Finally, commercial exploitation requires special attention to application stability and reliability as well as user interface design fine-tuning and support for multiple platforms. For example, during the Multimodal Infotainer development some difficulties were experienced when trying to use pure Java for mobile devices: Parts of the required functions and application design are so novel that there is no full support for these with the devices and Java implementations available on the market. The discovery module required device-dependent development that was based on C++ language for Symbian version and C# for Windows Mobile platform; this especially applied to communication features including Bluetooth discovery and wireless networking through a WiFi interface. These and many other issues, including both technical and user evaluation aspects, are considered in Chapter 8. Additionally, commercial exploitation requires also careful analysis of business models and marketplace dynamics; these needs are the starting point for Chapter 9.

7.11 Acknowledgements

The following individuals contributed to this chapter: Péter Boda (Nokia, Finland), Nermin Brgulja (University of Kassel, Germany), Stefan Gessler (NEC, Germany), Giovanni Giuliani (HP Italiana, Italy), Johan Koolwaaij (Telematica Instituut, The Netherlands), Miquel Martin (NEC, Germany), Dario Melpignano (Neos, Italy), Jean Millerat (Motorola SAS, France), Rinaldo Nani (Neos, Italy), Petteri Nurmi (University of Helsinki, Finland), Pekka J Ollikainen (Nokia, Finland), Petr Polasek (UNIS, Czech Republic), Mateusz Radziszewski (BLStream, Poland), Marcin Salacinski (BLStream, Poland), Göran Schultz (LM Ericsson, Finland), Michael Sutterer (University of Kassel), Dari Trendafilov (Nokia), and Libor Ukropec (UNIS).

References

[1] Bluetooth SIG for Medical Devices: Online: www.bluetooth.org/.
[2] ContextWatcher portal: Online: www.contextwatcher.com.
[3] European Data Format (EDF): Online: www.edfplus.info/.
[4] Eurostat: Online: epp.eurostat.ec.europa.eu/.
[5] Flickr: Online: www.flickr.com.
[6] ISO8601: Online: hydracen.com/dx/iso8601.htm.
[7] Kurvinen E. (ed.): 'Results of Service and Application Evaluation'. IST-MobiLife Project Deliverable D14 (D1.9), December 2006. Online: www.ist-mobilife.org.
[8] Open eXchange Data Format (OpenXDF). Online: www.openxdf.org/.

8

Trials and Evaluation for Acceptance

Edited by Esko Kurvinen (Helsinki University of Technology, Finland), Renata Guarneri (Siemens, Italy) and Jukka T Salo (Nokia, Finland)

In earlier chapters, an architecture for mobile services (Chapter 3), several enabling technologies (Chapters 4, 5 and 6) as well as different reference applications (Chapter 7) have been described. However, if one wants to reach mass markets, one needs a good understanding of the needs and requirements coming from the users and other stakeholders (as discussed in Chapter 2), as well as continuous evaluation and validation of the intermediate and final results.

Large R&D projects, covering heterogeneous areas from enabling technologies to end-user applications, need a very strong process in order to enable evaluation and validation of the results from different points of view, requiring therefore expertise in different areas. The key areas of the evaluation are as follows:

- *Evaluation and validation from the end-users' points of view*, using the approach and methodologies typical of the user-centred design process (see Chapter 2).
- *Technical evaluation*, looking at the key technological aspects, highlighting the major innovative aspects of the technical results, and analysing them against the requirements and specifications set at the beginning.

User evaluation is an integral part of a process to ensure the user centricity of the overall process, which in turn will result in better user acceptance and better assessment of the market

Enabling Technologies for Mobile Services: The MobiLife Book Edited by Mika Klemettinen

potential. This is a key element for large R&D projects, in order to identify – in the early phases of the development – the key application areas of the innovative technologies developed.

The technical evaluation that naturally complements the user evaluation will then have the objective of validating the overall innovative concept proposed, and, in terms of the actual implementation, will first of all evaluate the alignment with the specifications, then also the impact that the innovation will have on the existing technologies. At the same time, the technical evaluation process will also identify gaps and steps to be undertaken to actually deploy the new concepts (novel applications, novel architecture, new enablers) in the 'real world'.

Section 8.1 introduces the case study referred to in the user and technical evaluation parts. Section 8.2 discusses the user evaluation part, while Section 8.3 reviews the technical evaluation aspects. Section 8.4 is a short conclusion.

8.1 User and Technical Evaluation: A Case Study

As referred to earlier in this book, the goal of the MobiLife project [23] was to bring advances in mobile applications and services within the reach of users – focusing especially on families – in their everyday life by innovating and deploying new applications and services based on the evolving capabilities of third-generation (3G) systems and beyond. The research challenge of MobiLife was to address the multi-dimensional diversity in end-user devices, available networks, interaction modes, applications and services. To deal with this complexity and reach its strategic goal, MobiLife researched context awareness, privacy and trust, adaptation, semantic interoperability, and their embodiment in novel services and applications that match key use scenarios of end-users' everyday life. These results are reviewed in earlier chapters of the book.

Methodologically, the project followed the interactive approach, introduced in Chapter 2, which acknowledges that the user requirements must be learned partially from experiments with novel service and application prototypes illustrating the end-user value and possibilities of new enabling technologies.

MobiLife was concerned with the needs of family members juggling multiple roles. These roles included their roles in the family, at work and at school, and in other social contexts. To better understand the communications and group dynamics among the family, MobiLife decided to focus upon families undergoing some kind of a transition period. Therefore two target groups were selected:

- young/growing families (e.g. family with a child going to school for the first time); and
- older families with children moving away (also referred to as 'third age' families).

Before the project start, MobiLife organised and co-organised three workshops in Finland, France and Germany to collect requirements and views from different stakeholders (end-users, manufacturers, operators, service providers, regulators, etc.) to guide the work in MobiLife. These requirements have been summarised in Chapter 2.

At the beginning of the project, review of the overall user and business environment, including an overview of modern family life and communication in families, the role of mobility in everyday life, the expected evolution of the marketplace, and the legal and regulatory context for mobile applications and services was conducted. Scenarios (see the Appendix) were developed based on an analysis of the user tasks and behaviours found in an extensive review of scenarios from other projects and from within the MobiLife technical work areas. These results were organised into 15 categories of user tasks and behaviours presented. Within each category, the high-level user requirements and guidelines for further work were extracted.

Regarding the *user evaluation*, the following characterises the evaluation rounds carried out:

- There were four rounds of evaluations: Scenarios, Mock-ups, Probes 1 and finally the combined round of Probes 2/Integrated Applications, which was mainly in the form of real-life trials.
- During the evaluation activities, approximately 210 persons were addressed, mainly in Finland and in Italy: in Scenario evaluations 17 families with 61 persons; during mock-up evaluation during six sessions with approximately 85 persons; during Probes 1 evaluation with approximately 20 persons; and during the final combined Probes 2/Integrated Applications round (consisting of six activities) with 40 persons. During the last evaluation round, there were altogether four real-life trials performed; two in Finland and two in Italy.
- The number of real evaluated applications was 20 (including the scenarios as one application) due to the development of applications during the project.The user evaluation results and settings have been described in more detail in [18] and [20].

Regarding the *technical evaluation*, the five evaluation dimensions presented later in this chapter (Section 8.3) were used as a basis for the technical evaluation. The specific aim of the MobiLife technical evaluation was to provide guidelines on the different aspects of the Mobile Services Architecture, on the components therein, as well as on the overall approach towards the key features of the system.

The evaluation questions identified and outlined (Section 8.3.4) were further detailed, specifically addressing the implemented components, the interfaces, the selected protocols, etc. From the answers provided to such detailed questions, the evaluation data was collected and then analysed, in order to provide feedback to the parties involved, but also to extrapolate more general conclusions towards the shareholders of the project.

The technical evaluation was performed during the installation and integration of the developed software components, services and applications on the server platforms located in the two *Integration and Validation Centres* (IVCs), in Finland and Italy. These components, services and applications were made available to the project for the integration activity as well as for the trials that were performed at the end of the project and that constituted an important source of information and data also for the user evaluation.

The technical evaluation results and settings have been described in more detail in [14].

8.2 User Evaluation

This section includes a review of the key learning from the user research related to both enabling technologies and prototyped reference applications and services. These learnings are collected from a real-life case study [18,20] described in Section 8.1, which involved more than 200 end-users over two years. In this case study, the iterative user-centred design (UCD) process introduced in Chapter 2 was applied. Each round of the user evaluation activities focused, in addition to the application/service prototypes, also on a subset of research themes and enabling technologies.

Section 8.1 briefly describes the case study setting, while Sections 8.2.1–8.2.4 concentrate on the technological themes and their evaluation findings. Finally, Section 8.2.5 includes remarks on the user evaluation, for example related to uncertainty in user research and social interaction as a design driver.

8.2.1 Context Awareness, Personalisation and Adaptive Systems

The idea of context-aware computing has been around for a while. In the age of mobile devices, context recognition and context awareness are seen to have great potential. First, from the usability point of view, as the connectivity of devices, the number of applications and their complexity increase, context awareness can help people in managing this complexity. Second, context recognition can serve as a foundation on which altogether new types of mobile applications and services can be built.

A related term, *personalisation*, refers to a variety of processes ranging from adaptive changes produced by the system to adaptability, where the user produces the change [24]. For example, Fink *et al.* [11] report a system that automatically inserts shortcut links for web pages that are frequently visited by a given user. My Yahoo displays weather forecasts as a function of the postcode of the customer [22], and News Dude [5] presents news stories to the user, who then rates the articles according to whether they are interesting or not. To predict whether its user will be interested in a new story, News Dude forms a content-based profile, where similarity to other articles is based on co-occurrence of words appearing in the stories. These are all cases of adaptivity, the system being responsible for producing the changes.

8.2.1.1 Adaptation vs. User Control

Both context awareness and personalisation are fundamentally about machine-made recommendations and proactive UI behaviours based on machine reasoning. Both concepts take that systems can act on behalf of the user when they know what the situation and/or person are about.

While there is nothing wrong with the core idea, known user research findings imply two types of challenge.

1. It is Very Difficult to Build a System that can Interpret the Context of the User

This is partially, but only partially, due to lack of context information (e.g. sensors). For example, in MobiLife one was not able to build a system that could always (or often enough) provide the location of the user with the detail required to have location-aware services. Cell-ID, GPS or WiFi positioning all have their limitations. While building radio-based positioning covering indoor, outdoor and 3D needs is technically possible, there is no business incentive to drive such efforts. This is a chicken-and-egg problem; without context-aware services there will be no need for infrastructure – and without a proper infrastructure, it makes no sense investing in service creation.

While coordinates (x,y,z) are unambiguous yet so difficult to obtain, challenges related to situation or activity recognition are far more complex. This is because 'situations', as treated by humans, are inherently equivocal, deliberately left open for multiple interpretations, based on whatever the participants see fit for their current purposes. In this respect, there may be an irresolvable discrepancy between mainstream developers' way of seeing 'context' and the way humans actually see it [10].

2. Even if One had a Machine that could Correctly Tell what the Users Need, the Users may not Like it

While liking technology that eases everyday activities [7], the need for control is stressed repeatedly by users:

> 'I want to decide for personal things.'
> 'I won't let it coordinate my time.'
> 'It can propose changes but the user must be able to authorise these. Technology must not be the master.'

One should not think that decision-making is something that people would like to actively avoid. Free will – and the urge to express it to others – is an essential part of being human. People want to have the control, or at least the feeling of being in control of what happens next. One should avoid the so-called *prosthesis approach* – observing how people go about decision-making and trying to replace this activity with machine inference and behaviour. Instead, the aim should be to support human reasoning and for this purpose the context-sensitive systems themselves need not be very intelligent. They could, for example, merely display context-related information to the user, in a digestible format, letting her make sense of it together with the people the information relates to [21].

In systems that involve adaptive behaviour, one should provide also the conventional way of accessing the same information (e.g. organised alphabetically or according to fixed categories), so that users can themselves decide which approach they like the best. This would also enable proper testing of the reasoning approach.

What people want is generally a very difficult question to answer. However, correctness or accuracy of the recommendations is not always an issue. For example, the user may adopt a playful or exploratory attitude, or, instead of being focused on a task, she may be just killing time.

Even when the recommendation matches the current needs of the user, before deciding what to do, she may want to consult a complementary information source or she may want to know more about the item being offered. This is not only practical and useful for the activity at hand, but also gives the users the feel of control they desire.

8.2.1.2 Collecting Context Information

As already noted, one challenge relates to acquisition of context information. Dey and Abowd [9] provide a definition of context that is frequently cited within computer–human interaction (CHI):

> Context is any information that can be used to characterise the situation of an entity. An entity is a person, place, or object that is considered relevant to the interaction between a user and an application, including the user and applications themselves.'

It should be noted that the definition has two components. First, there is some information about the situation. Second, there is the issue about relevancy of that information to the user.

While the former has gained lot of attention, there is little discussion on what the items one should be measuring are. Developmental work at the information gathering side has been driven by the availability of sensor technology, not by the in-depth understanding on what should be measured in order to capture relevant aspects of human activity. The question is: Are the right variables being measured, or just those that are conveniently available at the moment?

Technology developers have built systems that utilise, for example, location information, noise level, room temperature, etc. Whereas sociologists would argue that *language* is the key instrument by which humans make sense of the world and maintain structure and orderliness in various situations and society at large, product developers have not taken this as a design challenge. Instead, the typical approach is to pick thermometers and other components that are readily available at electronics suppliers, or utilise features of current devices, such as GSM cell information. This is all done in trusting that, regardless of how unintelligent the individual data sources are, once one has enough of them, they can capture the relevant aspects of human conduct.[4] This is not to say that one needs speech interpreting systems, but to use exclusion of language as an example to point out that the selection of measured context variables has not been rational – and that such technology-driven selections really matter for the outcome.

8.2.1.3 Abstracting and Applying Context Information

Bits and pieces of context-related information (e.g. room temperature) are not very useful. That is why one typically aims at abstracting the information from various sources. So far there have not been commercially successful applications that use high-level context information. It may be that collecting context information and interpreting it is technically more complicated than expected. Furthermore, the key challenges are not so much technical, but relate to practical usefulness and social organisation as well as the interpersonal relations the context-aware systems affect.

In addition to the *context*, one typically – at least implicitly – assumes that there are some rules of application: the context affects the way people behave, and therefore the way systems should behave. Technical papers often use 'meetings' as examples of high-level context descriptions. In a prototypical scenario, during meetings mobile phones go automatically to a silent mode [32]. Accordingly, to illustrate how rules are assumed to work, one can extend the list and add rules that could apply to 'meetings':

'In meetings I want to have my phone on silent.'
'In meetings I want to receive only SMS, not MMS.'
'In meetings I don't surf the web.'
'In meetings we talk about work issues.'
'In meetings the chairperson steers the conversation.'
'In meetings I take notes.'

Some of the rules are directly transferable to computer code, whereas some of them have impact on the people through human interactions. However, any office worker who has attended meetings must have experienced how the above types of rule for meetings are broken by the participants. For example, when taking a phone call during a meeting, one can *ask to be*

[4] Try imagining culture without verbal interactions or language of some kind. Contrast this with some culture without thermometers, which is quite a feasible scenario.

excused, perhaps because the call is *very important* or *very short*, or if it can be arranged so that it *does not disturb the meeting*. This means that rules, as treated by people, are inherently open-ended and negotiable. In addition, people not only recognise and follow, but constantly create and maintain, contexts and related rules in concert with other people.

It should also be noted that what mobile IT product developers see as *meetings* or *rules that apply to meetings* may not match with how other people see them. For example, the social gatherings that teenage girls might call meetings are substantially, structurally and rule-wise very different from meetings organised by white-collar people at some company.

A system engineering solution to this problem would be to add another layer of information so that there are *types* of meetings: *internal work meetings*, *business meetings* and *meetings with friends* that have different sets of rules attached to them. This, however, does not take away the fact that there always will be some sort of mismatch between the ontological structure embedded in the system and the way people would like to name things. Alternatively, one can enable users to adjust these rules or design systems that seek to learn what the rules are. However, while the former is not feasible from the usability point of view, the latter is not possible in the near future, if ever (because of reasons discussed above).

One should not design away the fact that people are constantly breaking, bending and playing with labels and rules that relate to situations. People do all this for good reason. This is how they display social competence, manage social situations and steer them toward some preferred direction.

In the case of system-generated labels, some minor detail can have a significant effect on user acceptance. In the ContextWatcher (Chapter 7), the trial users were especially annoyed about the way the system suggested that people being inside the same cell-ID cluster were having 'a meeting'. Some quotes (F26, F24, M27 and F27 refer to quoted trial users' gender and age):

> 'We've had a meeting for 157 hours.' – *F26*
>
> 'The meeting reasoning is very bad. According to this I'm always having a meeting with F24 even though she quit from the test over a month ago.' – *M27*
>
> 'Meetings are more comical than useful . . . It would arouse interesting questions if one didn't know the context. I've spend the last three hours with M27 instead of my husband'. – *F27*

In this case, the problem was merely in the selection of terminology, or mismatch between the fidelity of the context measurement and the abstraction used to represent it. While 'meeting with' in CW meant merely 'in the neighbourhood of', people took it quite literally.

Aside from system-made inferences, one can leave room for the users themselves to reason about context data [21]. For example, the *ContextContacts* [25] application extends social awareness by providing contextual information about people appearing on the contact list of a mobile phone. The information is not interpreted in reference to some ontology, but summarised from raw data into a visual and easily digestible format.

The extended senses approach takes into account that if the system does the inference on behalf of the users, it may take away the very thing that is meaningful for them. Kurvinen *et al.* [21] discusses how collaborative sense-making of sensor data during a football game helped the participants learn more about each other better and provided a joint perspective for viewing and experiencing the game. Also, as discussed in [19], as people *initiate contexts themselves,* instead of some system doing it for them, they often do this step-by-step, in subtle and socially considerate ways, taking into account possible conflicting agendas or interests of the other participants.

8.2.1.4 Reflexivity of Human Behaviour

While we are still a long way from systems that can tell what situations are about, one can nevertheless hypothesise about their possible implications. Using a system that helps in managing complex everyday situations might lead into new kinds of problem, for example in the following ways:

1. The user is using a system that tells her what the situation is about. The system directly or indirectly *helps her to better manage the situation*, by for example being considerate, knowledgeable or humorous.
2. Other users may *perceive that she is being helped by the system*. For example they can see or otherwise infer she is using some application for *getting the right answers*.
3. The other users assume that she is *socially incompetent*, because she uses or needs devices to manage things that ordinary members of the society accomplish without.

For example, a person asking tricky questions and presenting detailed information at meetings is not considered an intellectual if it appears that she is 'googling' the material with her laptop. Similarly, a person telling a lot of jokes is not considered a humorous person if he is reading them aloud from a joke book.

The problem of reflexivity poses tremendous challenges for systems that seek to interpret situations on behalf of the participants. This is of course not so relevant for systems solving *marginal* or *non-existing problems*, such as *silencing your mobile phone at meetings*.

8.2.1.5 Correctness vs. Usefulness of Context Inferences

Although the *naive positivistic approach* (seeking to abstract and label situations) can be heavily criticised, it could prove useful in many ways, for example in *proactive resource management* [27].

In usability, a standard benchmark is that user/system tasks need to be 95% successful or error-free [26, p. 45]. Thus, on theoretical grounds only, one can expect that a system interpreting the context of the user with less than 95% accuracy will be perceived as unusable. This is, of course, not a static measure, but dependent of the application, the amount of intrusion or other problem caused by the misunderstanding.

On the other hand, when the system *is seeing things right*, it may be that the situation is already so obvious that there will be nothing the system can help the user in. Systems interpreting social situations and taking part in the interactions might be perceived as a 'friendly idiot', trying very hard to be helpful, but being constantly slightly out of focus and just a little late.

'The friendly idiot' might also turn out to be useful (funny, exciting, etc.) in unanticipated ways. For example, in the FamilyMap trial (Chapter 7), it was discovered that the participants liked also POIs that did not match their interests or needs. Although these mothers with small children were not into rock climbing, these pieces of information were considered *nice to know*. Already during the early mock-ups stage, the study participants approved push-ads on their mobile – when presented in a non-intrusive way. It seems that the way recommendations are presented to the user is at least as important as their correctness or match with the user preferences. The more complex the reasoning is, the more important it is to prepare for failures.

For example, it is strategically better to aim at application domains where damage done by system-made misinterpretation is minimal.

8.2.1.6 Standards vs. Design-driven Approach to Context Awareness

In mobile and ubiquitous environments, the synchronisation of information across clients, services and components is a true challenge. Context awareness and personalisation add another layer of complexity into the picture. In addition to exchange of data, systems should be able to exchange information about how they infer the context of the user.

The typical approach among large initiatives is to aim at standards or frameworks that aim at maximum compatibility across platforms, applications and application domains. One should, however, not expect that standardisation solves all challenges or is the only way to go. It is also possible that if someone comes out with a compelling design-driven application family (for context awareness, multimodality, personalisation, security, etc.), this can become a *de-facto standard* – regardless of the limitations such a combination may have.

Instead of relying on extensive structures/ontologies of the world, one can utilise context information that is relevant for a particular application and combine that with some structured understanding of how people use it. Contextual information and reasoning (if needed) would then be implemented in this limited domain, making things considerably easier, because the system would not need to account for all possible cases and variation across domains (as the ontology approach implies).

8.2.2 Multimodality

Multimodality refers to human–computer interaction utilising a combination of modalities or senses. For example, a PC input interface may consist of a combination of a keyboard, mouse and speech recognition. Similarly, output can be a mixture of visual and auditory feed. Extending the modalities from finger-tapping based input and visual output offers possibilities for improved usability in non-traditional environments.

The study of a mock-up of the Wellness-aware Multimodal Gaming System (Chapter 7) showed that seeing sensor data of a player can turn into an important conversational resource between the spectators: a discussion starter, a way to start watching the game together rather than everyone on her own, a tool for comparing individual players, getting a fuller picture of the game situation, and so on. Presenting sensor data on a device can therefore be useful also in social settings, not only in individual use.

Figure 8.1, from the FamilyMap mock-up trial (Chapter 7), shows some situations that are related to multimodal interaction. In these situations, the users are occupied in doing something during which the display is not pointed towards them. Based on the number of such instances in the data, *in location-aware information systems, multimodality should be benefited in achieving more robust communication rather than in increasing the amount of information communicated to the user*.

The findings about the difficulties in device usage are not specific only to a setting of mothers, babies and baby prams. Actually, urban contexts are information-intensive and ergonomically challenging even without a task of pushing a pram and monitoring wellbeing of a baby. For instance, simply walking in a crowd causes a situation in which a user cannot easily stop to investigate contents on a mobile device without disturbing others and being disturbed. The

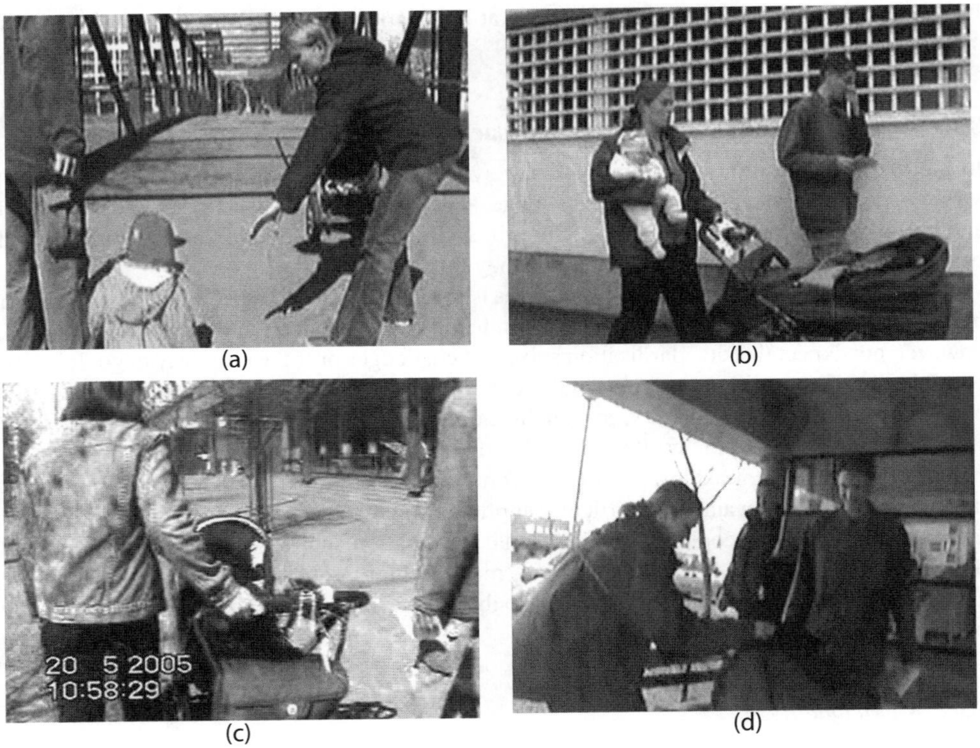

Figure 8.1 Situations in which the participants could not attend to the PDA display.

findings presented here point out the importance for considerations on multimodal aspects in designing any kind of device for urban contexts.

The study of a mock-up of the FamilyMap application showed that the interface feature design for a multimodal application must adopt one of the two strategies below:

- *Amplify* the same message from the device to the user (e.g. when a mobile phone rings, it creates a sound, blinks the screen and vibrates), thereby increasing the probability of making itself understood by the user.
- *Enrich* the message by providing complementary information (e.g. when watching a mobile map, the screen shows a map and the audio channel explains the different sights in the map), thus providing more information but making the communication more vulnerable to errors (due to low lighting or noise).

The latter approach provides also for increased expressiveness by increasing the bandwidth of information. In mediated environments, this increases immersion and sense of presence. For instance, video calls are able to provide a stronger sense of presence than normal telephone calls, because of the additional visual communication modality.

Currently multimodality is associated with virtual reality and other immersive environments. In contrast to such laboratory environments, achieving advanced multimodality in wireless mobile settings is far more challenging. Aiming at immersion by extensive coverage of senses is, however, not the only way to do successful multimodal applications.

Technology developers typically speak of modalities in terms of human senses (sight, hearing, touch, taste, smell), or their corresponding equivalents in terms of hardware/software needed to input/output the information at hand. From the user perspective, this definition is, however, insufficient, because there are different modes also within senses and these modes can co-exist in the same 'feed'. For example, the audio track of a movie can contain ambient noise (crowded street), speech (the actors are having a conversation) and music – all at the same time. The combination is not problematic for the audience, but in fact favourable for the immersive effect movies aim at. The implication is that to be multimodal, the UI does not have to address multiple human senses. Instead, one can take just two or even one channel, and utilise its possibilities to the maximum.

8.2.3 Privacy and Trust

The issues of privacy and trust are pervasive in applications and services. If the consumers do not trust the new context-aware services to protect their privacy, they are not willing to use them. Alternatively, context-aware features are switched off and the added value is lost.

> Privacy is the claim of individuals, groups and institutions to determine for themselves when, how, and to what extent information about them is communicated to others. [3]

As the definition above states, revealing one's information to others is the key issue in privacy protection. The always present technology, new kind of sensory data and personal devices increase this amount of information significantly. Moreover, this information needs to be shared as the context-aware applications and services are largely dependent on it. These technologies put the user to a completely new kind of situation, where personal data that have been previously only inside the user's mind can be shared with others and with different services.

When asked about privacy and trust, for example in interviews, people distinguish between different types of group. With trusted groups like own family, more information is shared than with a less trusted group of, for example, soccer team-mates. When context is changed from personal to professional, the key issues are the same; controlling of when, how and to what extent information is shared with different stakeholders. Figure 8.2, summarising one set of interview studies conducted in MobiLife, gives an overview of participants' willingness to share context information with their phone book contacts.

While people are able to come up with assessments of the type described above, without much effort, when asked in more detail and when using examples they soon add several reservations and exceptions. There are many imaginable situations that contradict the general pattern; for example, when location information would not be given to a spouse, or when it is acceptable to give the information to strangers.

Privacy and trust-related assessments are clearly dependent on the activity at hand, as well as on earlier experiences with the system or other analogous systems. The problem from the system development perspective is that people do not behave rationally or consistently to the extent that can be modelled, for example in terms of rules. For instance, the study by Acquisti and Grossklags [2] on web privacy shows that, while people have privacy concerns and good understanding of what is required to protect oneself, they are nevertheless often willing to compromise their long-term privacy for short-term insignificant benefits. Similar

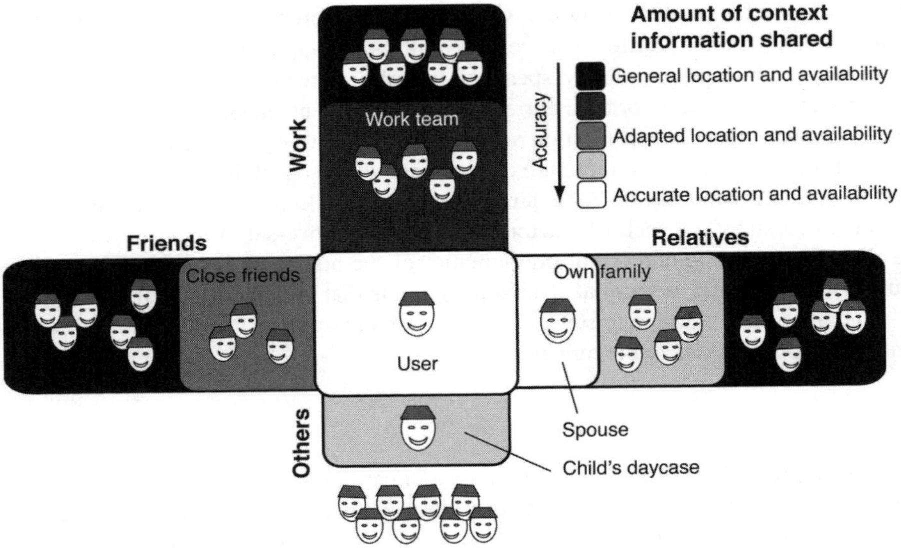

Figure 8.2 Typical categories of people based on a phone book exercise.

balancing between effort and benefit was seen in the MobiLife user evaluation participants' orientation towards the trust and privacy components. People understand the idea of centralised trust and privacy management, but anticipate that it means extra work they are not willing to invest in. Related to privacy, in both ContextWatcher and Personal Context Communicator (Chapter 7) initial interviews, there were many concerns related to tracking each other's whereabouts. However, after the users had tried out context sharing as a part of their everyday lives, their attitudes towards it became much more positive – but still there was the definitive need to stay in control of the data revealed.

Another important observation is the interdependence between privacy rules and the content they apply to. In the case of the exercise above, people did not impose yes/no types of rule, but wanted to decide on the amount of details shared across groups. This has the technical implication that privacy rules should not be considered independent from the data, but they should carry along information, for example, on how to down-sample, modify, disguise or provide previews of the content. Furthermore, the rule structure should also be able to trigger dialogues of different kind before granting access and support logging of information shared so that the user can herself assess how she is doing in terms of privacy.

Extensive trust and privacy management features can work against their original intention. An early version of MobiCar (Chapter 7) included a detailed profiling of passengers. During the user evaluation, the overwhelming presence of trust and privacy-related concepts made people utterly concerned; instead of trusting the application more because of these features, they now trusted it less. First, trust and privacy features made unwanted events appear more frequent than they actually are (car pooling and hitchhiking are still relatively safe). As put by one of the participants:

> 'Looks like the person who designed this has at some point been very disappointed in humans.'

The participants also anticipated that user profiling and trust rating would enable manipulation of the system for unintended, even hostile purposes. Common sense and instinct cannot be replaced by trust-managing technologies.

8.2.4 Group Management and Group Coordination

Several mobile applications and services, including the reference applications in Chapter 7, include a set of features for group management and group coordination. These concepts are similar in that 'group management' stresses what a user can do as her own initiative in relation to other people that are part or are going to be part of a certain group, and 'group coordination' addresses more the communication flows that allow people to align their schedules or in some other way reciprocally to adapt their actions.

People's understanding of groups relate mainly to the activities they are associated with. For example, being a member in a Flickr [12] group means (until further notice) only that one has some vague interest related to the group's topic or people within. Preference order between groups cannot be decided in abstract. Even relatively stable or formal groups, such as 'families', become alive only in interaction. For example, when the Wellness-aware Multimodal Gaming System (Chapter 7) was tested with a teenager football team and their parents, the boys interacted only minimally with their parents. They, for example, largely ignored the comments of their parents. In the presence of team-mates of the same age, it was not in their primary interest to be associated with their mum and dad. Instead of strengthening the family bond, the application increased the coherence of a looser group, the parents and the coaches beside the field. They were now following the players keenly and talking with each other more than ever [21].

On the other hand, in another test setting (TimeGems, see Chapter 7; and Tourist Info System), critical or confused reactions were prompted by the users on group management features mainly because the nature of the implied social relationship was either out of place in the presented context or not properly specified. It is one thing to propose an evening out to someone who is already in contact, maybe through some online interest-based community (e.g. fans of a specific movie genre in a city), but a completely different thing to imagine that a user is willing to initiate contacts with strangers in a foreign country (or on the street), even if some benefit is at stake (e.g. a group discount ticket for the same movie).

Another aspect that seems still under-addressed is the support for users in their personal relationships. In everyday life people engage in activities with each other not only because they belong to certain groups but because they share something personal together. Turning this shared thing into computer code is practically an impossible task: there are too many different ways in which people relate to each other.

Communication applications have solved the question in the following ways (for example):

- *Buddy lists.* This approach lets people define the most important people to them, while leaving open the question of the nature of these relationships.
- *Situated or ad-hoc lists.* In other applications, it has been found useful to let the user define the most important persons every time separately.[5]
- *List of most recently used contacts.* This is a standard approach in call logs.

[5] In MobShare picture-sharing system, the user first takes pictures with a phone camera, and then creates a gallery for publishing the pictures on the web. Since the user is asked to list the entitled persons separately every time when she creates a gallery, there can be different access rights for each gallery [28].

The approaches could be combined. For instance, being able to use the same personal buddy list in different applications would be an improvement to the current systems where each system requires the user to maintain a separate buddy list.[6]

Due to the variation in types of personal relationships, ontological modelling could not be very structured. A blackboard approach, where the user could associate different pieces of information to each person, would be a better approach.

Group management features could also explore the potential of 'tagging', the practice of adding free textual descriptions to a certain element that has gained popularity with social software-based services such as Flickr (for photo sharing) and del.icio.us [8] (for bookmarks sharing). These 'loose metadata' systems (sometimes called 'folksonomies') rely very much on the potential of distributed, informal description features of very similar items [33,34].

8.2.5 Remarks on the User Evaluation

8.2.5.1 Tolerating Uncertainty

How to treat ambiguous and conflicting user research results and how to use these results is a key question. Developers often ask for clear answers and immediately useful findings. However, there are occasions when such answers are not available.

Kirk and Miller [16], in their overview of qualitative research, present a well-known allegory of the blind men and the elephant. The blind men approach the elephant from different angles; one grasps the trunk, one touches the tusks and one its ears. The blind men soon end up quarrelling about their irreconcilable reports. The 'vulgar positivistic' view would be to compensate for errors in measurement and conclude that the elephant is a formless blob covered with wrinkled skin. Equally unfruitful would be to adopt 'pure perspectivism' – treating all viewpoints as equally valid, denying possibilities to mediate between them. In contrast to these approaches, if the blind men (the field researchers) can come to a consensus that all of them only feels one part of some huge object, contradictory reports are no longer an issue. Such reports are not non-objective, nor should they be treated as global features of the item being analysed. Rather, they are partial descriptions of a complex whole [16].

Research on the potentials and caveats of complex future technologies must tolerate some degree of uncertainty. This involves taking contradictory and ambiguous information that research generates. When dealing with complex phenomena, seeking simple answers is likely to be useless, even damaging.

In many cases, a clear discrepancy can be identified between how people talk about technologies and how they actually use them if given the opportunity. For example, the reference application FamilyMap evaluation (Chapter 7) – the fourth round of the Big Loop introduced in Chapter 2 – pointed out the differences between focus group findings and observations from the field trial. Similarly, the mock-ups stage evaluation of the reference application Wellness-aware Multimodal Gaming System (WAMGS; Chapter 7) – the second and third rounds of the Big Loop – describes how people, when introduced to the heart-rate sensor technology in the context of amateur sport, offer partyline stories of how they would have little use for the application and how the coaches could use the heart-rate information to improve the performance of the individual players for the ultimate benefit of the team. This is in clear contrast to how

[6]See a similar direction emerging also from the well-known cases of open or commercial Internet instant messenger such as Jabber and Trillian.

they then actually used the system; for getting to know everybody, for making collaborative sense of the HR data, and for social chit-chat and joking beside the field [21].

Discrepancy between findings should be taken as indication that the technology at stake is challenging as people have difficulties in coming to terms with the new capabilities. It also means that people are likely to behave inconsistently when facing such technologies. For example, the study by Acquisti and Grossklags [2] on web privacy shows that, while people have privacy concerns and good understanding of what is required to protect oneself, they are nevertheless often willing to compromise their long-term privacy for short-term insignificant benefits. Similar balancing between effort and benefit was seen in the MobiLife user evaluation participants' orientation towards the trust and privacy components. People understand the idea of centralised trust and privacy management, but anticipating it means extra work they are not willing to invest in. Related to privacy, in the reference application ContextWatcher (CW; Chapter 7), initial interviews – the fourth round of the Big Loop – many concerns were received related to tracking each other's whereabouts [17]. However, after the users had tried out context sharing as a part of their everyday lives, their attitudes toward it became much more positive. However, there was still the definitive need to stay in control of the data revealed.

It is sometimes better not to aim at true/false statements, but to seek to capture the phenomenon in all its richness and then seek ways of taking into account this complexity in the developmental activities that follow. In case of the WAMGS mock-up version, it was argued that the instrumental view – although 'false' when looking at the actual doings of the participants – could be used to legitimate the technology at the outset, while there is some indication that in the long run the sensor technology could be used also for some other purposes.

8.2.5.2 Content Creation for Ubiquitous and Context-aware Services

The technology enablers introduced in this book are designed to tackle the complexity of the anticipated future service landscape. Enabler components – like context awareness, personalisation, multimodality, and privacy and trust – all address particular types of challenge in the hypothetical world that consists of:

- multiple actors and types of role in the areas of service creation and distribution;
- many different types of mobile and ubiquitous service;
- large amounts of content that can be accessed through these services.

Logically, the user-related problem then is that, because there are so many services, the user is incapable of finding the one(s) that are most relevant for her at the moment or for the activity at hand.

The problem is then solved by context-aware systems that find the most relevant services, contents or resources and offer them to the user (see the Proactive Service Portal in Chapter 7 for an example of such a system).

In the area of context-aware, ubiquitous applications and services, developers' focus is typically on technologies of content distribution and access, not in the production of content. It is taken for granted that content that does not exist today will be produced by someone in the future. Furthermore, it is often assumed that this content will be produced more or less following the traditional top-down model; i.e. by some commercial party. Recent experiences with, for example, the failure of WAP or narrow offering and slow adoption of 3G services, indicate that within mobile media content producers and consumers don't meet so easily (more discussion of new business models and changing roles are included in Chapter 9).

In MobiLife, even if a UCD process was applied, a relatively traditional technological approach was followed in the sense that the primary focus was on researching and developing technologies and enablers for novel applications and services. Also the user research questions were mainly related to what components should be like and whether people can understand them and find them useful. However, a finding of the project was that it is perhaps more important to create conditions in which the components are needed.

Aside the favourable circumstances, one should design tools for creating such a variety of services and contents in an environment that is too complex to be controlled by any single actor alone. For providers of future context-aware ubiquitous applications and services, the earlier chapters (from Chapter 3 to Chapter 6) present architecture/framework for ontologies, semantics, annotation, etc. Also applications with different approaches to content distribution have been developed (Chapter 7) and user tested. The next step needed is to support content creation and maintenance. For example, the trend is that users create content, but there are no easy means to create context-aware content; i.e. how to geotag, annotate, apply semantics, etc.

Context awareness and ubiquity bring along a new dimension to content creation. Whereas *mobility*, to some extent, is about freedom of space and place, *context awareness* re-establishes them. It is not about the user connecting to *anyone*, *any service*, *anywhere* in the world (e.g. via mobile Internet), but that she connects to people, services and items locally available and locally relevant for her activity at hand. The key challenge here is not about technology, but about content: Who are the actors who will create content for ubiquitous services, and under what circumstances would they be motivated to do so? What are the tools they are using? If there are only a small number of ubiquitous services available to the user, there will be no need for context-aware filtering or sorting of these services. Instead, a basic static list, split into make-sense categories or merely in alphabetical order, will suffice.[7]

On theoretical grounds, and based on existing services, two opposite approaches can be outlined:

- *Top-down.* Commercial service providers produce the content for the users. Users' mobile devices are able to acquire this information, and possibly integrate multiple such sources within each application.
- *Bottom-up.* Users create the content by themselves, distributing and exchanging it through dedicated channels and platforms. On the Internet, this is already taking place in forums like SlashDot [30] and Wikipedia [35].

However, as will be argued below, these two approaches should not be thought of separately. Instead, they should be made to complement each other. Firstly, the users can help in the creation and maintenance of the top-down offered content. On the other hand, the availability of top-down provided content can inspire interpersonal communication. Both production and consumption of content should be seen as social processes that technology needs to support – not vice versa.

[7]Furthermore, in contrast to adaptive lists, static lists are good from the usability point of view, because of their consistency.

8.2.5.3 Social Interaction as a Design Driver

A simple but overlooked design driver or recommendation is to enable people to interact with each other. When developing technology for privacy and trust, multimodality or context awareness, for example, developers tend to focus on human–computer interaction; that is, on an individual person who interacts with the computer system.

However, people use technology together with other people, or they use technologies to communicate with other people. Even in the case of lone interaction with a computer system, people most often have goals or objectives beyond human–computer interaction that relate to other people. For example, a person using a word processor does not type at the keyboard just for the fun of it, but in order to communicate something to her (human) audience.

Especially related to mobile, ubiquitous and context-aware technologies, the developers should not think of an isolated person, but groups of people who use the technologies together or with an eye to the other participants. Studies reported here show that this approach brings about several benefits.

First, think about the initial moment when a consumer faces an application, service or appliance that has many novel and innovative features. It is likely that she is puzzled about the product, its features, potential benefits, and unaware of its limitations. However, the user is not alone; she can explore the usefulness of the product in co-operation with other users. If it is a communication-related product, it is quite natural to try out its features and wait for the responses of the other human party. Even if the product offers no such features, the user can turn to other users – or perhaps technologically oriented acquaintances – asking for consultancy on the matter.

Second, in the case of ubiquitous information services, it is especially important to make people interact with each other. When this is done properly, one can improve the quality and relevance of the information offered. People can collaboratively edit the content, report bugs and errors etc. (cf. Wiki). While doing so, they are also appropriating the information to better match the interests and needs of their peers. In contrast, following a strictly top-down content production scheme, one can only offer rather generic information that then either is or is not useful for those who access it. Even when the information is largely useless, people can make some use of it (see the WAMGS trial in [21]).

Third, by involving larger groups of individuals (or other actors, such as organisations) into a content production scheme, one can ensure that there will be enough location-sensitive and context-sensitive information to be filtered or explored using the technologies envisioned.

Fourth, the effect of social interaction can reach beyond the core features of the application/service at stake. For example, when adding their own Points of Interest (POIs) in the FamilyMap trial, the users occasionally mimicked the formats of public information and commercial ads. Since the user-generated POIs were mostly targeted to the other participants, one can argue that top-down provided content inspired interpersonal communication. Therefore, instead of relying on extroverts to fill in the *tabula rasa* of location-aware/ubiquitous information space, top-down offered information services can provide support (advice even) for larger masses of consumers when they start to explore what this type of content could be.

Fifth, social interaction by definition builds social pressure to use these systems. For analytic purposes, one can recognise several layers of such pressure. At the top level, there is the idea of technology adoption as a social phenomenon; if increasingly more people subscribe to some products or services, this builds expectations towards late adopters and laggards. Then there is the notion of reciprocity; seeing close ones, peers or meaningful others using the system (e.g. contributing to content creation for the benefit of the peer group) builds expectations that one should also participate. Finally, on the level of concrete actions, people can build sequential expectations as to what should be done next and by whom. On such occasions, missing the right type of responses is noticeable, sanctioned even. For example, POIs can be formulated as questions, teases or puzzles that are pointed to a particular participant. All these levels require that the user – one way or another – sees the other humans doing something with the same system she is using.

8.2.5.4 Services for a Vacuum?

During the ContextWatcher trial, the so-called *Quality of Service* (QoS) problem was faced. Python was selected as the prototyping platform because it enabled rapid versioning, component-based development and distribution of work across different developer teams. This, however, means that it becomes difficult to decide at the outset what the guaranteed performance level of the individual components should be. This is especially because the components can later be used to provide some unanticipated application feature. While architecture can be demonstrated with software that suffers performance problems, from the user's point of view such problems can be fatal.

Interestingly, it was also observed that achieving a relatively high QoS measure may not be enough. People are benchmarking the software built to analogous services and functionalities found elsewhere; in the web or in designated smart appliances. For example, the maps offered in FamilyMap and ContextWatcher were compared to navigation software and navigation devices, and map services found on the Internet. The mobile search features in TimeGems were compared against 'googling' with a PC – and the camera capabilities of novel smart phones and PDAs provided a benchmark against which to compare the rather modest picture-taking and sharing features of ContextWatcher and TimeGems.

The methodological implication is that as (mobile) technology matures, it becomes increasingly difficult to test individual new features apart from the service landscape, the myriad of cutting-edge features already in the marketplace. Applications and services are not developed in a vacuum.

In addition, as the complexity of the application increases, one should no longer think of developing 'a mobile application', but a service that is a combination of mobile and web-based features. For example, in the ContextWatcher trial, the users appreciated the Web Services provided by Google Earth and Blogger, although they were not part of the original application and although there were several problems in interfacing them with ContextWatcher. This approach enables the developers to, at least partially, tackle the QoS problem. One should aim at well-balanced and complementary combinations of mobile and web-based features. Complicated and resource-consuming features can be offered via the web interface instead of a mobile one, so that the mobile client can be stripped down to features that are really needed when on the move.

8.3 Technical Evaluation

Successful evaluation of the results of an R&D project depends on the availability of sufficient and valid feedback from all concerned stakeholders. The user-centric approach and the user evaluation guarantee that the target user groups play a central role in the design of the service scenarios.

The technical evaluation complements this approach by involving, in addition to the end-users, also other stakeholders. In this view, key stakeholders to be considered are network operators and service providers, but also manufacturers, software developers, third-party service and application providers, etc. This is to guarantee that the entire value network is part of the process, so that valid and profitable business models can be supported by the innovations proposed.

User and service trials are the ultimate tool to provide such results; however, initial tests will be run in the laboratory environment, where the system is first integrated and then tested. In the case study described in Section 8.1, Integration and Validation Centres (IVC) were used for this purpose. The integration and testing of the system is an iterative and interactive process involving the different actors in the development process. Actually, the integration process by itself can be seen as a technical evaluation tool, where conformance to the specifications, interoperability among the different parts of the system, as well as the documentation provided with the components, are evaluated.

In the case study referred to with examples, the main objective of the technical evaluation was to test the viability of the Mobile Services Architecture and the new enabling technologies for developing and delivering novel mobile applications and services in the areas of context awareness; multimodality and personalisation; and privacy, trust and group awareness. The technical evaluation covered five different dimensions of analysis:

- ease to implement applications;
- exploitability of the components' functionality;
- testability;
- impact on network architecture and terminals;
- evolution.

After a discussion of related work, this section outlines a general approach and methodology for the technical evaluation (Section 8.3.2). Planning and implementation of the technical evaluation are introduced in Section 8.3.3, and the five evaluation dimensions are presented in detail in Section 8.3.4 with remarks and examples from the concrete case study. The joint conclusions from the work and technical evaluation results are given in Section 8.4.

8.3.1 Related Work

The integration and testing of systems, services and applications is becoming increasingly important as their complexity increases. A typical commercial software-intensive system is developed as a composition of software components, electrical components and/or mechanical components each of which uses different modelling techniques. In addition, the individual requirements of each system component must be consistent with each other and finally result in a coherent realisation of the overall system requirements.

Testing and evaluation of systems, services and applications as a part of the product development process has been much studied in the literature. A comprehensive approach to defining a methodology for evaluation can be found in [1], where a methodological approach to the evaluation of an expert system is described.

A step towards a better integration of test development into the system development process has been taken by [6]. According to this approach, test components are developed in parallel with the system components as soon as the interfaces and the overall architecture of the system to be developed are defined. TTCN-3 – the Testing and Test Control Notation developed by ETSI and also adopted by ITU – has been used as a specific technology for the specification and implementation of test systems [13]. TTCN-3 is a powerful test specification and test implementation language, which addresses the testing needs of modern communication, software and embedded systems technologies. Typical areas of application are protocol and service testing, component and system testing, testing of embedded systems, testing of communication-based distributed systems, etc. The standardised test language has a similar look and feel to a typical programming language.

However, since developers still use different languages for system modelling and for test models, there is a gap between system and test development. Further work has thus been done to improve the integration of system and test development by adopting a common specification language. A consortium at OMG defined a UML2 testing profile (U2TP [4]) in order to support the specification of test models. This UML profile provides particular means for test modelling; i.e. for describing test objectives, test procedures and test data. The same language (i.e. UML) can now be used in order to specify system and test models. This eases the early integration of system and test development.

The SysTest project (see [31] for details) aimed at the improvement of validation, verification and testing processes through the application of cost-of-quality and scheduling models during software development.

Technical evaluation in R&D projects does not attempt to verify that the implemented prototypes of the target systems, services, functionalities, protocols, APIs, etc., are error-free. The focus of such activity is instead to verify on a higher level that the proposed technical solutions can meet the requirements set for these new systems, provide valuable information on system behaviour to be used in further R&D work, and to make proposals for improvements, or even new research topics, if needed. Technical evaluation of research results has normally been handled only within each research project separately. Re-inventing the wheel in each project is a waste of resources, however, and should be avoided.

8.3.2 Generic Approach and Methodology

The main objectives of the technical evaluation are:

- to test the viability of the proposed service architecture and functions;
- to assess the fulfilment of the requirements with regard to the capabilities of the terminals and networks;
- to evaluate tools, interfaces and protocols for developing and delivering new mobile services and applications.

As indicated in the introductory remarks to this chapter, evaluation of a complex system and architecture such as the one described in this book requires a holistic approach. As noted in [15]:

> . . . it is important to understand that there is no such thing as the architecture of a system – that is, there is no single artifact that one can definitively point to as the architecture. There are, however, many relevant and important views of an architecture depending on the stakeholders and the system properties that are of interest.

Further, considering that the evaluation is to be performed for a system that may not be fully implemented, reliable and consistent, results from the evaluation process imply that different system aspects have to be covered and different methods used for collecting and analysing the available information and data.

The selected methodology must allow evaluation of the viability of the architecture and technologies under consideration, and must cover aspects and elements that allow extrapolation from the specific results towards considerations at higher level. The approach and methodology to be adopted for the technical evaluation will therefore include the following:

- *Collecting experiences achieved in the design, implementation and testing of the system components and applications.* The data can be collected from the experts involved in the specification and implementation phases of the work through interviews and questionnaires.
- *Collecting experiences from the team responsible for the integration and testing of the system components and applications in the Integration and Validation Centres.* The same methods for collecting data as in the previous item can be exploited. In exploring new areas, the target of research projects is not to produce error-free code, but to develop implementations to test such new concepts and ideas; errors in the code may influence the opinion of people responsible for testing.
- *Testing the viability of the architecture also with regard to interoperability with existing solutions.* Backward and forward compatibility is an important element in the evaluation. Existing or developing standard reference architecture should be considered in performing such testing.
- *Testing the system components and applications in different configurations.* In this activity, special attention has to be paid to the interfaces of the components, the functionality available through them, and the 'consistency in behaviour' when re-using components, either by the same application or by other applications. Also, it is important to assess whether the distribution of functionalities between the different components is reasonable.

8.3.3 Planning and Implementation of the Technical Evaluation

It is important to plan the technical evaluation of the project results in a very early phase of the R&D project work. As soon as the first specifications of the system and components have been created, people responsible for the evaluation can have first ideas on what shall and can be evaluated, and the necessary actions for collecting data through different methods can be initiated. Also, it is important to exploit the early evaluation results, and even the comments and feedback from the evaluators to the technical specifications, to increase the viability of the to-be-developed system and components.

Specific steps in the technical evaluation will include the following:

- *Preparation of the test plan including identification of the items to be evaluated and the potential methods for collecting respective data.* It is important to communicate this plan within the project, and make it clear what is expected from project people at different phases of the work.
- *Identification of specific questions to be answered for each item to be evaluated and selection of methods for collecting data for each question.* It is important to check beforehand that questions are clear and understood by the involved people, and that there is no room for

different interpretations. The preparation of the final questionnaires is likely to follow an iterative process.

- *Outlining of configurations of the Integration and Validation Centre(s).* In a big research project, where there are many items to be evaluated, it may be important to share the approach to, and feedback from, the evaluation activities between all the IVCs involved. This is even more important in international collaborative projects, where developed applications may be country-specific, for instance in terms of used language. In the integration and testing phase of the components and applications it is important to share experiences between the IVCs, and log-keeping is an important tool enabling this action.
- *Collecting and analysing data at different phases of the project work.* As already pointed out, early data collection and analysis will result in more benefits towards the actual research and development work. It is also important to organise group discussions on the analysis results in order to remove potentially unclear items.
- *Extrapolating and drawing final conclusions and making recommendations towards exploitation of project results.*

8.3.4 Evaluation Dimensions

As mentioned earlier, one of the difficulties in carrying out technical evaluations is the ability to draw generally applicable conclusions from what often is a partial implementation of the system architecture. A vectorial approach looking at different aspects needs to be applied, where final conclusions and recommendations come from a combination and cross-analysis of the key aspects. A vectorial approach enables efficient and parallel implementation of the evaluation work. Each of the vectors in the evaluation space, called an *evaluation dimension*, addresses a specific focus area in the evaluation of the system architecture. This multidimensional approach has been used for the evaluation of complex systems, and it has been applied to several and quite diverse fields; examples can be found in [29] and [36]. For the purpose of performing a consistent and reliable evaluation of the system architecture described in Chapter 3, and the new mobile services, applications and service enablers, the dimensions presented in Table 8.1 were used.

The dimensions will now be presented and discussed in detail. For each dimension, the scope and objectives are introduced followed by a more detailed description of the evaluation

Table 8.1 Evaluation dimensions

Dimension	Focus area
Ease to implement applications	The effectiveness of the offered API to implement services and applications
Exploitability of components' functionality	The efficiency of the architecture and the components' functionality in supporting the creation of new services
Testability	Testability of the developed system and applications
Impact on network architecture and terminals	The impact of the new services on the operator's network systems and terminal devices
Evolution	The evolvability/expandability of the developed architecture and components

questions. For each of the questions, a short description is provided with selected case study examples and discussions.

8.3.4.1 Dimension 1: Ease to Implement Applications

The purpose of this evaluation dimension is to understand how easy it is to develop new applications on top of new service architecture. This is why the evaluation questions focus both on how understandable the system is with its components, and how usable the Application Programming Interfaces (APIs) are for the development of new services by the service providers or third parties.

A theoretical approach would require the identification of several application and service scenarios, and testing the system accordingly. However, within the course of the actual evaluation activity, it is more realistic to expect the components to be tested for a few service scenarios only. The component APIs, which may well fulfil the requirements of those scenarios, may therefore lack functionalities required for other, potentially more complex service scenarios.

As a minimum requirement, the following aspects must be considered:

- How do the proposed architecture and APIs support the development of new services and applications by third parties?
- Can the components' functionality be easily accessed and exploited by the new services and applications via the APIs offered by them?

Evaluation Questions to be Addressed Within this Dimension

1. *Is the new architecture clearly structured for developing new applications?* This can be investigated based on hands-on experience in developing such new applications. The application developer has to understand, first of all, the role and functionality of each component (through the technical documentation) and to identify which ones are required for the targeted application.
2. *Are the APIs of the new components general enough and simple to use?* The focus of this question is on the APIs of each system component with a target to understand whether they are clear, easy to use and with the appropriate level of detail to access the functionality they are supporting.
3. *Are the APIs of the new components well documented?* Since the usability of the new architecture and its components is very much dependent on the availability of the appropriate documentation, this question focuses on the quality of descriptions made available about the architecture and its components. In general, there is often a strong correlation between the quality and viability of a new system for the purpose it has been designed for and the documentation, especially user guides, illustrating it.
4. *Are the APIs of the new components offered with appropriate protocols?* The capacity and performance of the system is dependent on how the system components have been distributed between the system elements, and on the used protocols between the distributed components. The purpose of this evaluation question is to investigate the system from such a perspective, and to make proposals for potential improvements.
5. *Are there tools (programming environments, application containers, third-party libraries) that are highly recommended, required or preferred to integrate new components?* The system and its components can normally be run in different hardware and software

environments, and the applications can be developed with different tools. This is especially true when considering the hardware and software environment in end-user devices. The purpose of the evaluation question is therefore to understand whether there are specific limitations related to these environments and tools, and which features or functionalities require more attention.

Case Study Remarks

One practical problem specific to this dimension related to the simultaneous availability of all the components during the development and integration phase. This problem was solved through the local installation of the majority of the developed software at the Integration and Validation Centres and through the use of the Web Service paradigm to access the remote components.

A critical point was identified at the client side, where a wealth of incompatibility issues was highlighted during this phase. In fact, it is just about impossible to create a single client that would work with all the possible user devices, especially when smart phones are to be considered. This was not a new issue, and the mobile device manufacturers are well aware of this problem, although differences across devices are exploited by the manufacturers themselves as they can be seen as unique selling points. Within an experimental and innovation project such as MobiLife, however, this problem was directly linked with the issue of portability and introduced some limitations with regard to the devices that could be used. While the use of Java technology, in principle, carries portability[8] as an embedded feature, this principle often fails when mobile terminals, which constituted the target MobiLife user devices, are used. JWTI (Java Technology for the Wireless Industry) defines a number of minimum requirements for an application to work on different devices; however, often such a set of interoperable features is not sufficient for the advanced services and applications which were dealt with within the MobiLife project (see a subset of them in Chapter 7).

8.3.4.2 Dimension 2: Exploitability of Component Functions

This dimension analyses the components of the new architecture, with respect to the specific functionality they are able to provide, and how such functionalities can be used as enablers for the current and future services and applications. When doing this analysis, the key requirements identified during the early stages of a research project must be kept in mind. The purpose of this evaluation activity is therefore to assess the level of compliance to the requirements of the new component implementations. Additionally, the evaluation assesses the extent up to which the new components can be integrated and used as part of service platform architectures other than the specific architecture developed within the project.

As a minimum requirement, the following aspects must be considered:

- matching of the capabilities of the new components with the requirements stemming from the need to support new services and applications;

[8] An application, or a component of a solution, developed in Java can be moved from one platform to another and it is still able to work. The issue of portability with regard to J2ME is a subject considered by JTWI (Java Technology for the Wireless Industry), whose objective is that of creating a standard environment for running J2ME applications, in such a way that all applications JTWI-compatible are able to work without problems on any device compliant with this standard. JWTI defines a number of minimum requirements for an application to work on different devices.

- use of standard communication links, interfaces and protocols, and interoperability (this includes access to external resources – local sensors, remote service enablers – as well as access from, for example, external service providers);
- portability of the implementations to different platforms – experience from the prototype implementations and recommendations for future implementations.

Evaluation Questions to be Addressed Within this Dimension

1. *Are the components useful and adequate (functionality-wise) for creating new context-aware services and applications for the mobile users? Are the functionalities of the components responding to the requirements coming from the services? How well do they fit the service needs*? The purpose is to investigate whether the new components include all the functionalities required for developing new context-aware services and applications. The main reference documents to be used here are the scenarios illustrating usage of the future services and applications, and the requirements document generated from the scenarios.
2. *Can the functionalities be used within the context of other architectural models/approaches (e.g. IMS)*? This question investigates whether the proposed new components and their functionalities could be implemented on top of the existing, state-of-the-art service delivery system, like the IMS system currently. As the deployment of new telecommunications systems and services often follows an evolutionary path, it is important that the research projects mirror their research results against the state-of-the-art systems.
3. *Does the implementation conform to the requirements of the specifications? Does the implementation behave as expected in the specification*? The role of this question is to verify that the implementation conforms to the specifications, which is a prerequisite for most of the overall technical evaluation activities.

Case Study Remarks

By its nature, this dimension is in itself multidimensional and addresses directly the core of the Mobile Services Architecture, analysing the components envisaged for the application areas, so that detailed questions were developed addressing the different components related to personalisation, group and privacy management, and context awareness.

The overall scope of evaluation activities under this dimension was described as follows:

- Analysis of the capabilities of the UI components to support the implementation of multi-modal user interfaces for distributed applications.
- Analysis of the Mobile Services Architecture trust and privacy models and architecture to enable mobile individuals to join trusted groups and to interact with other members, with user-controlled privacy policies.
- Analysis of the functions and capabilities for the creation of ad-hoc mobile groups and communities.
- Analysis of validity and relevance of context-aware information in different contextual situations of users, including the related reasoning mechanisms.
- Analysis of user profile scalability for dealing with changing user roles in different contexts.

As the Mobile Services Architecture concept is quite innovative, the whole development was experimental and significant improvements would be required towards commercial development. However, the responses to the questionnaire prepared indicated that all the issues relating to this dimension were clear, and that specific restrictions had to be made within either the specification, the implementation or the integration, in order to meet requirements coming from the user side, including providing users (and this was particularly true for the trials) with fully functional applications that would allow collection of significant data for the user evaluation.

8.3.4.3 Dimension 3: Testability

The purpose of this evaluation dimension is to test the viability of the proposed new architecture from the viewpoint of how easy it is to assess the new system and new services. The evaluation analyses potential hurdles in service deployment from the installation of a new service to the overall service management.

This dimension covers evaluation aspects such as 'How well do the system and its components meet the configuration and customisation requirements of the new service?', 'How does the architecture help in the overall understanding of the technical relations and dependencies between different components?', and 'Does the system and each component provide the necessary interfaces for testing purposes?'.

More specifically, the approach and methodology to evaluate dimension 3 is based on the service components and application integration, as well as on the experiences and findings from service management and operations in the testing centres. In the technical evaluation process, the key approach is to focus on what problems are encountered and how the issues can be resolved when new services based on the new components are integrated and tested.

Evaluation Questions to be Addressed Within this Dimension

1. *Are the new services and applications able to be easily tested in their deployment environments, and by whom (e.g. network operator, third-party service provider)*? The objective is to collect views on the testability of new components, separately and as a part of the full system. This will give information on how they should be further developed for enabling the boost of new services by the wider service development community.
2. *How can sufficient Quality of Service be reached for commercial take-up of the service*? The aim is to evaluate whether the system provides the required monitoring, event tracing and management capabilities for testing a new service, and to understand what can be considered a sufficient level of provided facilities.
3. *What problems (if any) were encountered in the research project when integrating and testing the new system and services*? What were the related solutions? The purpose is to complement the understanding of the system and services testability through practical experience gained when integrating and testing the components for the demonstration purposes of a research project.

Case Study Remarks

The focus within this evaluation dimension was to test the viability of the Mobile Services Architecture from the overall system testability point of view. The assessment analysed potential hurdles in service deployment from the installation of a new service to the overall service management.

The dimension addressed evaluation of the capability of the system and its components to meet configuration and customisation requirements of the new service, and how the system and component capability provided the required interfaces for testing purposes.

The approach and methodology to evaluate these aspects were based on service components and application integration activities in the MobiLife project, as well as on the experiences and findings from service management and operations in the Integration and Validation Centres.

Despite the lack of some instructions and component implementations, the component architecture was considered a major asset for deploying a new service.

8.3.4.4 Dimension 4: Impact on Network Architecture and Terminals

The purpose of this part of the technical evaluation is to assess the impact of the new services and architectures on existing and near-future network and terminal architectures and capabilities. The following aspects have to be addressed in the assessment:

- The impact of the new service architecture and new capabilities on the existing and near-future capabilities in the terminals; the capabilities required from the underlying terminal platforms; and the potential changes required to the developed new architecture to enable commercial deployment of the new services.
- New capabilities required in the terminals for maximising the positive user experience stemming from the new service features – like better displays, keypads and voice recognition features.
- The impact of the new architecture on the existing and near-future capabilities in the networks (like service enablers in IMS). Or should the new architecture be changed to make commercial deployment of the new services and applications viable on top of current systems?
- The required supportive capabilities from the underlying networks (e.g. connectivity set-up time, data transfer speed and access types).
- The existence of reference points in the new architecture to support a multi-vendor approach in implementing services and applications.
- The distribution of functions from the perspective of the system performance. Can the current and near-future terminals and network systems provide the needed computational resources for running the new service features?

Evaluation Questions to be Addressed Within this Dimension

1. *How do the new service architecture and new service enablers affect the existing terminal devices?* This addresses the availability of the interfaces to the auxiliary devices (e.g. a GPS module), as well as terminal capabilities and resources such as displays, keyboards and software development environments in order to enable development of attractive future services and applications, and portability of their implementation among the different user devices.
2. *How do the new service architecture and new service enablers affect the existing network capabilities?* The purpose is to understand the required capabilities in the underlying networks, and to assess how they support the connectivity requirements between the terminal and network sides of the new services and applications. The focus is on the existing and near-future cellular and wireless technologies. It is important that the key technologies under evaluation are actually available in the evaluation centres.
3. *How does the new service architecture affect the existing service provisioning architectures and systems (like IMS as an example of a state-of-the-art system)?* The main purpose of this question is to assess how the potential exploitation of the new service architecture would

affect the capabilities and interfaces of the existing service provisioning systems. The key issue in the exploitation is how easily the new capabilities can be integrated into existing systems, and what changes should be designed to the developed service architecture to make it more adaptable. The focus of the analysis is on the state-of-the-art service provisioning systems, like the IP Multimedia Subsystem (IMS). The identification of interfaces and reference points is seen to be important in enabling complementary roles for the old and new systems.

4. *What are the most appropriate software environments for implementing the new services and applications?* This addresses the capabilities and support of the different software development tools for creating new services and applications. Both the terminal and network side implementations must be taken into account.

Case Study Remarks

PDAs and advanced mobile phones available at the time of the case study were able to support relevant network access and interfaces for external devices, such as GPS modules. For most sophisticated services and applications, voice recognition and touch-screen capabilities would permit friendlier user interfaces, and should be offered as input modalities in new end-user terminals.

With regard to client development, most was conducted using Java and it became very clear during the project that the 'write once, run everywhere' concept familiar from the desktop environment is not yet valid for mobile devices. Java support is still lacking, and there is no one solution that would be available on every platform.

On the network side, the following remarks can be made:

- The underlying 3G networks are considered sufficiently fast in terms of the connection set-up time and data transfer. In addition, HSPA technology will further improve the data transfer rate.
- A mapping done with the IMS specification showed that many of the Mobile Services Architecture components could be mapped into the current functionality of the IMS entities. Additionally, Web Service interfaces at all IMS components (e.g. HSS, SLF, MRPF) would be needed to support the complete interoperability.

8.3.4.5 Dimension 5: Evolution

The purpose of this last dimension in the technical evaluation is to understand how the new architecture can be extended with new components and functionalities, and what the consequences of such extensions would be to the existing components and the overall consistency of the architecture.

Evaluation Questions to be Addressed Within this Dimension

1. *What is the impact of inserting a new component into the architecture?* The purpose is to understand how the insertion of a new component may impact the existing architecture. It is also important to understand whether, from the architecture simplicity and consistency point of view, it is more feasible to add new components or to extend existing components with new functionalities.
2. *What is the impact of new components on the existing ones and on the applications? How easy is it to integrate new components with others and with applications?* The focus is on

the technical aspects to be solved when adding new components to the system. In addition to having implications for the architecture, adding components may also affect existing components and applications. The priorities in running the processes (components) may change, affecting overall performance.

3. *How does the addition of the new functionalities affect the existing ones?* New functionalities can be added to the system by specifying new components, or by extending existing components with new functionality, as stated above. Extending the functionality of existing components potentially leads to changed performance of the component and the overall architecture. This last evaluation question investigates the evolutionary capabilities (evolvability) of the system especially from the performance point of view.

Case Study Remarks

The main objectives addressed the capability of the Mobile Services Architecture system to evolve in terms of supported services, context sources, supported mobile groups, etc. The approach therefore focused on an analysis of the scalability of the system in terms of new Context Providers, addressing in particular how the addition of new CPs is taken into account within the reasoning system, and how removal of CPs affects the system. Finally, attention was paid to the extension of the basic context awareness support to mobile groups.

In general, the impact of inserting a new component into the architecture depends on the specific component and where it is located. Specific analyses were made only with regard to adding new CP components. Both qualitative analysis and simulations were used. The simulation allowed consideration of load balancing when replicating more CPs. Replicating CPs allows support for more context interrogations, which in turn means more concurrent usage of applications. The qualitative analysis looked at the addition of new types of CP. In this case, a full implementation of the Context Broker (CB) is a key to making the system more scalable, also reducing the impact on requirements for client updates, as well as for the server side of the application, where the new context information is accessed through the broker without a need to have a direct link between each application and the CP.

The design of the whole architecture lets new components be well integrated with existing ones and, as a consequence, there is minor impact on the development of the application, without dependencies on the number or the specific characteristics of the single components. Of course, addition/removal of Context Providers may impact on the quality of the inferred context data and on the related applications.

Further work has to be done for a smooth transition in the application of the concept of context awareness from an individual to a group of people. A specific function, the Group Awareness Function (GAF), has been implemented and partly used in the TimeGems application. One evolutionary path could be extension of the GAF, while another could be implementation of special Context Providers joining pieces of context information from multiple sensor sources. The current architecture is open to both of these solutions.

8.4 Conclusions

In this chapter, the user-centred design (UCD) process has been applied to perform a user evaluation. Aspects of key enabling technologies have been highlighted with additional remarks about user evaluation. An approach to performing technical evaluation as an integral part of the UCD process was introduced and applied in Section 8.3. The dimensional approach, developed in generic terms, is applicable to large research projects and has the merit of allowing more general conclusions to be derived from analysis of specific project results. Both user and technical evaluation were applied on a concrete case study project.

Evaluation of research results is an integral part of collaborative research projects, which involve different types of organisation from big and small industrial companies to academia. Technical and economic assessment of the evaluation results provides a basis for exploitation in commercial product development, training programmes, etc.

In order to keep the research project on track through the project lifetime, it is important that the evaluation takes place at different stages of the project and involves representatives of the target users of the project artefacts. Here the user-centred design process with a 'Big Loop' and its four rounds gives a solid framework to follow. In general terms, one could even simplify the process into two main steps:

- concept evaluation; and
- reference implementation evaluation.

Both of these steps includes several sub-steps, so it is finally compatible with the 'Big Loop' approach introduced in Chapter 2. The concept evaluation is based on storyboards, mock-up implementations and simulations of the different components in the service architecture and the underlying service platform. This will give the project an important opportunity to learn about the socioeconomic impacts of the new service concepts and about requirements for future services, enabling technologies and user devices. It is important to pay attention also to the multicultural differences between people. Feedback from the concept evaluation can be used to further improve the service and feature specifications prior to implementation.

Evaluation of the reference implementation is based on using the different components on top of state-of-the-art service platforms, and on simulation of such elements, which are not within the focus area of the project work. It is important that usability experts review the prototype implementations and give their feedback to fine-tune the functionality, performance and user interfaces of the services [26]. When all needed improvements have been implemented, the final and most comprehensive evaluation round with users can be initiated. This evaluation round is often implemented through service trials, where users can access the system and services in their genuine usage environments.

When the research project focuses also on new enabling technologies and processes, it is vitally important that developers of the future systems and services participate in the evaluation of the research results from a technical perspective. This can be seen as an industrial assessment of the usability of the developed engineering, management and provisioning methods. It will give the project internal and external stakeholders a clear picture of the potential provided by the developed methods, concepts and technologies.

8.5 Acknowledgements

The following individuals contributed to this chapter: Agathe Battestini (Nokia, Finland), Luca Galli (Neos, Italy), Renata Guarneri (Siemens SpA, Italy), Annakaisa Häyrynen (Elisa, Finland), Mika Karlstedt (University of Helsinki / Nokia, Finland), Esko Kurvinen (Helsinki University of Technology, Finland), Harri Lehmuskallio (Helsinki University of Technology, Finland), Kari Lehtinen (Elisa, Finland), Mia Lähteenmäki (Nokia, Finland), Rinaldo Nani

(Neos, Italy), Pekka J Ollikainen (Nokia, Finland), Marcin Salacinski (BLStream, Poland), Nicoletta Salis (Telecom Italia, Italy), Jukka T Salo (Nokia, Finland), and Antti Salovaara (Helsinki University of Technology, Finland).

References

[1] Adelman L.: *Evaluating Decision Support and Expert Systems*. ISBN 0-471-54801-4, John Wiley & Sons, New York, 1992.

[2] Acquisti A. and Grossklags J.: 'Privacy and Rationality in Individual Decision Making'. *IEEE Security & Privacy*, Vol. 3, No. 1, 2005.

[3] Anderson C. et al.: 'Enabling Autonomity for the Mobile Internet using the mCrowds System'. Proceedings of the Second IFIP 9.2, 9.6/11.7 Summer School, 2003.

[4] Baker P., Dai Z. R., Grabowski J., Haugen Ø., Lucio S., Samuelsson E., Schieferdecker I. and Williams C.: 'The UML 2.0 Testing Profile'. Proceedings of the 8th Conference on Quality Engineering in Software Technology 2004 (CONQUEST 2004) in Nuremberg (Germany), September 22–24, 2004, pp. 181–189.

[5] Billsus D. and Pazzani M.: 'A Hybrid User Model for News Story Classification'. Proceedings of the 7th International Conference on User Modelling, Wienna, Austria, 1999.

[6] Born M., Schieferdecker I., Kath O. and Hirai C.: 'Combining System Development and System Test in a Model-centric Approach'. RISE 2004 International Workshop on Rapid Integration of Software Engineering Techniques, LNCS 3475, pp. 132–143.

[7] Conaty G. (ed.): 'Initial Scenarios, Requirements and Guidelines: User-centred approach for the design of future mobile services and applications'. IST-MobiLife Project Deliverable D6b (D1.1b), February 2006. Online: www.ist-mobilife.org.

[8] del.icio.us social bookmarking: Online: del.icio.us.

[9] Dey A. K. and Abowd G. D.: 'Towards a Better Understanding of Context and Context Awareness'. Proceedings of the Conference on Human Factors in Computing Systems (CHI'00) Workshop on The What, Who, Where, When, Why and How of Context-Awareness, The Hague, Netherlands, April 2000.

[10] Dourish P.: 'What We Talk About When We Talk About Context'. *Personal and Ubiquitous Computing*, Vol. 8, No. 1, 2004, pp. 19–30.

[11] Fink J., Kobsa A. and Nill A.: 'Adaptable and Adaptive Information Provision for All Users, Including Disabled and Elderly People'. *New Review of Hypermedia and Multimedia*, Vol. 4, 1998, pp. 163–188.

[12] Flickr: Online: www.flickr.com.

[13] Grabowski J., Hogrefe D., Rethy G., Schieferdecker I., Wiles A. and Willcock C.: 'An Introduction into the Testing and Test Control Notation (TTCN-3)'. *Computer Networks Journal*, Vol. 42, No. 3, 2003.

[14] Guarneri R. and Salo J. (ed.): 'Technical Evaluation Report'. IST-MobiLife Project Deliverable D43 (D6.3), December 2006. Online: www.ist-mobilife.org.

[15] Jones L.G. and Kazman R.: 'Software Architecture Evaluation in the DoD Systems Acquisition Context'. *News@SEI Interactive*, Vol. 2, No. 4, December 1999. Online: www.sei.cmu.edu/news-at-sei/columns/the_architect/1999/December/architect-dec99.pdf.

[16] Kirk J. and Miller M. L.: *Reliability & Validation in Qualitative Research*. Beverly Hills, CA. Sage, 1986.

[17] Kurvinen E. (ed.): 'Results of Mock-ups Evaluation'. IST-MobiLife Project Deliverable D8 (D1.3), May 2005. Online: www.ist-mobilife.org.

[18] Kurvinen E. (ed.): 'Results of Service and Application Evaluation'. IST-MobiLife Project Deliverable D14 (D1.9), December 2006. Online: www.ist-mobilife.org.

[19] Kurvinen E. and Oulasvirta A.: 'Towards Socially Aware Pervasive Computing: a turntaking approach'. Proceedings of IEEE PerCom 2004, Orlando, Florida, 2004, pp. 346–351.

[20] Kurvinen E., Häyrynen A. and Klemettinen M.: 'MobiLife UCD Process'. IST-MobiLife Deliverable D1.4c, January 2006. Online: www.ist-mobilife.org.

[21] Kurvinen E., Lähteenmäki M., Salovaara A. and Lopez F.: 'Are You Alive? Sensor data as a resource for social interaction'. *Journal of Knowledge, Technology and Policy*, Vol. 20, No. 1, 2007.

[22] Manber U., Patel A. and Robison J.: 'Experience with Personalisation on Yahoo'. *Communications of the ACM*, Vol. 43, No. 8, 2000, pp. 35–39.

[23] MobiLife project: Online: www.ist-mobilife.org.

[24] Oppermann R. and Simm H.: 'Adaptability: User-inititiated individualization'. In Oppermann R. (ed.): *Adaptive User Support: Ergonomic design of manually and automatically adaptable software*. Lawrence Erlbaum and Associates, New Jersey, 1994.
[25] Oulasvirta A., Raento M., Tiitta S.: 'ContextContacts: Re-designing SmartPhone's Contact Book to enhance mobile collaboration and awareness'. Proceedings of Mobile HCI 2005, September 19-22, Salzburg, Austria, ACM Press, 2005.
[26] Rubin J.: *Handbook of Usability Testing: How to plan, design, and conduct effective tests*. John Wiley & Sons, New York, USA, 1994.
[27] Salovaara A. and Oulasvirta A.: 'Six Modes of Proactive Resource Management: a user-centric typology for proactive behaviors'. Proceedings of the Third Nordic Conference on Human-Computer Interaction, Tampere, Finland, 2004, pp. 57–60.
[28] Sarvas R., Oulasvirta A. and Jacucci G.: 'Building Social Discourse Around Mobile Photos: a systemic perspective'. MobileHCI 2005, Salzburg, Austria, 2005.
[29] Shamo M.K., Dror R. and Degani A.: 'A Multi-dimensional Evaluation Methodology for New Cockpit Systems'. Proceedings of the 10th International Aviation Psychology Symposium. Columbus, OH, 1999.
[30] SlashDot: Online: slashdot.org.
[31] SysTest: 'Developing Methodology for Advanced System Testing'. Online: http://www.incose.org/secoe/0105.htm.
[32] Wagner M., Luther M., Hirschfeld R., Kellerer W. and Tarlano A.: 'From Personal Mobility to Mobile Personality'. *Telektronikk*, Vol. No. 3/4, 2005, pp. 155–164. Online: www.telenor.com/telektronikk/.
[33] Walker J.: 'Feral Hypertext: When hypertext literature escapes control'. Proceeding of the 16th ACM Conference on Hypertext and Hypermedia (HYPERTEXT '05), ACM Press, 2005, pp. 46–53.
[34] Weiss A.: 'The Power of Collective Intelligence'. *netWorker*, Vol. 9, No. 3, 2005, pp. 17–23.
[35] Wikipedia: Online: www.wikipedia.org.
[36] Xiangzhu G., San M. and Lo B.: 'Multi-dimensional Evaluation of Information Retrieval Results'. Proceedings of the IEEE/WIC/ACM International Conference on Web Intelligence, September 2004 pp. 192–198.

9

Marketplace Dynamics and Business Models Framework

Edited by Ulla Killström (Elisa, Finland)

In earlier chapters, different key components – from architectures and enabling technologies to technical viability and end-user acceptance – behind potentially successful mobile applications and services have been described. However, a good technical solution or fit to the end-user needs does not guarantee success; a complex and dynamic set of value-added creation, marketplace dynamics, business models and socioeconomic factors, as discussed in Section 9.1, affects the puzzle.

This chapter introduces a dedicated business model framework suitable for mobile applications and services with its seven components (user/customer, products and services, earnings logic, resources, suppliers, organisation and processes), and describes four generic business model alternatives ('Survival Battle', 'Old Rules', 'Wind Turns' and 'Eye of the Tornado') for new mobile services in terms of design choices for each business model component (Section 9.2). As discussed in more detail in this chapter, it is hard to predict today how economy, society and technology will develop in the coming years, and how they will shape the future marketplace. This is why, in Section 9.3, various future scenarios are presented and the business models defined are mapped to two rather probable scenarios, 'Old Rules' and 'Wind Turns', and the challenges of each business model are discussed in Section 9.4. Before the conclusion in Section 9.6, socioeconomic aspects affecting the adoption of new mobile applications and services are covered in Section 9.5.

Enabling Technologies for Mobile Services: The MobiLife Book Edited by Mika Klemettinen

9.1 Introduction

Companies in various industries can take advantage of enabling technologies for mobile services to improve their current way of doing business. They enhance the opportunity to build their value proposition around the users' need to stay informed and to respond quickly. Future communication environments, new mobile applications and services, and a multitude of available networks, devices and technologies give many possibilities in composing the actual service. However, in order to capture value for the company, the impact of the technological evolutions to business has to be considered.

While business models are useful in analysing how technological developments impact ways of doing business, they are perhaps the least understood terms in management and technology innovation. Scientific and practitioner publications have advanced understanding of various modelling approaches in different business contexts.

It is not easy to provide a clear picture of users' needs, the competitive situation, or the state of technology development today. It is even harder to predict how the economy, society and technology will develop in the coming years, and how they will shape the future marketplace. The degree of change can vary widely. Although moderate changes affect business models in a predictable way, very dynamic changes in the marketplace have a nonlinear, less predictable impact on business models, and market boundaries may become blurred. Figuring out the business value of technological innovations is therefore challenging.

The great strength of a business model as a planning tool is that it focuses the attention of planning efforts and managerial decisions on how all the elements of a business fit into a working whole. The business model is a set of basic assumptions about the components that describe the way of conducting business. The business model components – users/customers, products/services, earnings logic, resources, suppliers, organisation/architecture and processes – profoundly influence the success of a company in the marketplace. However, understanding the opportunities and threats related to defining those components is very important. Table 9.1 gives an insight into a few of them.

This chapter describes a framework for the components of a typical business model. It illustrates the modelling task in applying the business model components in four generic models. In the first of these, technological changes have a clear impact on other modelling topics. The second meets the challenges of users' unwillingness to pay and therefore starts from the earnings logic component. New mobile business possibilities could simply extend existing businesses, and in the third generic business model the adaptation of new services to existing user segments, service promises and processes are key questions. The fourth modelling approach describes the typical situation where a customer can buy some content by ordering it from a service provider.

As the four approaches each start with changes in a different business model component and then assess the impact of design choices in the other components, this set of models provides a comprehensive illustration of how the component-based business model framework may be used in practice.

Success depends not only on the choice of a certain business model but also on the future environment in which the model is deployed. Some business models may have a better 'fit' with an assumed future environment than others. The four generic business model alternatives are described in terms of design choices for each component. The assumptions made for business modelling are highly related to the two main environmental drivers that shape the

Table 9.1 Opportunities and threats on the marketplace

Business model component	Opportunities	Threats
Users/customers component	• Users emphasise usefulness and convenience. • The role of communities is increasing. • Management of the customer relationship portfolio is still a useful tool. • Enables flexible and mobile workplaces to be taken into account.	• Customer needs difficult to understand but important to identify. • Launch of new technology before it is reliable and the usability is reached. • Cultural and national specificities have impact on business success. • The risk of unwanted development and side-effects.
Products/services component	• Voice and messaging services still important. • To devise innovative ways to extend services well beyond currently available. • Empower user content production, elaboration and delivery. • Digitalisation process of any media extends marketing and advertising free service models.	• The delivery of content and applications over different platforms. • Limited interoperability negative impacts on user experience. • Devices are often too complex and unfriendly. • End-users are not aware of new services/potentially interesting solutions. • A lot of uncertainties in the market.
Earnings logic component	• New business roles and revenue models become feasible. • Digitalisation of content continuous. • Subscription models, margins for providers are available. • Value-based pricing depending on perceived value of the services. • Possible to sell bundles of services.	• Division of roles, investments, costs and revenues becomes complex and difficult. • The share of revenues to each actor in the value net. • Valuation of contribution of partners is challenging. • Monitoring and visualising the performance of mobile services and business models is needed.
Resources component	• Convergence continues: fixed-wireless, mobile media, home technology. • Unlicensed frequencies make wireless access. • P2P environments utilise resources by the participating nodes. • Proximity technologies ease identification. • Standardisation continues.	• The application technology innovation needed for mainstream market. • Diversity of terminals makes deployment challenging. • Difficulties in service discovery, initiation of service prevent services from becoming widely used. • Negative aspects of openness; spam and viruses. • Unidentified core competences of potential players.

(Continued)

Table 9.1 Continued

Business model component	Opportunities	Threats
Suppliers/actors component	• Mobile service industry transforming from static value chain to more dynamic net, new and more players. • Transition from supply driven to user-centric demand-driven value creation.	• The value nets are more complex, including several industries and interdependences. • The power of actors depends on their resources.
Organisation/ architecture component	• Governance in the business network, orchestration of nodal actors and roles in relation to the services are dynamically related. • Interest for mobile data services service has impact on new service lifecycles. • Management of resources, digitised services, and interoperability done in the service oriented architecture.	• Industries are blurring. Co-operation between a lot of different stakeholders, increasing need for management. • Organisations can capitalise a technology only if they pay attention to the service architecture, information and money flows between actors. • Technological standards and interoperability play important roles in co-operation.
Processes component	• Being able to define effective processes is a way to enhance business performance. • Roles of customers are changing, Customer relationship management emphasises customer self-management and control.	• Increasing amount of more complex networks and services calls for automation and optimisation of business processes. • The firms need to interact with process flows defined by external parties. • Processes are not nearly as flexible or adaptable as resources unless defining early enough.

future marketplace: 'business growth potential' and 'changes in business logic and industry structure'. Each business model has to be adapted to the marketplace, and the social issues define the business context further.

The business modelling approach provides insight into how to define the components for a viable business model in different future environments. However, an understanding of management and the concrete decisions they make related to the business modelling components are of crucial importance in explaining the success or failure of individual business models. Therefore, these aspects conclude the discussion of modelling the business benefiting from enabling technologies for mobile services.

9.2 Business Models Framework

9.2.1 Framework for Modelling the Business

Business models have become a popular subject for research. A business model is essential to every organisation, whether it be an Internet business or in a traditional sector. The definition of a business model brings clarity to and helps the performance of the business.

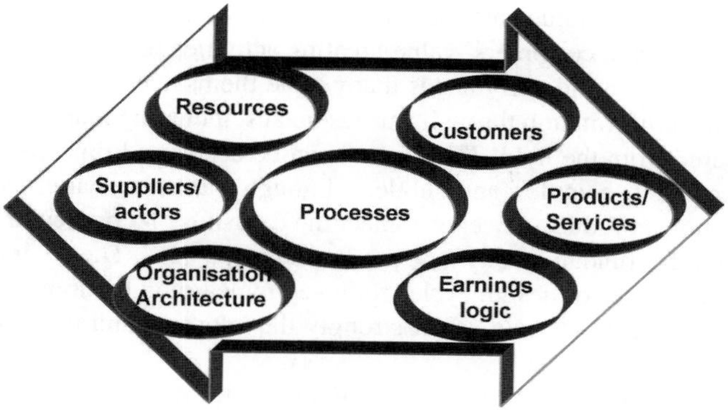

Figure 9.1 A framework for a business model.

Several authors, management researchers, economists and organisational theorists have invoked the concept of a business model in their search for answers to a broadening range of questions. However, there is no generally accepted definition of what a business model is [7, 18, 25]. The notion of a business model is frequently applied in the information and communications technologies literature [13, 19, 31]. However, despite its vagueness, the business model concept has become a central notion in the managerial vocabulary [28].

The main motivation for using a component-based approach comes from the strategy literature. All components included in a business model are also used in the strategy definition of a company. The framework presented in the following sections naturally represents an archetype. The components included in the framework are: *users/customers*, *products/services*, *earnings logic*, *resources*, *suppliers*, *organisation* and *processes* (Figure 9.1). However, the business model or components presented in the theory cannot be directly adapted to a company as such. It is important to notice that each business model is centred on a particular firm. However, all these business model components are also necessary for the business definitions for products or services provided by a network of companies.

The functioning of a business model becomes visible in managerial decisions and actions [28]. Managers view and make decisions based on their understanding. The company's strategy gives meaning and direction to the development of the company's business model. In addition, the future environment in which the business model will be used guides managerial decisions as well.

9.2.1.1 Users/Customers Component

The key question in building a business model is what types of customer benefit a company should seek to provide in five, ten or fifteen years. It is important that a company learns about the precise dimensions of customer demand and required product performance.

A company must do much more than be customer-led [12]. A company's relationship with a customer is an access channel to the customer's ongoing value-creating activities. Any customer, whether another business or an individual, uses a wide range of inputs in order to create value. A company's offerings have value to the degree that customers can use them as inputs to leverage their own value creation. In this respect, then, companies do not profit from customers, they profit from customers' value-creating activities [22].

In order for companies to identify needs that people themselves may not recognise, companies need to collect information through various means, including market research and user information gathered from the field. The process can be costly and time-consuming because people's needs are often complex and subtle. Although market research can be helpful in fine-tuning well-known product concepts to meet the demands of a particular class of people, it is seldom the spur for fundamentally new product concepts [12]. The company must enable people to complete a series of design cycles followed by learning by doing. Lifestyle, values and past usage experiences seem to affect strongly the adoption and usage of products and services.

Customer's expectations are not stable but changing. Consumer attitudes change most at the beginning of the use of a service or product. These changing priorities, and the way in which they interact with new competitors' offerings, are what trigger, enable or facilitate the value migration process [21].

The management of the customer relationship portfolio is one of the most crucial aspects in a company's business model. Customer relationship management processes are key tasks that address all aspects that identify customers, create customer knowledge, build customer relationships and shape their perceptions of the organisation and its offerings [8].

As people organising themselves *virtually* becomes more widespread, companies must recognise communities as part of the value delivery system and respond appropriately in their strategies [31]. The business model may offer more value to a discrete group of customers or it may completely replace the old way of doing things and become the standard for the next generation of entrepreneurs to beat [18]. The most profound aspect of interaction in the virtual model is the emergence of electronic customer communities. These communities are information-gathering and information-disseminating conduits.

9.2.1.2 Products/Services Component

The products/services component defines which customer value the company intends to offer to its customers. The company has to develop products that deliver optimal customer value. In technology-enabled businesses, for example telecommunications, the evolution depends on the technology adoption lifecycle [26]. Technology causes shifts in alignment among the various strategies. The technology-based products pass through multiple phases of adaptation during which the market (i.e. customers) behaves in different ways specific to each phase. This means that the market and customer priorities change, so competition between alternative technological approaches changes too. The company may have to change its business model to meet new customer expectations.

The concept of a business model should guide the product development process. In this context, product development can be seen as one value-creating process. The central issue is the creation of value for buyers, either in the form of differentiated product or in one produced at lower cost [24]. The key process of product management includes offering development and testing of products, identifying and managing internal relationships, and developing and maintaining linkages with external co-operation.

Innovation strategies and their impact on product development have also to cope with the challenges of an increasingly networked world. Interdependent players won't switch to new products unless they believe other players will. For instance, a bank would carefully consider other banks' choices if a new transaction-processing system is proposed. A set of directions has been identified to address this issue [2].

9.2.1.3 Earnings Logic Component

The earnings logic component in a business model defines how the company gains revenues and how it will charge (or subsidise in the case of advertising supported models) for the services provided. In addition, it describes how the company will divide revenues among the actors in the value-creating network.

In the business model literature, especially in the literature on e-commerce, several taxonomies of basic business models are available that describe the position of a company within its transactional environment, together with its revenue model [10].

The traditional vision on earnings logic in the mobile business is based on the situation of having solely operators and voice services. Consumers and business users act as the revenue sources and competition between providers is much on price. However, the future value nets will involve more players in more complex value networks. This requires more co-operation and partnering as well as new ways of sharing and distributing revenues [27]. Each of the business actors must gain sufficient benefit from participating in the network of actors.

Mobile and wireless devices have the promise to enable organisations to conduct business in more efficient and effective ways [9], for instance by empowering an organisation's sales force or remote workers. The number of companies that implement mobile and wireless solutions is increasing. Operators no longer play a central role; rather, specialised service providers and system integrators benefit from the adoption of mobile services [27].

The companies in the value net can choose between three different pricing strategies. The first is *cost-based pricing*, which involves setting a price such that costs are covered with an acceptable margin. This approach has – especially in the case of a value net – a fundamental problem in that allocating fixed costs depends on sales volumes that are intrinsically linked to prices. The second approach is *competitive pricing*, whereby prices are set based on similar offerings in the marketplace. The danger is then that the entire industry is vulnerable to price wars. The third possibility is *value-based pricing*, where the price level depends on the consumers' perception of a service's value.

Mobile advertising can serve as a new revenue source. Advertising companies paying to get messages to recipients is a promising development towards new high-potential revenue streams [20]. Utilising user profiles and user context information, such as location information, will enable contextual advertising. Mobile information and entertainment services that are linked with contextual advertising are currently a developing area. Under certain conditions, including easy opt-in and opt-out possibilities, consumers are willing to accept advertising in exchange for free services. Another aspect are subsidies which could be provided by organisations that see value in providing a mobile service to customers for free or for a mitigated fee – for example, subsidies for services, such as tourist services, for the general public.

Several industries show an interest in mobile services to improve their current way of doing business. The content industry considers the mobile channel as an additional distribution channel, which may increase overall content sales. In the telematics industry, personal navigation is moving from stand-alone navigation products to navigation systems embedded in a service offering including several mobile data services. In the healthcare industry, mobile services are considered to lower costs, for instance by empowering a mobile workforce, and to improve quality. The general interest in (monitoring) personal health and 'wellness' creates opportunities for mobile services, for which people are willing to pay. Mobile advertising provides interesting opportunities for advertisers and service providers, as individualised and customer-oriented marketing becomes a more desired form of interaction with the customer [11].

9.2.1.4 Resources Component

The means to create value are defined in the resources component of the business model. It includes assumptions on how resources should be configured, and how skills and technologies can enable the company to provide a particular value proposition to customers.

Resources required for viable approaches to generating value include more than technology alone. A key question in the definition of a business model is the allocation of three types of resources: *human*, *technological* and *financial*. Resources can also be divided into *tangible ones* like human assets (people), equipment, technologies and cash, and *less tangible ones* like product designs, information, brands and relationships with suppliers, distributors and customers. These tangible and less tangible assets all create and strengthen the company's possibilities to perform.

The resources are either owned or controlled by a company by various kinds of agreements and arrangements. In the latter case the company needs to carefully manage the resources in order to maintain control.

Technological change affects competitive advantage and industry structure. Whereas sustaining innovations make a product perform better for mainstream customers, disruptive innovations create entirely new markets. When considering a response to disruptive innovation, managers must take into account the unique resources/processes/value framework that defines a company and shapes its capability to change; this means changing the business model [7]. Technology alone, not embedded in an effective business design, is no longer a viable approach in generating sustained value growth [25].

Capabilities define the company's ability to exploit and combine resources through organisational routines in order to accomplish its targets. Creation of value depends on the competencies and ability to provide and deliver efficiently benefits that are important to the customer. In addition, these capabilities are related to the business processes and enabling technologies. They call for agility and are created and acquired over time.

As various economic players participate in value nets in order to create and capture value, it is critical to identify core competencies of potential players in the future mobile services domain. The following is a fairly comprehensive list:

- terminal design, hardware and software design (specifically in developing intuitive user interfaces with minimal interaction by maximal utilisation of context and personalisation), and integration, manufacturing and logistics (mobile terminal vendors, consumer electronic vendors);
- service platform and application development (IT companies);
- subscriber management: identity, payment, helpdesk (operators, merchants, banks);
- content production (media companies, advertisement agencies);
- content distribution (broadcast companies).

In addition, the use of innovative mobile technologies described in this book has several implications on the resources required (Table 9.2).

- Personalisation and context management requires intimate knowledge of the customer. The *capabilities* required include (a) technical capabilities – automatic analysis of large amounts of event data in order to understand user behaviour; and (b) organisational capabilities – finding partners and organising co-operation by applying knowledge on user behaviour for various types of services. A large information base of personalisation data covering users from different groups and locations, as a *resource*, is required having a critical mass for personalised applications.
- Trust and privacy are needed for developing strong and intimate customer relationships. When achieved, this means strong brand and added perceived value. The *capabilities* required include (a) technical capabilities – development of secure solutions; and (b) organisational capabilities – secure operation, fast response to incidents, and strong brand management.
- If a service can get communities (groups) involved, valuable content is created by active group members with no extra cost to the service provider. The content attracts new users, and thus an audience for advertising, and a potential customer base for charged services. This is an example of how users–customers ('prosumers') can participate in creation of a *resource* (data) for the service provider. The *capabilities* required include (a) technical capabilities – development of solutions that are easy to manage and use for non-professionals (communities); and (b) organisational capabilities – brand management in order to attract communities and media type of competence in order to create an attractive media for advertisers.
- The ability to manage multiple devices and interfaces requires (a) competencies in several user technologies, and (b) orchestration capabilities for defining required interfaces in co-operation with other companies.

9.2.1.5 Suppliers/Actors Component

In the suppliers/actors component of a business model it is important to define how the actors co-operate. As the resources as outlined in the previous section are typically not available within the boundaries of a single company, managing an efficient customer-driven supplier net and the ability to integrate and coordinate the value activities of each net member are essential. An actor taking a central role in a value net needs to be able to mobilise other actors to form a tightly coordinated way to act. The coordination capability requires information and management systems that combine business processes of each actor and monitor efficiency of production, logistics, and customer delivery and service.

The creation of viable networks of actors to develop novel products and business concepts demands several complex capabilities. Network-orchestration capability at the right end of the value-creation continuum refers to an actor's capacity for influencing the evolution of a whole new business network. Orchestration presupposes the ability to envisage the emerging business field – which may be very complex like the convergence of the ICT field suggests – and its key actors, and to identify potential trajectories [20].

Getting the right suppliers and building sustainable value nets with the right actors and roles for mobile services with respect to business modelling are key aspects. Putting together a network of companies to develop the set of capabilities necessary becomes a major challenge.

Table 9.2 Meaning of the technology aspects for modelling the business

Technological aspect	Key features (innovativeness, functionality)	Meaning (needed capability) for the business model
Personalisation technologies	Adaptation to user preferences either automatically or based on user control	(a) Technical: automatic analysis of large amounts of event data in order to understand user behaviour; (b) Organisational: co-operation by applying knowledge on user preferences for various types of services.
Context-awareness technologies	Adaptation to user context	(a) Means to analyse user context (sensors, software algorithms); (b) Co-operation by applying knowledge on user context for various types of services.
Group awareness technologies	Adaptation to social environment	(a) Development of solutions that are easy to manage and use for non-professionals (communities); (b) Brand management in order to attract communities, media type of competence in order to create an attractive media for advertisers.
Trust and privacy	Intelligent combination of trusted platform technologies with intuitive trust-enabling user experience	(a) Development of secure solutions; (b) Secure operation, fast response to incidents, strong brand management.
Multimodality	Ability to manage multiple devices and interfaces	(a) Competencies in several UI technologies; (b) Orchestration capabilities for defining required interfaces in co-operation with other companies.
Service platforms	Advanced terminal and server platform components; fast and flexible network connectivity	(a) Complex set of UI, data management and connectivity software and hardware development skills; (b) Software and hardware integration, definition and maintenance of development interfaces.
Service platforms	Advanced terminal and server platform components; fast and flexible network connectivity	(a) Complex set of UI, data management and connectivity software and hardware development skills; (b) Software and hardware integration, definition and maintenance of development interfaces.
Service platforms	Advanced terminal and server platform components; fast and flexible network connectivity	(a) Complex set of UI, data management and connectivity software and hardware development skills; (b) Software and hardware integration, definition and maintenance of development interfaces.

(*Continued*)

Table 9.2 Continued

Technological aspect	Key features (innovativeness, functionality)	Meaning (needed capability) for the business model
Service lifecycle	Making services available to customers, including service creation, provisioning (deployment, usage, retirement, operational management), and lifecycle management	(a) Complex set of IT skills; (b) Advertising, cost-efficient (automated) help desk operation.

Smart combinations of technological developments is driving the emergence of real innovative mobile services. The static value chains are transformed to more dynamic value nets with new and more players. Therefore, innovating and combining emerging technological capabilities to create innovative and attractive mobile services and related business models could be a key factor for success of mobile services. However, that calls for strong relationships with suppliers and key partners that are crucial in adding value to the offering. In addition, the end-users and user communities are aquiring more central roles in these value creation processes.

A *supply chain* has traditionally been defined as a network of facilities and distribution options that performs the functions of procurement of materials; transformation of these materials into intermediate and finished products; and distribution of these finished products to customers. A supply chain essentially has three main parts: supply, manufacturing and distribution [11]. However, the traditional thinking about supply chain management is moving to business design that uses digital supply chain concepts to achieve both superior customer satisfaction and company profitability [5]. It is not a sequential chain but a dynamic network of customer/supplier partnerships and information flows. Contrary to the traditional supply chain where a company manufactures products and pushes them through distribution channels, the value net way of satisfying customers' actual demand begins with customers and allows them to self-design products [5].

The characteristics of the supply markets have changed as creating networks on virtual markets has become easier. These virtual markets combined with the vastly reduced costs of information processing allow profound changes in the ways companies operate and how economic exchanges are structured. Thus, the ways value is created and the business models are modified is being challenged [1]. As an electronic network with open standards, the Internet supports the emergence of virtual communities and commercial arrangements that disregard traditional boundaries between companies along the value chain. Business processes can be shared among companies from different industries, even without any awareness of the end customers. As more information about products and services becomes instantly available to customers and as information goods are transmitted over the Internet, traditional intermediary businesses and information brokers are circumvented and the guiding logic behind some traditional industries (e.g. travel agencies) begins to disintegrate. At the same time, new ways of creating value are opened up by the new forms of connecting buyers and sellers in existing markets (re-intermediation) and by innovative market mechanisms (e.g. reverse market auctions) and economic exchanges.

The value nets for emerging mobile services are more dynamic and complex than the state-of-the art ones. They have more, as well as new, actors and include actors from different industries – such as IT, telecoms, consumer electronics and media – that are increasingly operating in each other's formerly separated markets. Each of these actors has one or more roles to perform in the value net. Various economic players may participate in the value nets (e.g. companies, families, public bodies, non-profit institutions) by taking responsibility for one or more activities or roles. They may also participate in more than one value net.

Actors can be more or less powerful in the value network depending on their resources and capabilities. Hawkins [14] identifies three basic types of partner in a value network: *structural* partners, which provide essential and non-substitutable (in-)tangible assets; *contributing* partners, who provide goods and/or services to meet specific network requirements; and *supporting* partners, who provide substitutable, generic goods and services to the network. Structural partners in principle are better positioned to exert control over the network than supporting partners.

Actors differ with respect to the strategy and goals they pursue with the collaboration. Collaboration requires partners to share information and give insight into their ways of working. However, strategic interest may induce partners to act against what is agreed upon, hide the truth, or try to extract confidential information from their collaboration partners. Trust between partners is an important condition for an open and constructive collaboration. Collaboration gives rise to complex interdependencies between organisations because no single partner has formal authority over another partner. In order to govern the collaboration, actors need to agree formally and informally on how to divide and coordinate the value activities and how then to monitor them.

In the future service creation, it is all about customer *and* network value creation. When designing and developing a mobile service, a company has to take into account that all important actors – telecoms operators, content providers, but also new players from industries like IT and consumer electronics – in the future mobile service value nets should be able to capture value. The business model has to include the user-centric, multi-industry view with multi-actor business models and multi-actor value networks.

The typical traditional roles with only minor changes in value drivers and earnings logic (e.g. network provider, application developer, device manufacturer, content provider and advertiser) are still used for new mobile services. However, some changes in some current roles are also demanded, such as when a user participates in creating content (e.g. in health monitoring). In essence, the service concepts to be developed are mostly globally applicable. However, because a lot of future mobile services are context-dependent, also local organisations (like museums, restaurants and public transport companies) may become increasingly important actors of future mobile service value networks at least as information providers for – or as providers of – products or services related to context-dependent mobile services being used by end-users. Also, new roles are expected to emerge, such as:

- *personalisation provider*: manages personal profiles and adapts the service offering to them.
- *identity and trust provider*: manages users' and third parties' identities, gives users and third parties a possibility to set and update appropriate trust levels.
- *context information provider*: provides and distributes context information (e.g. location, situation) of users.

• *group awareness provider*: offers community functionality that supports having peripheral awareness of each other.

The most important question in the entry of these new roles is whether there is realistic earnings logic for them. For example, continuous flow of context information increases costs which have to be covered by revenue sources. The new roles can be performed by existing companies like telecoms operators, IT and other Internet companies, or new types of actor who create their business through the new roles.

9.2.1.6 Organisation/Architecture Component

The number of relationships between organisations in mobile service value networks is growing, and they are increasingly complex. In this environment organisations can capitalise on a technology only if they define a business model that pays attention especially to the service architecture, information, knowledge and money flows between all needed business actors, including understanding the roles and potential benefits for the various actors. The organisation/orchestration component in a business model defines those key topics that combine the actors in a value net together. It enables the provisioning of the future mobile applications and services.

Interorganisational co-operation includes the definition of governance systems (network governance systems), which describe how the inter-company coordination is organised and what kind of informal social and formal contractual relationships are needed between the companies [16]. Strongly related to network governance mechanisms are concepts like orchestration and organisational arrangements used to organise collective action.

Different roles in provisioning mobile services need to be fulfilled by one or several actors. Each of these actors has different requirements and strategic interests, which need to be constantly taken into account in order to organise or orchestrate collective action [29]. The creation of value for the end-users and the capturing of value for participating organisations are crucial for successful co-operation.

The organisation/orchestration component describes those structural mechanisms that provide the necessary backbone and stability for different actors. The network form of governance carries with it special problems in adapting, coordinating and safeguarding exchanges, because it relies on autonomous units operating in a setting of demand uncertainty with high interdependence, owing to customised, complex tasks [16].

The business models, and therefore also the value networks, need to be continuously revised when new challenges and opportunities emerge. This means that the governance system has to be dynamic and externally aimed in order to be able to repeatedly reconfigure the needed resources. Intercompany-level governance systems based on rules and discipline mechanisms play an important role here.

The use of technological standards and concepts like open Service Oriented Architectures (SOAs) play an important role in this context (Figure 9.2). In the networks of actors providing mobile services, each market actor has one or more roles to perform. A role represents a set of functions that enables making a mobile service and delivering it to the environment. From the SOA point of view, the roles form the Business Interface needed to provide services to the user. The governance mechanisms of the business are defined at this level. The definitions include the responsibilities of actors, compensation to each actor, descriptions of service

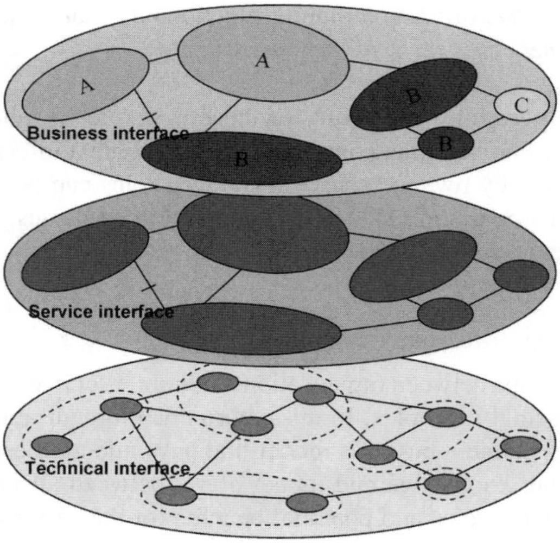

Figure 9.2 Service-oriented architecture as interfaces.

levels and confidentiality of the information used. The management of resources is defined at the Service Interface of the SOA. This interface includes the data and interface definitions, controls which ask what and who manages the value chain and different functionalities. The main questions asked in defining the Technical Interface focus on how to execute the services, how the negotiation and communication between interfaces as well as interfaces between different technologies are done.

A relatively high level of standardisation and interoperability will be needed for future mobile services. Every reasonable business model can be considered in two dimensions from the separation and distribution of the mobile system components point of view.

- In an *open model*, different companies (e.g. mobile application provider and mobile service enabler provider) operate the functional components. Common reference points/interfaces between the different domains are needed in order to specify their interworking.
- In contrast, in a *closed model* one company operates and provides all mobile applications and service enablers. The development of those components might still be done outside this company.

Combinations of the two options are possible and the degree of openness of the business models in real life will lie between the two extreme views described above. The new roles needed for future service provisioning and the potential business models have a strong impact on the specification of the architecture. That underlines the importance of the openness and scalability as well as modular approach to service definitions in customised value creation.

Facing the revenue questions included in the earnings logic component description, the mobile architecture has to support operational management functions beyond the lifecycle management, including accounting features. To be able to share revenues in value networks, system and application monitoring is needed to support billing systems for collecting, reporting and sharing the revenues. Accounting features have also to be offered for the mobile service enablers, so that attaching billing systems is also possible there.

Business Interface, Service Interface and Technical Interface in Service Oriented Architecture play important roles in the inter-organisational collaboration between different actors. The new roles in future service provisioning have a strong impact on their specifications. The various architectural domains and careful architectural specifications are the key enablers and thus necessary for future mobile services.

9.2.1.7 Processes Component

Processes bind the business model components together. Improving the efficiency of processes has always been a way to enhance business performance. The importance of well-defined processes grows even more when products and services are no longer developed and produced by single companies working in a stable business environment. When value nets become more and more complex and at the same time product and service lifecycles shorten, it is not possible to handle the business without properly defined processes.

A process is a group of repeating functions being dependent on each other in timing and/or functionally, which produces a certain outcome. The purpose of a business process is to describe (1) what actions will be done, (2) why they are done, (3) what preconditions are needed, and (4) what the actions will be produced. The processes of a business consist of the activities and procedures with which a business procures resources, adds value, and produces and sells products and services. Some processes are formal, in the sense that they are explicitly defined and documented. Others are informal: they are routines or ways of working that evolve over time. Both types of process typically occur simultaneously in a company.

The basic idea behind conceptualising and categorising business processes is to identify and design repeatable processes that have enough elements of consistency (e.g. clearly identified inputs and outputs) to justify developing a common, aligned process for the organisation. The key elements consist of sequentiality, routinisation and organisation of activity chains, input–output and material flows, organisation of work, and the cross-functional and inter-organisational nature of the organisation [28].

In a business model, the processes should be concretely described in order to provision new mobile services. The services developed and the roles of different actors in the value net have impact on the process description and thus it has to be done separately for each business model. Processes between different actors are slow to change and it is very important to start analysing them at early stages of applications and services development in order to be able to take into account all needed aspects. The challenge is to connect the new applications to the existing processes of the company.

Customer Relationship Management (CRM) is one of the important processes in mobile service delivery. The purpose of CRM is to enable organisations to better manage their customers through the introduction of reliable systems, processes and procedures for interacting with those customers. The principal business drivers for having CRM include increasing customership lifetimes, reducing costs and improving efficiency. The main obstacles in becoming customer-centric are poor processes and practices. Business Process Re-engineering (BPR) is often an unavoidable consequence of the introduction of CRM. Unless the process is changed across the business, however, CRM will simply be an expensive new initiative with no benefits. Linking processes will also ensure a more consistent experience for the customer [23].

The roles of different actors and roles in mobile business models call for new requirements for process definitions. It is important to move from the older customer care or service orientation to a customer relationship management regime that emphasises customer self-management and control, increasing the value customers contribute to the service and the use of customer-related information in order to customise and personalise the services. The information flow should be friction-free across actors and roles, applications and devices. Properly defined processes will boost the operational work and thus impact on cost efficiency of a company.

The resources used in future mobile services across technologies (i.e. application, computing and network) have to be managed by integrating the processes. Networks and services form complex structures and the rhythm for new technologies escalate. At the same time there is a greater need for automation and optimisation of business processes.

The roles taken by different actors are changing. The operator focus is moving from managing the network technologies to managing the end-user service experience, which demands different management solutions and processes. In addition, future mobile services are provisioned in a network of companies interacting with each other and the process flows are very often defined by external parties. This calls for efficient and interoperable tools in the service creation, deployment and assurance areas. In the changing markets the processes have to be easy to modify when needed.

9.2.2 Generic Business Models

This section describes four generic business models that give practical insight into the modelling work described above.

- The *technology-based model* describes a situation where a company meets new technological challenges. In order to benefit from these challenges all the other business model components have to be defined accordingly.
- The *advertising-based model* uses contextual information, taking into account the possibility that a user is not willing to pay for the services.
- The third model describes a *mobile extension to existing business of a company*.
- The final model is based on *content delivery*.

The focus has been on the business model components that are most challenging to solve. All the models include the components needed for modelling a business. The value mechanisms are included and pointed out in the figures describing the roles needed for the business model and the service and financial flows between these roles.

The *value for the user* means benefit. In a business model, it is always a question about money and the source of revenue is indicated as *value as money*. The *value proposition* has to be given to the user and the role responsible for this task is also included in each figure. The *value creators* – all of them – have an impact on the value proposition. They all have clear roles and they all change somehow the proposition which could not be done without those roles. Finally there are several roles needed for the provisioning of the service but these roles do not change the value proposition and can been seen as *costs* for the service.

The following sections discuss also how mobile architecture should support business modelling, including the relationship between business roles and architectural components.

9.2.2.1 Business Model A: Technology-based Model

Technological developments drive the emergence of real innovative mobile services in the marketplace. New technologies generate changes in business models. Also the actors needed for provisioning the service may be different. Additionally, the use of new technology may provide a quite extraordinary value proposition to the customer compared to the competitive alternatives of the existing technology. If a company aims to benefit from the new technological possibilities, it has to reconsider the definition of the resources component in the business model. It also has to consider what kind of impact the changes in the resources component has on the other components of the business model. As an output of these analyses, the company defines and applies new business models in order to take a good competitive position in the marketplace.

This section describes a generic business model that is based on new technological challenges. The convergence of industries provides a possibility to sell and distribute some mobile applications to consumer and enterprise customers as software, either separately or as a part of the terminal or service package. Typical examples are general-purpose communication and collaboration applications, groupware applications or media players. This modelling perspective could thus be applied to, for example, the reference applications ContextWatcher (Section 7.2), Multimedia Infotainer (Section 7.5) and MobiCar (Section 7.9).

Users/customers

The applications in this model may be part of quite a new service bundle or add new features to existing services. The definition of the user benefits may be based on the experience of existing services. However, if the complete package or service offering is quite new, the precise dimensions of customer demand and required service performance have to be defined in order to describe the user component for the business model.

The customer may be a consumer or enterprise (acquiring a mobile service package for its employees). Segmentation is done for the entire package (e.g. working people, teenagers, media enthusiasts), where each segment needs to be sufficiently wide.

For consumers, ease of use and entertainment are key factors. For professional usage, productivity (using the applications for work processes) and manageability can be regarded as most important. In both cases the services should not be launched before technology is reliable enough and usability has reached the appropriate level.

Products/services

The product is defined as a seamlessly integrated solution: terminal (e.g. terminal intended for a certain segment like mobile workers, media enthusiasts), network and service applications. Integration with various Internet services (web, email etc.), and with office systems in enterprise systems, is very important.

The motivation behind this business model is to sell the entire new package, or the purchase of a particular terminal model. For example, an enterprise might purchase mobile terminals and associated connectivity and other services because of a mobile email solution, or a consumer could buy a new terminal because of a music application. The value proposition is based on practical or innovative usage possibilities.

Mobile applications based on technological development can integrate with legacy systems, such as legacy calendar systems, and offer additional functionality. They can provide context-based reminder and navigation services as well as personalised communication support within and between the groups. Through serving the individual user this application also supports different types of groups.

Earnings Logic

The application is sold by the service provider (providing the entire mobile service package including the terminal) to the consumer or enterprise customer. The customer either pays some extra for the application when acquiring the service package or the price is not separately visible. In the latter case, the application is part of an enhanced service package and motivates the customer to buy this combined product.

The service provides extra value to the customer and it is assumed that s/he is willing to pay for it. The division of roles, investments and costs in this specific business model is clear and there is no need for definition of metrics for actual service usage.

Resources

The definition of this business model starts from the resources component. It conceptualises the resources, capabilities and technologies that enable the company to provide the application to the customer. For the service provider, it is important to have the following resources: efficient, reliable server equipment operated with sufficient security, and a Customer Relationship Management system with a cost-effective helpdesk.

New technology like an enhanced calendar system is provided as part of the service offering. This technology by itself is owned by the companies designing and providing the related software (terminal software provider, services software provider). The service provider needs to have a certain control over the technology (e.g. defining requirements or features and ensuring interoperability) and the capability to utilise the technology to offer benefits to the users. It also needs to have capabilities for service lifecycle management.

Suppliers/actors

The *service provider* (see Figure 9.3) provides the entire service package including the terminal and the necessary applications. The application could also be sold separately by the *terminal software provider* or *terminal manufacturer*, but this option is not considered here. The general nature of the application and suitability for a large target group justifies this model.

In addition to providing the application-related service for its customers, a single company, acting as a complete service provider, is assumed to cover the roles of *network provider* and the *provider of trust services, identity services and group awareness services*.

For the value creation, needed roles are *terminal software provider* (terminal application software) and *service software provider* (service application). The terminal application software and the service software are provided by *terminal manufacturer* and *systems integrator*. These roles do not change the service promise and are as such included in the value net as costs.

Organisation/architecture

It is assumed that in this business model, a single company (e.g. a mobile operator) is acting as a complete service provider, responsible for the application services provider, connectivity provider and trust services, identity services and group services provider roles. Orchestration is focused on managing relationships with the terminal manufacturer companies and services software companies for creating a robust end-to-end offering for the customers. Standardisation (de facto) may be necessary for this mobile service. Joint marketing is useful in launching the

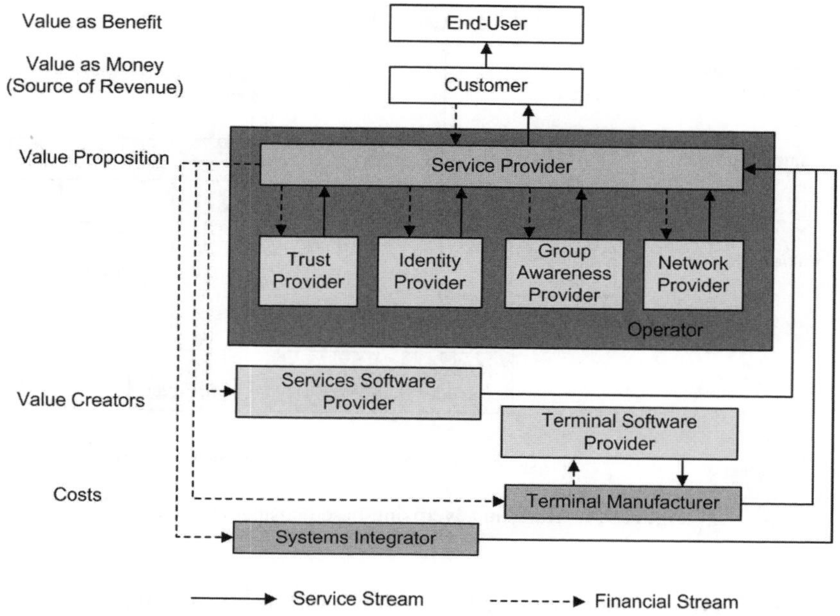

Figure 9.3 Roles in the technology-based business model.

services. The software provider could benefit from having the company (brand) name visible to the user.

A nodal company could be, for example, a company acting as the terminal software provider and services software provider. The governance of these relationships is based on contractual relationships between the actors and no additional definitions are needed.

Processes

Effective processes are needed for business performance. The service provider needs to have in place processes for Supplier Chain Management (SCM) and Customer Relationship Management (CRM). For suppliers, especially important is the management of terminal software releases and service software releases (testing, interoperability issues, delivery). These have to be designed early enough in order to have a common structure.

The CRM includes processes for software update deliveries and helpdesk operations. Since the creation of awareness among users of all the extra services in devices is slow and old routines are very strong, practical help in using the services is important.

9.2.2.2 Business Model B: Contextual Information Applying Advertising

Context-related services benefit from technological development and thus the resources component definitions are important also in this business model. However, the main challenge for services using context-related information – and thus also for the modelling task – is the

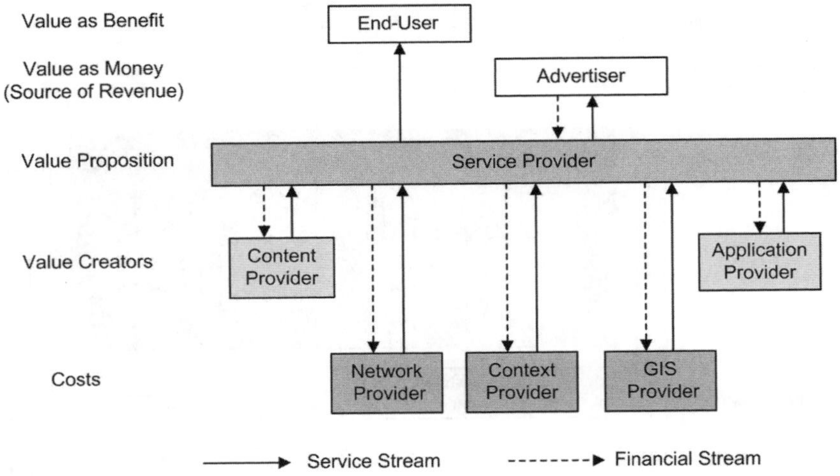

Figure 9.4 Roles in advertising-based business model.

earnings logic. Familiar revenue models have to be reconsidered since the user is not necessarily willing to pay for the services.

The design of an advertising-based business model (see, e.g., Figure 9.4) is related to the definitions made in the earnings logic component of the business model. This model is useful, for example, when the user is planning on how to spend leisure time, taking into account personal profiles and explicit preferences and relevant context information. The earnings logic in that service is based on a context-aware advertisement revenue model. From the reference applications of this book, FamilyMap (Section 7.7) and TimeGems (Section 7.8) could be mentioned as application examples.

Users/customers

In order to attract advertisers, the user base for the service needs to be large enough and reach a sufficient portion of potential customers of products of each advertiser. This may involve segmentation based on social groups and geography. One segment could be people interested in socialisation through online services, people with some well-defined, specific interest about (e.g. a certain type of movie or sport – the application would help you in finding buddies with a similar interest), and travellers, especially independent travellers. The paying customer in this model is not the user but the advertiser. The most promising advertiser categories include event organisers and ticketing agencies as well as retail stores and cafés.

Products/services

Interest-based customisation (including recommendations) already has a central role in many Internet-based services (e.g. the recommendation system of Amazon and the advertisement model used by Google). Having such services in a mobile, personalised, context-aware environment could prove to be valuable as well.

By matching the profile and status information of users with the services available, content providers and end-users are getting matched in such a way that end-users can easily enjoy the most appropriate services (e.g. an interesting exposition in a nearby museum, a concert, a movie, mobile content) in a given context. Content providers and advertisers are able to 'automatically' come in contact with valuable prospects.

The service should optimally utilise context and group functionality. There needs to be balance between the commercial exploitation of personal information (user preferences, location, agenda, etc.) and the privacy of end-users.

Earnings Logic

The revenue of a successful Internet company like Google is primarily based on contextual advertisements based on search terms given by the end-user or other type of service-specific content. A contextual advertisement revenue model may also be the most appropriate revenue model for a mobile service which offers suggestions on how to spend leisure time, taking into account personal profiles and explicit preferences, availability and other relevant context information of the users. However, in order to get enough revenue based on advertisements, a service needs a large customer base. Of course such advertising will only work if sufficient relevant context and personal information can actually be reliably gathered, interpreted and used against acceptable costs.

Charging for the advertisements can be implemented in the following way. When the service provider shows potential interesting content services to customers via the application, each time the end-user sees and/or selects such a content service the content provider pays a small fee to the service provider. Also for each time a content service is being 'consumed' via the leisure recommender service, an additional fee may be paid to the service provider (a sort of commission, such as for amount of tickets sold, amount of digital content sold).

Resources

The resources component in this business model definition includes several technological enablers. The service provider has to provide an intelligent matching system between the priorities of its customers (e.g. via preference profiles and usage of context information like location information) and available content services. This forms the basis of the value proposition of, for example, a recommender service. Therefore, enablers like personalisation technology, context-awareness technology, and, when offering group-based recommendations, group-awareness technologies also play important roles. Of course, the content as provided by the content providers can be regarded as important resources.

Suppliers/actors

The *advertiser* invests directly or indirectly in the service to promote its products and services and provides advertisement content to the service.

The *service provider or content aggregator* composes the service for the end-user. This party packages content from content providers and advertisers, matches the content to the personal profile of the user, and delivers it to the user. The *content provider* provides content other than advertisements that is included in the service offering. The *application provider* provides the application that the service provider needs in order to offer the service.

The *network provider* offers connectivity that enables the service provider to deliver the service to the end-user. The context provider provides information about the context of the mobile user to the service provider, such as location information. The *GIS provider* provides geographical services – needed to map the location of the user to the advertisement content – to the service provider.

Organisation/architecture

Due to its central position in the network, it is expected that the service provider will play an important role in organising the collective activities. The service provider acts as an intermediary between content providers, advertisers, the application provider, network providers and (potential) customers, and has to create and manage relationships with the diverse group of content providers and advertisers and has to intelligently market their services to the end-users of the network. Possibly, this role will also involve the highest risk, as the service provider needs to reach two critical masses to make the service work: it needs to acquire a sufficient number of customers to provide value to the advertisers, and it needs to mobilise a sufficient number of advertisers to have a viable service. When the service provider fulfils this role effectively, it may ultimately gain power over the other roles in the network and become the nodal actor in the value net.

So, the role of service provider will probably be crucial to the success of the business model, as the actor fulfilling this role will be practically in charge of the value net. It remains still to be seen which party will actually be the service provider. In theory, this could be a network provider, an Internet start-up company or an established company like Google that already has experience with targeted advertising. The role can also be fulfilled by a large advertising agency, but only if they have a sufficient volume of advertising content to be distributed.

Processes

The service provider needs to have processes in place for the matching and customisation process needed for matching the available content services to the right users of the service based on information and context-aware information like location and time. Then there need to be processes for getting the advertisements and a process for counting and collecting the advertisement fees.

9.2.2.3 Business Model C: Mobile Extension to an Existing Business

Mobile technologies and their potential in creating new forms of business dominate the description of this business model. The main challenges are related to existing business processes. These processes describe what actions the company is doing today, why they are done, what preconditions are needed and what the results of these actions are. Adding mobility features to existing services differentiates them from the competitors, increases the quality of service and gives new possibilities from which to get some benefit. However, the modification of a business model to include also these features (see, e.g., Figure 9.5) has to be done with care to avoid the loss of existing customers.

The possibility to digitise almost everything, including text, sound, speech, film, graphics, animation and music, increases the possibilities to benefit from mobility in many existing businesses. In order to simplify the business model description, a solution designed for sharing wellness information among selected individuals or groups is focused on. This model could thus be applied to, for example, the Wellness-aware Multimodal Gaming System presented in Section 7.6. This solution integrates fitness monitoring features and comparison capabilities into one package. It supports the user as well as family life by sharing wellness, environment and location information. Furthermore, it can be included in healthcare businesses as well as in small sport clubs. The following discussion will explain the underlying logic according to the business model components.

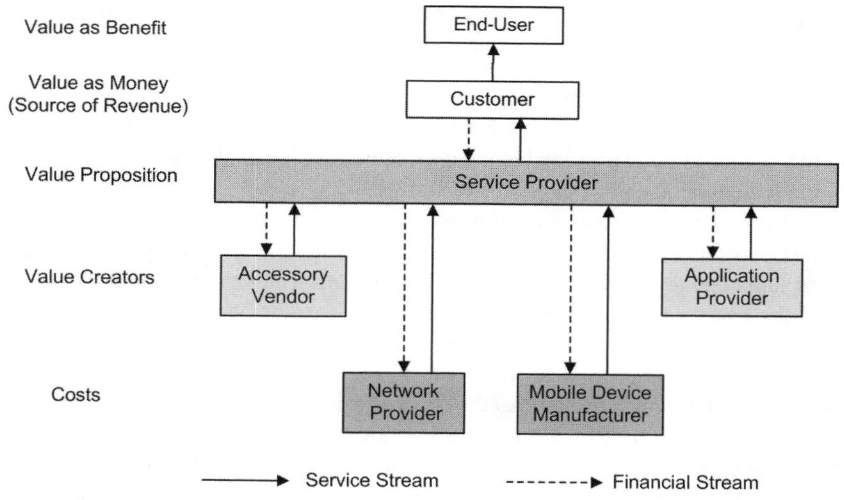

Figure 9.5 Mobile extension to existing business of a company.

Users/customers

In this model, users and customers are those that are already using and paying for the existing basic service, such as of a wellness or healthcare company. The customers may be individual consumers, enterprises buying services for their employees, hobby clubs acquiring services for their members, and government or municipal organisations providing services for citizens.

The definition of the user/customer component includes the issue of segmentation and deep, structured understanding of end-users and customers. Existing service providers (e.g. healthcare and training centres) are in the best positions in the market for having a sound knowledge of their customers, as they already have an established relationship with them and are most likely ready to profit from the new learning allowed by this service extension.

Products/services

Applying mobile solutions for enhancing the existing service adds usefulness and convenience for the end-users and efficiency for the customer. For example, a healthcare company can use mobile technology for appointment reservations. The information generated about training centre customers can be useful for planning proper health-related activities, increasing the service level and sophistication, augmenting the organisation knowledge base. When the service is provided to existing users/customers and it can be integrated with standard devices and platforms (mobile phones, PCs, web services, etc.), it will be distinctively convenient.

Members of a training centre will experience an improved and better-than-average service, because wellness monitoring and data sharing add something new and specific to the standard training service. In other words, the existing business leverages innovative mobile technologies to distinctively shape and differentiate its offering.

The definition of the service component in this business model may also be motivated by the reducing of costs. For a healthcare company or public healthcare institutions these new enabling technologies would mean an opportunity to offer the service to multiple prospects thanks to their access to an existing customer base.

Earnings Logic

The service is a mobile extension to an existing business. The earnings logic is thus strongly related to the service already provided to the user. However, it can also be defined as a premium service – the added value can be charged separately.

A similar logic is applied to costs and investments that are allocated according to business opportunities, risks and roles of different actors in the value net providing the service.

Users are supposed to be willing to pay for the services. The service is added to the existing service and the usage is paid at the same time. In the case of healthcare services, either the final user is willing to pay it as a private service or the service is included in the service of public healthcare providers.

Regarding costs, it is worth noting that the company providing the service to end-users could benefit from some economies of scope. These innovative services are included into existing offers according to a bundling logic; wellness monitoring is something already performed in a more traditional way at wellness centres and as such it just extends an existing activity. A healthcare company could enlarge its customer base without investing more in human resources.

Resources

Human and technological resources needed to put this service in place are likely to be most successfully aggregated by a specialist company acting as a service provider. For example, extending a healthcare or wellness service with an advanced solution requires a sound understanding of a vertical market typical of a specialist business; combining the service with other ones means also better access to complementary technological resources.

Using standard technologies (such as mass-market mobile terminals) makes it possible to utilise the innovations at acceptable cost and without disrupting the available technical infrastructure.

The extension of existing services with mobile features requires the capability to exploit and combine resources through organisational routines. These valuable capabilities are created and acquired over time with several actors. This calls for a strong coordination capability, as service provisioning requires information and management systems that combine the processes of each actor. The service quality has also to be constantly monitored.

Suppliers/actors

The *service provider* is any company or organisation that uses a mobile service extension of its existing activity, such as a healthcare company or a training centre.

The *application provider* is a technology company focused on designing and implementing the software solution; some or even most of the components needed could well be provided by third parties, with the application provider taking advantage, perhaps, of local market expertise, vertical knowledge, strong integration and innovation capabilities or other factors. The application can be provided as software or as a service. The latter option would make deployment easier for customer companies, whose main expertise is to provide vertical services like healthcare or wellness, not running network services.

The *accessory vendor* focuses on designing and implementing the hardware components related to the particular service, such as sensors for body information monitoring.

The *mobile device manufacturer* provides devices with advanced capabilities and standard interfaces. The *network provider* provides necessary connectivity services.

Organisation/architecture

The definition of this business model – especially for a healthcare company and public healthcare provider – has to pay attention to the service architecture, information and knowledge sharing between different actors. This basic requirement for capitalising the technology as part of the existing business architectures, service architectures and technological architectures calls for a high level of standardisation and interoperability. The company would like to keep the premises of existing services stable. However, the mobility provides characteristics that require dynamism also at the architectural and at the service provisioning levels.

Rules and responsibilities between different actors as well as privacy issues are of extreme importance in the case of healthcare services.

Processes

The biggest challenges in modelling the business as an extension to an existing business are met in the main processes of service provisioning. The company providing the service to the user has to modify at least partially its standard processes and procedures.

Modularity is a key concept in challenging the demand for customised healthcare services. Modular services, processes and knowledge architectures enable service providers to create greater service variety, introduce technologically improved or new products more rapidly, and lower costs of product creation and realisation. Modular architectures affect the processes inside the company as well as between different companies in value nets.

Similarly, the service provider will have to adapt itself as required by the integration of a technology-based business-to-business offer. In customer service this may require specific customer support, service level agreements with its key suppliers and careful management of new information flows. There might also be a need for a helpdesk-type of service provided for end-users.

9.2.2.4 Business Model D: Content Delivery-based Model

On the Internet and in different mobile services, it is quite typical to use a business model where a customer can buy some content by ordering it from a service provider either by making a separate order each time or by using a regular subscription. Digitisation is affording huge possibilities for this kind of business model when almost every type of content can be delivered in digital form.

Examples of existing services using content delivery-based business models are news services, music downloads and offers of ring tones. This model could also be used in some mobile applications and services where content is offered to users as a main service proposal or as part of some wider service package; the applications Multimedia Infotainer (Section 7.5) and TimeGems (Section 7.8) are good examples. Figure 9.6 shows a possible value network for the content delivery-based business model.

Users/customers

A content delivery-based business model can be used to sell very different kinds of information or entertainment. The business model itself does not set constraints, so segmentation has to be done according to the content itself. For example, financial information might be relevant for

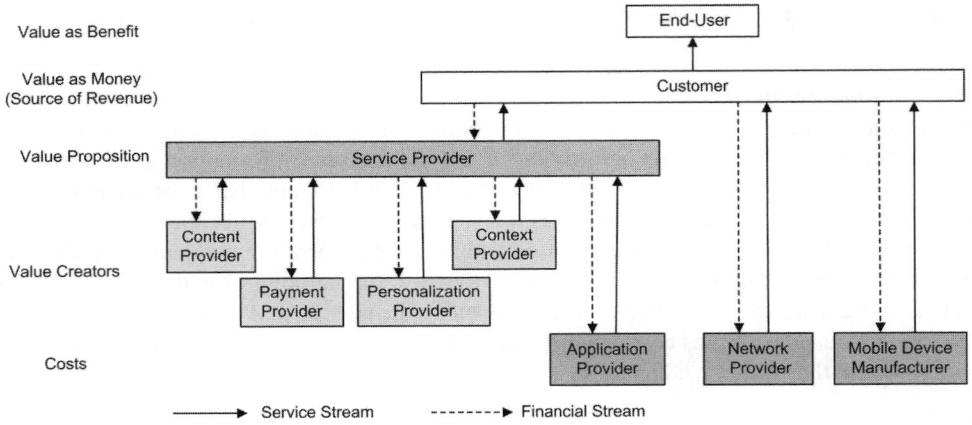

Figure 9.6 Content-based business model value network (example).

business people, and fans of a certain band are one potential segment for their music, videos, etc. Precise dimensions of customer demand have to be studied to be able to offer the right content to the right customers.

The user is quite often also the customer, especially in the consumer market. Thus, emotional and irrational needs are important in segmentation, and ease of use is essential. However, companies too can offer their employees access, for example to some business relevant information, in which case the company is the customer and the employees are end-users.

Products/services

A product or service promise is very much based on the offered content. The content can be news, music, videos, map, contact information, etc. There are lots of examples in the market already. The main point is that the content has to be relevant to the user and the user has to have a device through which she can use the content. In this context the devices are mainly mobile but they can also be a PC or a TV, or the service can support multimodality and multiple devices.

Technological enablers such as personalisation, context awareness, group awareness, privacy and trust, and multimodality have important roles in creating new distinctive service promises.

Earnings Logic

Earnings logic is very central in the content delivery-based business model. The customer can make, for example, a subscription to receive news automatically to her mobile device for a one-month period. Another possibility is to make each demand separately. The latter is extremely challenging for the service provider because the customer has to make a positive buying decision every time she wants to use the service. Ease of use and proper payment systems have a critical role here.

Consumers are used to free Internet access, so it is difficult to make them pay for mobile access to these services. Hence the service promise must be very attractive to be able to attract enough customers for a sustainable business. A content-based delivery model can be used, or there could be a combination of that delivery model and the business model supported by advertising. If part of the service is sponsored by advertising, the service can be offered to the user cheaper and thus it is easier to achieve larger user groups.

Resources

There has to be enough interesting content for each targeted segment to be able to make a sustainable business. There have to be appropriate end-user devices and access to a (mobile) network, as well as the availability of a proper payment system. Technology enablers offering added value in the mobile setting are personalisation and context awareness. There has to be a good balance between commercial exploitation of personal profiles and the privacy of end-users in order to avoid privacy and trust risks.

Suppliers/actors

Traditionally the *service provider* has quite often been the mobile operator, but separate service providers could also become more common; or, for example, the content provider could also take the role of a service provider in this business model.

The *content provider* has a central role because the actual content is what the end-users are paying for. Thus the power and importance of content providers in the value net are set to grow. *The payment system provider* has an important role since the customer pays to use the service. The payment solution has to be easy and convenient to use for all players in the value net, otherwise it might be an obstacle to service use. There also has to be a fair division of revenues between different actors in the value net. Content services do not currently use context information, but it might create added value to the service promise; for example, a user gets tourist information about places she is passing. The *context provider* role could be taken by the operator. Personalisation is especially important in mobile services, where displays are small and navigation difficult. The *personalisation provider* offers a service that makes it possible to use the end-user's interest profile.

The *application provider* supplies the application that the service provider needs in order to offer the service. The *mobile device manufacturer* provides terminals for using the service (directly to the customer, as in Figure 9.6, or through another route).

The *network provider* offers connectivity that enables the service provider to deliver services. In all mobile services the mobile network is essential, or the services cannot be used.

Organisation/architecture

The service provider orchestrates the whole service offering by acting as an intermediary between user/customer, content provider, payment system provider, technological enablers (personalisation and context information provider roles) and application provider.

The role of a service provider can be taken by a single company, operator or content provider. An example is Vodafone Live!. Vodafone has positioned itself to directly influence and profit from the customer's total wireless experience instead of simply providing a connection. Vodafone's role is to package, promote and sell the content, subscriptions and services offered by content companies. Vodafone also works closely with handset manufacturers. On the other hand, when customers are willing to pay for the content, the network itself does not bring so much added value. Thus the possibility for a content provider to become a leading actor in the value net grows if the network operator is not capable of taking on the service provider role.

Processes
The core process in a content delivery-based business model delivers interesting content to the users and also manages and shares money flows to the different actors in the value net.

In a changing environment, the viability of the business model calls for a common way to structure the processes. The processes between different actors are slow to change, and the delivery of interesting content requires more open value creation systems as well as open markets.

9.3 Marketplace Dynamics and Business Models

9.3.1 The Future Cannot Be Predicted

The mobile business sector is a turbulent environment [6]. To survive this turbulence, organisations must have the ability to develop rapidly new business models that fit changed circumstances. They have to take into account dynamics in the marketplace in terms of market offerings, customer demand, technologies and regulation [3, 4].

Scenario methods have been widely used in business research in general and in managerial practice. The scenarios encapsulate short-term challenges. Proper scenarios are more than informal projections of the future and should come with a certain warranty of quality. Scenarios should be internally consistent, plausible, and challenging enough to meet the intended purpose; and different scenarios should be clearly distinguishable. However, scenarios should not be seen as forecasts [3].

So, scenarios describe *plausible* future states, taking the present as a starting point. Extrapolating scenarios means that there is a *transition* from the current to a future state. The art of scenario building is to identify the relevant drivers of this transition.

The scenarios can be used as tools for further analysing the mobile technology-based generic business models. Success of a service in the market depends not only on the choice of a certain business model but also on the future environment in which the model is deployed. Some business models may have a better 'fit' with an assumed future environment than others. Also, a business model may be adapted or tuned to better fit an environment that seems unfavourable at first. Scenario methodology [30] is a powerful tool for thinking through the implications of business model choices. Rather than assuming a fixed (often implicitly assumed) 'official future', scenarios offer a range of possible outcomes used not as predictions or forecasts but more as 'wind tunnels' for plans or policies. The future cannot be predicted. However, using scenarios the uncertainty regarding the future marketplace can be harnessed, and business model choices can be evaluated and possibly adapted.

Four high-level scenarios are now examined to assess how the marketplace may develop. Subsequently the scenarios are used to discuss the impact of marketplace dynamics on the four generic business models.

9.3.2 Marketplace Dynamics: Scenarios for the Future Marketplace

This section describes some future marketplace scenarios. As the scenarios are generic, they could be used for testing the future viability of other types of business model. First, four high-level scenarios that are at the extremes of how the marketplace could develop are elaborated. Then, the two most interesting scenarios are selected and developed in more detail.

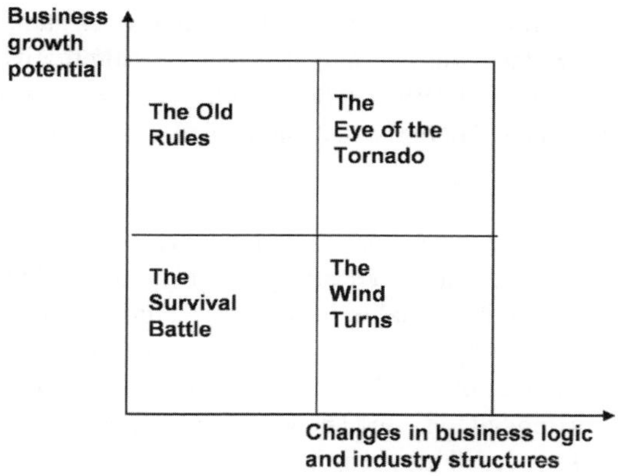

Figure 9.7 Macro level scenarios.

9.3.2.1 High-Level Scenarios

When developing future scenarios, forces that are both highly uncertain and have a big impact on the viability of business models should be taken into account. Two major forces meet both conditions: 'business growth potential' (the possibility for mobile services to take off in the future market); and 'changes in business logic and industry structures'. Four different scenarios provide unique combinations of extreme values in these two dimensions. The dimensions and the four scenarios derived from them are visualised in Figure 9.7.

- In 'Survival Battle', business growth potential and changes in business logic and industry structures are insignificant. Demand for services is decreasing and companies strive to defend their current position. In order to avoid unprofitable operations, they look for structural rearrangements and focus on rationalisation. Innovativeness is related to 'survival'.
- In 'Old Rules', business logic and industry structures still do not change, but business growth potential is significant. The gross domestic product (GDP) rates will be relatively high while inflation remains low. Domestic demand accelerates and consumer spending is continuing stable or even growing. Growth is sought mainly through traditional means – reconfiguring intra-industry arrangements and enhanced efficiency. Innovations are launched to expand the existing services, although investment activity is moderate. Traditional players and roles reassert their position in the marketplace in all industries.
- In 'Wind Turns', business logic and industry structures change. Interest rates will rise significantly. Rising fuel and labour costs will increase inflation, and unemployment rises. Industry structures are being crossed, driven by technological convergence. New players enter the market although growth potential remains low. The changes in consumption habits support demand. New companies develop new value-added services with new business models.
- In 'Eye of the Tornado', the growth of the global economy is rapid and it creates business growth potential. The changes in business logic and industry structures are significant. World trade is doing well, giving new possibilities. The networked and powerfully developing global economy and open technology standards build a new foundation for industry

consolidation and business reinvention. The triumph of new open standards provides an open playing field: easy to access new markets and easy to access alternative information sources. Innovations focus on new areas of services. Some of these make spectacular successes while others fail just as spectacularly. There is a constantly ageing but really well-off and culturally differentiated population.

Of the four scenarios described two are developed in more detail. In the moderately dynamic marketplace, changes occur frequently, but along roughly predictable and linear paths [8]. Typical for that market are relatively stable industry structures such that market boundaries are clear and the players (e.g. competitors, customers) are well known. The new services are added to the business in the context of the existing rules and routines. In this environment products and services are extended to include new functionalities and the actors form a vertical structure to provision the services. This scenario is described in the 'Old Rules' scenario.

In a very dynamic marketplace the change becomes nonlinear and even less predictable. Market boundaries are blurred, successful business models are unclear, and market actors are ambiguous and shifting [8]. The overall industry structure is unclear. In these markets development necessarily rely much less on existing knowledge and much more on rapidly created situation-specific new knowledge. Organisations rely on limited routines and they are looking for new ways to collaborate in modelling the business to meet the opportunities on the marketplace. This scenario is described in the 'Wind Turns' scenario.

Sections 9.3.2.2 and 9.3.2.3 discuss several environmental forces relevant for the two scenarios, focusing on the domains of economy, business, industry, society, users and technology. Indicators were sourced from several public sources, including characteristics of trends, reports from marketplace studies, published economic data by the European Central Bank, and academic studies. Each chosen indicator has to be assigned a value according to the two different scenario descriptions. The lists of indicators and values provides a basis to write narratives that describe the economic, business, industry, society, users and technology domains of the two scenarios in detail.

9.3.2.2 'Old Rules' Scenario

The economy is prospering and employment remains high. Inflation rates are assumed to remain low and economic growth continues at a steady level. One consequence of this prosperity is that shareholders are willing to invest more into companies, which increases the amount of capital available for companies in the marketplace. Unfortunately, only a small part of the capital is actually used to perform truly radical innovations based on mobile applications, as most companies are focused on simply extending their existing services and streamlining their existing working processes. Companies also seem reluctant to utilise the potential to digitise offline services and to bring them to the market. This lack of innovativeness explains the paradoxical phenomenon that the mobile data services market still shows little growth, although consumers generally have more money to spend.

When comparing today's mobile telecommunications industry to the industry in this scenario, one would notice little difference, if any. The same players are dominating the market as they are today. The position of these players, who are mainly mobile network operators and device manufacturers, is further strengthened following the large number of mergers and acquisitions that take place in the sector. Another contributing factor is the fact that convergence of mobile telecommunication businesses with other industries has ceased. The dominant players make some incremental innovations.

To remain competitive, device manufactures develop and produce more and more powerful terminals, while maintaining the current price level. Manufacturers are coping to keep the user interfaces at satisfactory levels while introducing more features and complexity. They stick to their existing products rather than introduce communication and interaction capabilities to entirely new types of device. Existing companies are enhancing capabilities of their voice, data and broadcast networks, and trying to find added value by using them together.

Applications and services can utilise multiple modalities to a higher extent. Personalisation and context awareness are slowly developing, while terminal and software vendors are adding corresponding features and existing service providers are enhancing their services by trying to utilise those features. Limiting personalisation and context to narrow application segments hinders wider acceptance. Services remain distinct. Service providers rather implement new features on their existing services than use services from other companies. Standard service interfaces are implemented to some extent but not deployed widely among distinct companies. Existing companies enhance their products to compete with disruptive technologies introduced by new entrants.

Users' active involvement in media and ICTs has been the hallmark of the Internet era and it was expected to be an equally strong driver for mobile and wireless applications and services. Nevertheless, the lack of innovative and open interfaces to include users in a constant and effective participation restrains their role to a limited interaction, mostly based on sophisticated but still restricted customisation and personalisation features. With few, big service providers consolidated in the market, users can rarely turn their content production or direct presence into new services platforms or open communication environments, as happened so remarkably with online communities, file sharing and blogging systems.

Companies differentiate by providing premium services. The improvement of solutions and services offers renewed room for old-fashioned but sound, well-proven business models based on direct payment from the users for what they consume. The thrust of the Internet fades, and even big online players differentiate more and more their services with very basic, free offers plus quality-driven, well targeted payment solutions. This sustains profits for the industry, even if they are not very distributed over a large range of players. The users' feelings about their freedom in the information and communication space are dampened by ever stronger security and privacy measures.

For the less favoured user segments, such as ageing or less educated people with lower incomes, there are no relevant improvements. The evolution of their participation in the information society is still blurred by a number of questions and contradictions.

9.3.2.3 'Wind Turns' Scenario

The economy is in what people on the street start calling a crisis, as interest rates, inflation and unemployment are on the rise. This makes shareholders less willing to invest in existing companies, and the only capital flowing into the industry comes from venture capitalists investing in start-up companies. Still, the number of innovations and the disruptiveness of these innovations are quite high, partly driven by increased digitisation and venture capital. Despite this, people are not willing to spend too much money on mobile data services, as they are worried about the economy and their financial status.

Players from various industries are benefiting from mobile services in business. The mobile industry is converging with other, traditionally separate industries. To mention a few, new entrants to the mobile services-based market include TV companies, music companies, retailers, open source players and phone manufacturers. As a result of these entries, competition in the market is increasing, threatening the position of traditionally dominant players. Furthermore, the radical innovations that are coming from the new entrants to the services dramatically change the rules of doing business in the sector, making resources and capabilities other than those evident today important. This further frustrates attempts of traditionally dominant players to strive for consolidation. The market is no longer dominated by a few actors, but new and small players are co-operating in strategic networks to offer innovative services.

New entrants present solutions that can function with less powerful terminals while still being attractive for customers. User interface development and innovations in that area are driving the competition. New types of device with embedded computation and interaction capabilities are introduced in the market, either by new companies or spin-offs of existing companies. There are innovative ways found to base all services on a single network technology, making other, possibly costly, networks useless.

Applications and services can utilise multiple devices to select the best device and modality at any moment. Required interactions between devices happen seamlessly. There are new ways found to collect personalisation and context information, and there are also new types of services that can enable use of this information by multiple businesses. There are new companies specialising in a certain service that satisfies a clear business requirement. The companies provide that service through well standardised and easy-to-deploy interfaces and business practices. Disruptive technologies are introduced by new companies.

Users' active involvement appears in the mobile and wireless arena. People's willingness to participate, driven by both technological and social reasons, produces a noticeable impact on the ways in which companies have dialogue with end-users. Mobile and wireless applications and services, managed on an expanding range of open platforms and services, are increasingly shaped by user dynamics.

Users have no or less money to spend on mobile-based services. However, a positive attitude towards novel ideas pushes the development of business models forward. These are characterised by highly innovative commercial proposals and complex and distributed business-to-business value networks enabling a rich offer of new services.

In addition, ethical and political aspirations towards a more inclusive information society push innovations. People with disabilities or elderly are recognised as niches of promising businesses; forward-looking entrepreneurs look at them to build highly personalised, context-aware mobile and wireless applications and services.

Security and privacy measures are still strong but don't produce uniform effects. The public debate over what is legal, ethical or sensible is active. Commercial companies, open source groups and consumer groups stand behind different and competing solutions. It is difficult to achieve a unique approach, which in turn gives opportunities to innovators in the field.

9.4 The Impact of Marketplace Dynamics on Business Modelling

The viability of the four generic business models presented in Section 9.2.2 can be considered in different future environments. In Section 9.3.2, different scenarios give an insight into the relationship between business modelling and future mobile applications. The ‘Old Rules’ and ‘Wind Turns’ scenarios are the two extremes and are as such the most descriptive. In this section, the four generic business models and their different components are analysed and a number of recommendations regarding the design of business models for mobile applications are given.

These business models are generic in the sense that they do not describe a particular service offering. Rather they show different ways of creating and capturing value from utilising ‘mobility’ and mobile technology. Collectively they provide a set of relevant business model alternatives for new mobile services and the various business model components. In addition, they provide practical insight into the business model design work. The attention in this section is focused on those components that are the most challenging to solve.

9.4.1 Wind Tunnelling Approach

The aim of the analysis is to understand how the four models behave in different future environments as described in the scenarios. The models are described in terms of design choices for each component. The scenarios are shaped by the outcomes of the scenario indicators as described in the previous section. The impact of the indicator outcomes on the business design choices guides the following considerations.

The analysis provides insight into design choices which can be regarded as robust – valid in multiple future scenarios. These may take the form of ‘future proof’ recommendations or design guidelines for business modelling. If certain design choices are found not to be robust, they may be changed in order to provide more carefully tuned business models that fit certain future scenarios in new environments.

9.4.2 Challenges for the Technology-based Business Model in Two Extreme Scenarios

The technology-based business model considers how the new technology can be leveraged in viable services. The way the technology is absorbed into new services is quite different in the two scenarios, as can be seen in Table 9.3.

The technology-based model can be regarded as viable in both scenarios. In the ‘Old Rules’ scenario, it is supported by users who are ready to pay for easy-to-use, secure service offerings. Mobile operators have a strong position to aggregate and offer such services. In the ‘Wind Turns’ scenario, more innovative services would be created, with multiple parties contributing to the complete offering using standardised interfaces. The leading actor is less clear in this case but would likely be one with an innovative technical solution and forming a value net around supplying that solution.

In the ‘Old Rules’ scenario, lack of innovativeness may hinder development of attractive enough technology. Here, the service provider should rely on its ability to supply complete, secure solutions and its strong position towards the suppliers: terminal manufacturers and software providers. It should utilise its current customer relationships and enhance current product offerings with enhanced features and try to get new customers with the current offering.

However, options exist also for established actors from device manufacturing, consumer electronics or software provisioning. For example, applications that provide for a seamless integration between PC and mobile device, and even CE devices, could be very fruitful. Still, such a solution would require some agreed forms of standardisation by the involved actors.

Table 9.3 The challenges for the 'technology based business model' in the two scenarios

Business model component	'Old Rules' scenario	'Wind Turns' scenario
Users/customers component	Firms focus on extending existing services for existing users by slightly enhancing the user experience, and widening the customer segment.	Users seek for new services for less money but instead putting in their own effort. Users are eager to use other than established sources of secure products.
Products/services component	Services and applications are based on the existing ones with small enhancements. Applications are sold as part of a package.	Use of more advanced technologies, e.g. context awareness and personalisation. Applications may be sold separately, with multiple parties contributing, using standardised interfaces.
Earnings logic component	Users are ready to pay and pay more for an enhanced service.	Providers cannot rely on users to continue to pay for the service.
Suppliers/actors component	Service provider has a strong position towards the suppliers: terminal manufacturers, software providers.	A service provider having an innovative solution builds the partner network around that innovation. Standard service interfaces are utilised to allow flexibility in choosing suppliers and partners. Utilising open source.
Organisation/ architecture component	A single company acts as a complete service provider, controls the suppliers.	Cooperation is based on standard, easy to deploy interfaces. Uncertain which player would have the role of service provider, and whether its power position would be sufficient for the service to thrive.

In the 'Wind Turns' scenario, a service provider should have an innovative solution and base its business on that. In particular, it should build the partner network around the innovation. Standard service interfaces should be utilised in co-operation to large extent, to allow flexibility in choosing suppliers and partners. The service provider could consider utilising open source. The earnings logic would also have to be reconsidered in this scenario. Other sources than paying customers could be required, such as advertising or cross-subsidies. In the latter case the revenues would come from selling or providing complementary hardware or services.

A situation in which the governance between service provider and systems integrator and terminal manufacturer is on a more equal basis could be more realistic in this scenario.

9.4.3 Challenges for the Advertising-based Model in Two Extreme Scenarios

The advertising-based business model is driven by the idea that end-users may not be willing to pay (enough) for innovative mobile services. The challenge is therefore in the earnings logic component to find new revenue streams (i.e. advertising) which support a sustainable business. It is therefore no surprise that the earnings logic component shows the most significant differences when this business model is compared in the two scenarios (Table 9.4).

Table 9.4 The challenges for the 'advertising based business model' in the two scenarios

Business model component	'Old Rules' scenario	'Wind Turns' scenario
Users/customers component/Earnings logic component	Users are assumed to have some spending capability, which relaxes the need for advertisements as a primary source of revenues.	Sponsoring and subsidies will be much needed revenue sources.
Resources component	Context awareness and group awareness technology is not developing enough, thereby hindering the application of the business model.	The technology will be sufficiently advanced to enable the service and its revenue source.
Suppliers/actors component	The network operator will probably fulfil the service provider role.	A new entrant will probably fulfil the service provider role.
Organisation/ architecture component	Complexity of orchestration will be somewhat lower.	Complexity of orchestration in the value network will be high because of differences in the level of standardisation.

The 'Wind Turns' scenario can be regarded as the more favourable for this business model. In this scenario the underlying earnings logic is more necessary and the required technology will be available.

It has to be noted that advertisements are just one source of revenue for mobile content services. If users have enough money to spend, as assumed in the 'Old Rules' scenario, there is less need to focus on advertising money and companies may keep the user in mind as the primary revenue source. If, however, users are similar to those in the 'Wind Turns' scenario, and do not have so much money to spend, advertising or sponsoring becomes important to companies that still want to make money with mobile services.

Still, in the 'Old Rules' scenario, established players like broadcasters and media players, with ample experience in advertising models, could team up with device manufacturers to provide advertising-based mobile content services (e.g. mobile TV). Such services would not provide the full potential of contextual advertising, but would be closer to traditional mass-market advertising. In combination with obvious economies of scale, providing such advertising sponsored services could be quite profitable.

Most current initiatives and trials with contextual mobile advertising are small in size and run by new players rather than by established players. This again suggests that a contextual mobile advertising business model, so familiar for the fixed Internet, is more likely to prosper in the 'Wind Turns' scenario. Success depends on the degree of personalisation and contextualisation that can be realised, as was argued earlier. It is dependent on the availability of advanced technologies, such as context awareness and group awareness, which are available in the 'Wind Turns' scenario. Large established players prospering in the 'Old Rules' scenario could buy smaller players to create some footprint in new lines of business.

The introduction of advertising as a revenue source in the earnings logic component has significant impact on other business model components as well. In the customer and product/ service component, not only does the end-user of the service need to be satisfied but the advertiser also. Success will depend on a careful balance between the provided customer value and the acceptability of advertising on the one hand, and the value for the advertiser on the other hand. Acceptability may depend on the relevance and unobtrusiveness of the advertisements. Value for the advertiser depends on the size of the provider's customer base, the richness of customer profiles and the targeting capabilities of the provider. These may be better in the 'Wind Turns' scenario.

The extension of the mobile value net with an advertising chain makes orchestration more complicated, especially in the 'Wind Turns' scenario. New modes of co-operation and governance are required in order to safeguard the activities in the value network.

Privacy problems may be problematic in both scenarios – not a surprise considering the potential intrusive character of mobile advertising. A suggestion regarding advertising could be that clear guidelines for a transparent application of mobile advertising are established by the industry. Such guidelines would, for example, describe how customers could opt-in or opt-out of advertising content, how providers would safeguard sensitive customer information or how much advertising a customer might expect. A similar recommendation can be given to regulatory and legal institutions, to make privacy legislation as clear as possible in order to aid development of the mobile advertising market.

9.4.4 Challenges for the Mobile-extension Model in Two Extreme Scenarios

The third generic business model adds 'mobility' to existing service offerings and solutions to existing user segments. The main challenges are found in the product/service component and the processes component. The key in this modelling task is to adapt new mobile possibilities to business and consider what kind of ties the service promises, user segments, processes, etc., have. Value creation is based on leveraging innovative services and efficient processes with existing customers.

The really generic nature of the business model under discussion (the extension of an existing business in the mobile domain) and its relationship with the specific industry at hand makes it very difficult to draw a final comparison (Table 9.5).

Nevertheless, it should be noted that the different indicators, which indeed play even in opposite directions, still sustain the business model potential in both scenarios, but with really remarkable differences in the impact or momentum that the model could have, both in the short and in the long term. The lack of standardisation – especially in terms of public, open standards – and interoperability hinder the model's potential in the 'Old Rules' scenario, while the same technology indicators, being pointed exactly in the opposite direction in the 'Wind Turns' world, provide the best premises for the entrance of new players in the market.

In the 'Old Rules' scenario, consumers' good spending capability gives room for premium extensions, but other factors (privacy and security measures, persistent divides, etc.) may limit their impact. In 'Wind Turns', users' eagerness to actively participate and the necessity to build new, open, multiple organisation models appear very much in line with the model principle of extending and renovating existing businesses.

The 'Old Rules' and the 'Wind Turns' scenarios depict quite different futures for companies and organisations interested in 'mobilising' their businesses and public services – in other words, adding mobile and wireless extensions to their day-to-day existing offers. In short, the main point is about who is taking the lead – in terms of increasing costs, investments and risks, but potential benefits and profits as well.

Table 9.5 The challenges for the 'mobile extension business model' in the two scenarios

Business model component	'Old Rules' scenario	'Wind Turns' scenario
Users/customers component	Users are open to mobile extensions offered by long-standing, big brand names. Consumers' good spending capability gives some room to premium extensions.	Users are more open to experiment with innovative products and services, setting new trends; new ways of doing things spread also for the new that they convey, beyond practical benefits.
Products/services component	Adding mobile dimensions extends existing products and services. Intention is to increase productivity in the business context and enhance customer service, access and experience.	Products and services are entirely redesigned as some dynamic players from differentiated industries embrace mobile technologies as an opportunity to radically innovate their offerings.
Earnings logic component	Dominant players take most of the profit out of mobile extensions, which would be likely marketed as premium, paid services.	The various industry players that engage with mobile extensions use differentiated earnings logic. Alternative to direct payment from the users, such as the ones based on interactive advertising and marketing, are commonly used.
Suppliers/actors component	Telecommunication service providers tightly manage suppliers; as some services go actually well there could be interesting opportunities for well-integrated, established suppliers coming from other industries.	Turmoil in the industry and disruptive technology innovations spur a lot of changes on the supply side, which shifts towards a more fragmented and dynamic structure.
Organisation/ architecture component	Current telecommunication-centric orchestration model wins. Especially operators are still in the driving seat; they are the ones coordinating existing products and services and extensions in the mobile domain.	Players from industries other than telecommunications and possibly even ICT take the lead in developing new businesses and acting as service providers.

Of course mobile and wireless solutions do not necessarily require expensive always-on mobile connectivity. As the boom in personal in-car navigation has shown, mobile services may be quite well based on devices with advanced capabilities such as large storage, processing power and GPS connection. This already provides opportunities for existing and new players to utilise capable mobile devices to 'mobilise' existing business.

In the 'Old Rules' scenario, for services utilising always-on connectivity, operators will likely be most prevalent, especially for service provisioning strategies and operations, which means that all the other key players in the network, namely manufacturers and all the other vendors and providers (of software platforms, custom applications, integration, etc.) will have to follow their choices. As industries and public sectors want to experiment with B3G technologies, the realistic option will be to turn to the operator as the only actor capable of covering all the needed critical roles. Of course one could question that operators would be able to provide these niche services, but actually some strategies could be of help here. For example, operators could establish separate but controlled entities that would more closely co-operate with other industries on specific new initiatives and offers.

The services realised under these assumptions might likely have some clear strengths and weaknesses. They will probably be managed in a sound way, have a limited but well-defined scope; possibly they could generate good margins – if not big ones; on the other hand, it could be difficult for them to become established.

As a consequence, extending an existing business into the mobile and wireless market would be a path to be followed with caution and somewhat contained expectations. The main point for a business would be engaging in the right consulting and outsourcing process, maybe with more chance of success than nowadays – a growing economy should help both businesses and consumers in developing and absorbing a certain degree of innovation.

In the 'Wind Turns' scenario, other industry leaders would jump in the driving seat, not being telecoms or ICT operatorst. Here B3G developments made of interoperable standards could be the driver. In the 'Wind Turns' scenario, one would have a marketplace of modular, easy-to-combine layers and components, so making mobile applications and services should be at least as manageable as most fixed Internet solutions are today. This means a lot of opportunities for start-ups and innovators. 'Mobilising' an existing business could be the play of people with smart ideas about new ways of doing traditional things, including the more mundane. There are many examples in the health and wellness domain as well as domestic arenas, from wirelessly sharing a recipe between some family members to the support of daily non-professional sport practice.

The 'mobile extension' would need to be well thought out and delivered by the business line (or the service line of a public body); in a sense, technology might be stolen from the hands of the technologists. As such, the change might be unstable, contrasted, and of limited impact in the end. In order for the mobile and wireless systems really to explode, a stronger convergence of traditional and new players is likely necessary.

9.4.5 Challenges for the Content-delivery Model in Two Extreme Scenarios

Digital-content delivery models are becoming increasingly important fuelled by digitisation of almost all content (text, images, music, video). Several reference applications presented in Chapter 7 could utilise such a business model to offer content as a separate business or as part of a wider service package. The fourth generic business model therefore describes a business model that is based on content delivery. The main challenge is found in the orchestration component. Value creation is based on leveraging digitisation and e-value chain co-operation.

Both scenarios are favourable for this model (Table 9.6). The content offerings as well as the service providers in the two scenarios differ, however.

The user has more money to spend in the 'Old Rules' scenario and the economy of scale in content production is important. However, the 'Wind Turns' scenario can provide a more tailored service promise for the user and thus the user can benefit according to individual taste and liking. The customer groups are smaller and thus business in this scenario may be more uncertain.

The content-delivery business model is used to sell very different kinds of information or entertainment. The services may be an enlargement of existing content services in digital and mobile form (the 'Old Rules' scenario) or radical innovations (the 'Wind Turns' scenario). In

Table 9.6 The challenges for the 'content delivery business model' in the two scenarios

Business model component	'Old Rules' scenario	'Wind Turns' scenario
Users/customers component	Firms focus on extending existing services that provide possibilities to reach a large number of customers.	The market is very fragmented and smaller customer groups are looking for innovative services.
Products/services component	Content services provided as mobile are part of a whole service package. Services are more or less the same kind as we see on the market today.	Innovative ways to use mobility to provide content services, even as a combination of several brands.
Earnings logic component	Shareholders are giving support to the large companies to invest in growth.	The money for new entrants and investments in innovative services comes from venture capitalists.
Suppliers/actors component	Content services provided by traditionally dominating players.	Content services provided by a value net composition, new entrants changing the rules.
Organisation/ architecture component	Still a lot of closed systems from standardisation point of view. Existing players dominating.	Standards for content creation and delivery are more open. It provides support for monitoring the service provisioning and thus sharing the money flow.

both cases the delay in information delivery decreases by the use of mobile connections that are independent of location. This feature provides service promise to companies that want to provide content to users, or, in case of company-specific content, to the employee. The living environment of the user changes with increasing technology/device integration (e.g. e-Learning while sitting in the train – content can be downloaded 'into your pocket').

The usage of content services, and thus the interest of companies to invest in content-based business models, depends on device development. The more capable the devices are to provide mobility features (context, personalisation, battery size, screen size), the more potential there will be in the development of innovative services in the 'Wind Turns' scenario, thus providing business possibilities. In the 'Old Rules' scenario, the market of mobile service-based content will develop slowly – phones with services cost too much. However, the availability of proper devices for accessing the content (e.g. mobile TV) is increasing and the users are changing their devices to new ones fast.

Consumers are used to paying for mobile connections. However, they are also used to free Internet services and thus it is difficult to make them pay for getting the content by using mobile devices. The usage of mobile devices increases and the user could benefit from getting the services she uses partly in digital form. The price of the service to the user can be included in the price of the main product. If new entrants are to provide the innovative content services, the service must be very attractive and generic enough to be able to receive enough customers for a sustainable business. If the user has to make a decision about using the service each time, the likelihood of usage decreases.

The earnings logic component is relatively simple in the 'Old Rules' scenario – it is part of the main services. In the 'Wind Turns' scenario it will be complicated to solve. Usefulness, ease of use and proper payment systems all have a critical role, but it is not clear who takes these roles. For new small companies it is necessary to earn enough from the narrow niches they are concentrating on, but the payment systems need a mass market in order to be productive. Content distribution sponsored by advertising is a well-known model (e.g. newspaper, news sites). This could also prove to be important for mobile content delivery.

Suppliers/actors co-operate in the 'Old Rules' scenario. In the 'Wind Turns' scenario the number of service concepts and content providers increases, value net type of collaborating increases and, in order to provision the services, the call for standardisation also increases. The Darwinist principles give guidelines for who will survive in these networks as content providers. Orchestration is the key in the development of this market. Players from content industry, consumer electronics, device manufacturers and enabling technology providers (e.g. DRM, payment) need to agree on standards for managed content services that provide users with flexible usage options.

9.4.6 The Impact of Scenario Context on Business Modelling

The two described environmental scenarios provide opportunities for the four generic business models. In the 'Old Rules' scenario the 'new services' will mostly be enhancements of current offerings, whereas the 'Wind Turns' scenario would support more 'radical' innovations. In both cases the increase in mobile services will benefit all the industries involved: the ICT actors providing services gain new business, and the businesses using mobile services may improve productivity. Consumers may receive a better mobile experience.

The 'Wind Turns' scenario supports business models providing mobile services that are carefully tuned to specific business or user segments. This is essentially due to the combination of more advanced technology and standardisation that enable stronger propositions (mobile-extension model, technology-driven model), new revenue sources like advertising (advertising-based model) and new forms of co-operation and efficient processes (content-delivery model). In the 'Wind Turns' scenario there would be a fragmented marketplace of modular, easy-to-combine layers and components, which implies opportunities for autonomous businesses to provide innovative applications as separate products. Spending capacity is lower in this scenario, but this may be compensated by other revenue sources such as advertising. Users' eagerness to actively participate in the service provisioning is important here as well.

It is interesting to note that, although services may be targeted to smaller segments, some of the involved 'components' (e.g. payment systems) will be successful only if they can reach sufficient scale; i.e. they need to be used by many providers and services.

In the 'Old Rules' scenario, one expects *packaged* selling of innovative applications and services, based on new technology that mainly extends current features of applications and services. The applications would be more general in nature, more secure and reliable and provisioned to larger customer segments, thereby benefiting from economies of scale. Providers would be the established players operating in much the same way they are used to. Businesses will adopt the enhanced mobile services if they increase productivity and support more efficient processes.

New revenue sources, other than paying customers, may be required to obtain viable business models for mobile technology-enabled services, especially in the 'Wind Turns' scenario. Advertising coupled to services and sponsoring of services is well known from the 'fixed' Internet, but they provide important opportunities for supporting mobile services as well.

Of course, conditions relating to consumer acceptability and advertiser value have to be met. The typical mobile technology enablers (personalisation, context awareness and profile management) can be of help here to specifically target customers with advertisements in an unobtrusive way. The industry, possibly together with regulators and consumer groups, could provide transparent guidelines for mobile advertising that help support consumer acceptability.

For advertising to work as a new revenue source in the 'Old Rules' scenario, established players from media and device manufacturers could set up a strategic partnership to provide, for example, advertising-sponsored mobile TV. Mobile operators would have to rethink their traditional revenue models, which are based on paying customers. Also mobile operators could adopt advertising.

All of the described business models involve more than just the traditional mobile services value chain. Whether it is a mobile extension to an existing business, an advertising model or a digital content-delivery model, multiple actors from different backgrounds need to co-operate in a value net. The essential difference between the two scenarios is found in who takes the lead. This difference very much determines the product/service component of the business model as well as the organisational components such as orchestration.

In the 'Old Rules' scenario the traditional players still lead. They provide extensions and enhancements to their current offerings in much the same way as they do now, building on current hierarchical governance forms to control the provisioning network. For companies providing enabling technologies or for SMEs developing solutions, the main guideline is to look for strategic partnerships to build and keep lasting supply relationships with established players in the mobile services field. This would mean partnerships first with operators but manufacturers as well. In short, the service developer positioning in this case is all about being a very close and trusted partner of some dominant or strongly emerging mobile player.

In the 'Wind Turns' scenario, new players rather than the traditional mobile players are in the lead. For example, we have seen that the service provider offering mobile extension types of service will most likely be a party from the business line rather than the operator. These parties bring their domain knowledge to create tailor-made solutions for specific customer segments. Services can be created from standardised, easy-to-combine layers and components.

In the 'Wind Turns' scenario, the technology providers and service developers could market their products or services as separate, easy-to-integrate products. The opportunity is to embrace public standards and interoperable technologies. Service developers should be prepared to co-operate in extended articulated value networks and promote their positioning as a stand-alone, autonomous business capable of partnering in an easy and efficient way with everyone that fits the model. Their main strategy could be going directly to vertical players; i.e. leaders in the various industries that are supposed to 'mobilise' their business. Operators may still provide certain enabling services, such as payments or authentication.

In the 'Wind Turns' scenario, new forms of governance are required to control the complex value nets. Rather than hierarchical structures, more open network forms of governance are required to support this horizontal modularised marketplace.

9.5 The Social Impact of Future Mobile Services

Mobile and wireless technologies do not develop in a void. Society at large, in all of its economic, ethical and cultural facets, is affected by them and vice versa. For this reason, the societal impact of future mobile services is analysed in the following pages, by focusing on some exemplary cases. This review provides a contribution to the business modelling discussion, with special regard to the four generic business models presented in Section 9.2.2. In doing so, the socioeconomic analysis of technology is taken as the privileged point of view; more precisely, this refers to the broader approach in which socioeconomics is understood as a meta-discipline that pursues a better understanding of the social and economic impacts of any product or service offering, new technology or market intervention [15].

As for the analysis of the cases at hand, the investigation is centred on the potential impact of a number of ideas about mobile applications and services on existing and emerging issues in the society and the economy as a whole.

Starting from the technology-based business model, it is important to remember that in this case the driver is given by the resources component, since the model describes a situation in which a company deals with technological change and innovation. The socioeconomic implications of the technology-based business model are discussed having in mind the example of applications founded on context-awareness mechanisms. Generally speaking, the major social implications of context awareness seem related to their role in strengthening already existing social ties, as they would act as potentially strong enhancement for the so-called 'connected presence' enabled by mobile technologies [17]. This is valid not only for consumer-oriented instances. Context-aware applications could help workplace mobility by sharing status information within work-related groups, especially with remote or commuting workers. Moreover, the usage of this type of application could encourage people who are at risk of exclusion to engage in a wider use of ICT, given its potential in reinforcing those same social ties – provided that economic and cultural barriers can be lowered.

In designing these solutions, developers and service providers should take into account that it would be sensible to include specific options to allow users deciding which elements of context should be shared with family members, friends and colleagues, in order to avoid conflicts regarding the separation of work and personal life, and maintaining privacy and trust. Furthermore, one should look at usability issues for disabled and elderly people. Policy experts and decision makers could play an important role here: clear rules for exposing intimate context information have to be defined, in order to prevent potential cases of misuse. Chapter 10 discusses these aspects in relation (for example) to personal data and privacy protection.

The second generic business model deals with users' unwillingness to pay for online services and therefore the main challenge is defining the earnings logic component. In this case, value creation is based on innovative services supported by new revenue sources, such as advertising.

Social media applications, like those aimed at group management, are very relevant here. By addressing, for instance, group interactions for organising leisure-time activities, one could exploit their potential as an innovative promotional channel. These applications, if deployed with this specific model, could also help lower-income user segments to have more access to mobile applications and services (as they would be sponsored by advertising money).

The third generic business model is centred on the idea of adding a 'mobility' dimension to existing service offerings and solutions. This model demonstrates the almost endless possibilities that private companies and public organisations could tap if they 'mobilise' their existing business and activities. One possible instance could be related to the case of an application devoted to health monitoring, in which healthcare, fitness and sport organisations would directly offer their services on a mobile device.

Several positive effects might emerge for both end-users and businesses or professional users of similar applications. As the appreciation of a healthy physical condition increases, as well as the related demand for information and monitoring, many people should appreciate the benefits of mobile technologies in this field.

Yet, there might be some downsides: in fact, personal healthcare applications bring a new wave of risks and worries. As previously noted about context information, participating organisations have to deal here with very sensitive data, whose protection is mandatory; yet, the lack of terms of comparison and empirical cases makes it difficult to determine exactly in which practical and legal ways this is supposed to be done.

In addition, taking the ICT inclusion perspective, there are other issues. The elderly and people with disabilities are indeed a very promising market, but they could be easily left behind if applications and services are too difficult to use or rely only on high-end devices or do not support backward compatibility with older, more low-end ones. As far as policies are concerned, it is worth underlining that personal context is also an area in which standardisation and openness are of strategic relevance, as public and private healthcare are too precious to be left only in the hands of pure market and technology dynamics.

Finally, the fourth generic business model is based on content delivery. Annotation and mobile location-based applications are considered most relevant here for the discussion. The most pressing point about similar services is that, as they let users share and comment freely on commercial content by using location and other context technologies, they raise the now much-debated issue of user-generated content to a new degree of complexity. 'Mobilising' user-generated content and making it context-aware could make it much more relevant and precise than today, as people would have better and more frequent opportunities to add a comment, share a piece of advice, or rate something on the spot as interesting or not.

On the other hand, there are some new phenomena that might be difficult to understand, and some challenges too. Relying on other people's published judgments in order to consider or not a certain place suitable for a stop-by is something new for mobile users. People might want to be able to get clues about the trust levels that they could concede to other users or to commercial information providers. Looking at the traditional media and even at the Internet, it is evident that building trust on a certain source is a difficult and sometimes lengthy process.

Policymakers and influencers of industry strategies might want to worry about the fact that these applications should be actually reachable by the widest sectors of the society. Poorer families, the elderly or people with disabilities have many information needs, sometimes stronger than average. Parents with disabilities might find concrete support in a context-aware content-based application that steer them around a city – provided that it does not require too costly devices and only standard keyboard and display-based interaction modalities.

To recap, mobile and wireless technologies, and the applications that embody them, are giving shape to a paradigm change in which all society, including businesses and public actors, not only communicate with their counterparts but manage ever richer and larger streams of data – including data that is manipulated by the users themselves. This is a far-reaching phenomenon whose importance has not been overlooked. Still, the specific socioeconomic implications are very much dependent on the concrete ways in which a certain business model is adopted and shaped in offering a certain product or service. More empirical studies are needed here – adding yet another dimension to the methodological challenges of combining user, business and technology activities in the same R&D framework.

9.6 Conclusions

This chapter has considered some business models for new mobile applications and services. It has described a business model framework and applied the framework to four generic models. These followed different modelling perspectives in order to provide new insights into how to define the components needed for viable business models. Marketplace dynamics analyses described as environmental scenarios provide opportunities and challenges for the modelling task. Parallel work in user-centric research and technological development gives added value to the modelling process.

Analyses of the behaviour of economic, technological, industry and social indicators have provided descriptions of two future environments. The variation of these indicators has impact on the four generic business models in these environments. Mobile services also have socioeconomic implications.

The managerial perspective in modelling the business is crucial. The business modelling aspects are thus synthesised to recommendations on what to take into account from the managerial perspective in running a company.

According to the research results presented in this book, some methodological topics are important for a manager to consider when developing business models.

1. Use the Component-Based Business Model Framework
The proposed framework for business models puts the essential parts of a business together. The four generic business models include different design perspectives and give insight into the relative importance of each component and the perspective taken.

2. Analyse Contemporary Marketplace Dynamics
Current marketplace dynamics have relevant impact on the managerial considerations. Assumptions of its impact on the company have to be made in designing a business model for a company or network of companies.

3. Use Future Scenarios to Assess Business Model Robustness
The viability of the business models should be assessed in possible future environments. As the future is hard to predict, the scenarios that describe what could happen in the marketplace increase the understanding of the modelling task. Applying the developed business models to the future scenarios can provide valuable insight into the robustness of design choices. The business model can be altered in order to fit to the different environments and to adapt to unanticipated changes in the marketplace.

Besides these methodological recommendations on how managers should develop new business models, the theoretical insights provide important hints to the topics that managers should address especially when developing business models.

4. Take Into Account the Need for Collaboration in a Value Net
Designing and deploying a business model calls for different kinds of resources and capabilities; also, collective action will be needed. The collaboration can be on an equal basis. However, the possible power imbalances between the actors in the so-called value net and its impact on the provisioning of the service have to be taken into account. The chosen form of value nets (e.g. strategic alliances, virtual organisations, supply-chain collaboration and value webs) has an impact on the relationships between actors.

Environmental development has an impact on the way the value nets will be defined. In the 'Old Rules' scenario, the dominance of existing actors shapes the form of organising the activities in the net, whereas the fragmented actors in the 'Wind Turns' scenario could bring about more co-operative ways of inter-organisational collaboration.

5. Implement Proper Governance Mechanisms to Provide Rules for the Collaboration in a Value Net

In the value net, the governance modes should be shaped in a way that they safeguard and coordinate the activities of the organisations and make them adaptable to changes. Governance modes can have aspects of loose coordination (market), hierarchical coordination (hierarchy) and co-operative coordination (network). Organisations should be aware that the governance mode they choose is tailored to the business context.

The new services will be embedded in larger social networks and new forms of governance for controlling cooperation are needed. Network and market governance may replace more hierarchical governance and control that is currently used. However, from a managerial point of view, the importance of common rules and dynamic organisational arrangements does not disappear but has to be agreed on.

6. Have a Broad View to Revenues and Revenue Sharing

Revenue and revenue division in a viable business model includes both relevant revenue sources and a logical way to report and divide the revenues. In addition, other revenue sources to sustain the business besides the paying customer, such as advertising or sponsoring, should be considered. Clear guidelines for a transparent application of mobile advertising is needed. Such guidelines are important in order to win over customers to accept mobile advertising in return for valuable mobile services. Guidelines would, for example, describe how customers could opt-in or opt-out for advertising, how providers would safeguard sensitive customer information, or how many advertisements a customer may expect to receive. A similar recommendation can be given to regulatory and legal institutions, to make privacy legislation as clear as possible in order to aid development of the mobile advertising market.

7. Use Service Oriented Architecture for Business Model Definition

The Service Oriented Architecture is a good way for organising collaboration and for making the inter-organisational interfaces more explicit. This makes the collaboration easier for the manager and thus for the company. Providers of mobile applications should be prepared to use functions (such as personalisation and context awareness) through service interfaces, in order to have the possibility to choose the most appropriate providers from the business point of view.

Orchestration is also a key in the development of the content delivery market. Players from the content industry, consumer electronics, device manufacturers and enabling technology providers (e.g. DRM, payment) need to agree on standards for managed content services that provide users with flexible usage options.

The following recommendations can be made on how to make mobile applications-based business models robust in different future environments.

8. 'Old Rules' Scenario Promotes Targets Related to Business Model Efficiency

The marketplace dynamics emphasises economic development. In the 'Old Rules' scenario, development is incremental and aimed at improving existing services. Thus, the main focus is in the efficiency of the business. Customers would be willing to pay a premium for extensions and enhancements to current service and product offerings, as long as they provide added value (e.g. increase productivity). Traditional players in the mobile value chain should develop and provide these services in much the same way as they do now, building on current hierarchical governance forms to control the provisioning network.

For advertising to work as a new revenue source in this scenario, established players from media and device manufacturers could set up a strategic partnership to provide, for example, advertising-sponsored mobile TV. Mobile operators would have to rethink their traditional revenue models, which are based on paying customers.

For companies providing enabling technologies or for SMEs developing solutions, the main guideline is to look for strategic partnerships to build and keep lasting supply relationships with established players in the mobile services field. This would mean partnerships first with operators but manufacturers as well. In short, the service developer positioning in this case is about being a close and trusted partner of some dominant or strongly emerging mobile player.

9. 'Wind Turns' Scenario Provides Fragmented User and Service Provider Markets

In the 'Wind Turns' scenario, the marketplace is modular including easy-to-combine layers and components. Carefully tuned end-user services have to be targeted to small user groups if advertising is used as a (secondary) revenue source.

The opportunity is to embrace public standards and interoperable technologies. Autonomous businesses may provide innovative applications as separate products. Providers of enabling 'components' (e.g. payments) should make sure that they can reach sufficient scale, for example by enabling several service providers.

Service developers should be prepared to co-operate in extended articulated value networks and promote their positioning as a stand-alone, autonomous business capable of partnering in an easy and efficient way with everyone that fits.

When applying mobile advertising, service providers should meet certain conditions relating to consumer acceptability and advertiser value. The mobile technology enablers can be of help here to specifically target customers with advertisements in an unobtrusive way. The industry, possibly together with regulators and consumer groups, should provide transparent guidelines for mobile advertising that help support consumer acceptability.

10. Take Into Account the Socioeconomic Dimension

The mobile technologies have implications on the socioeconomic dimensions, which can be included in the following twofold recommendation. First, there are a number of strong and well-identified social and economic issues regarding ICT that might be overlooked in the process of designing and developing new offers; for instance, the role of the elderly or of people with special needs that may not be properly understood. Second, companies and public organisations alike do not operate in a void but in the society at large. It would be beneficial for society if their services contributed positively to address some of the issues that the society itself feels as urgent and in need of intervention.

11. Take into Account the Legal and Regulative Aspects

The legal and regulative aspects are of importance in developing a business model. The legal essence of end-user communities and competence within them, as well as privacy and data protection law, intellectual property rights (IPRs), especially copyright, and contractual relationships including consumer protection are important. However, in different business models, legal issues may deserve different weights.

9.7 Acknowledgements

The following individuals contributed to this chapter: Luca Galli (Neos, Italy), Timber Haaker (Telematica Instituut, Netherlands), Olli Immonen (Nokia, Finland), Ulla Killström (Elisa, Finland), and Mark de Reuver (Telematica Instituut/TU Delft, Netherlands).

References

[1] Amit R. and Zott C.: 'Value Creation in e-Business'. *Strategic Management Journal*, Vol. 22, pp. 493–520.

[2] Ballon P.: 'Scenarios and Business Models for 4G in Europe'. *Info* (The Journal of Policy, Regulation and Strategy for Telecommunications (Special Issue: Mobile Futures – Beyond 3G), Vol. 6, No. 6, 2004, pp. 363–382.

[3] Bouwman H. and MacInnes I.: 'Dynamic Business Model Framework for Value Webs'. 39th Annual Hawaii International Conference on System Sciences, 2006, p. 43.

[4] Bouwman H., MacInnes I. and De Reuver M.: 'Dynamic Business Model Framework: a comparative case study analysis'. 16th Biennial Conference of the International Telecommunications Society, Beijing, China, 2006.

[5] Bovet D. and Martha J.: *Value Nets: Breaking the supply chain to unlock hidden profits*. ISBN 978-0-471-36009-4, Wiley, 2000.

[6] Camponovo G. and Pigneur Y.: 'Business Model Analysis Applied to Mobile Business'. 5th International Conference on Enterprise Information Systems (ICEIS 2003), Angers, 23–26 April, 2003.

[7] Christensen C.M. and Overdorf M.: 'Meeting the Challenge of Disruptive Change'. *Harward Business Review*, Vol. 78, 2000, pp. 67–76.

[8] Eisenhardt K.M. and Martin J.A.: 'Dynamic Capabilities: What are they?'. *Strategic Management Journal*, Vol. 21, 2002, pp. 1105–1121.

[9] Figge S. and Schrott G.: '3G 'ad' Work: 3G's breakthrough with mobile advertising'. Proceedings of the 8th International Workshop on Mobile Multimedia Communications, München, 2003.

[10] Galanxhi-Janaqy H. and Fui-Hoon Nah F.: 'U-commerce: Emerging trends and research issues'. *Industrial Management & Data Systems*, Vol. 104, 2004.

[11] Ganeshan R. and Harrison T.P.: *An Introduction to Supply Chain Management*. 1995. Online: http://lcm.csa.iisc.ernet.in/scm/supply_chain_intro.html.

[12] Hamel C. and Prahalad C.K.: *Competing for the Future!*. Harward Business School Press, 1994.

[13] Hamel G.: *Leading the Revolution*. Harward Business School Press, August 2000.

[14] Hawkins R.: 'Looking Beyond the .com Bubble: Exploring the form and function of business models in the electronic market place'. In Preissl B., Bouwman H. and Steinfield C. (eds), *E-life After the Dot.com Bust*. Berlin: Springer, 2003.

[15] Hovart M.: 'Mid-term Synthesis Report on the Integration of Socio-economic and Foresight Dimensions (SED) in FP6'. *Report to the European Commission*. June 2005. Online: ftp.cordis.europa.eu/pub/citizens/docs/sed_report_final_050720.pdf.

[16] Jones C., Hesterly W.S. and Borgatti S.P.: 'A General Theory of Network Governance: Exhange conditions and social mechanisms'. *Academy of Management Review*, Vol. 22, No. 4, 1997, pp. 911–945.

[17] Licoppe C. and Smoreda Z.: 'Are Social Networks Technologically Embedded?: How networks are changing today with changes in communication technology'. *Social Networks*, Vol. 27, No. 4, 2005, pp. 317–335.

[18] Magretta J.: 'Why Business Models Matter'. *Harvard Business Review* Vol. 80, No. 5, 2002, pp. 86–92.

[19] Mahadevan B.: 'Business Models for Internet-based E-commerce: an anatomy'. *California Management Review*, Vol. 42, No. 4, 2000.

[20] Möller K., Svan S. and Rajala A.: 'Network Management as a Set of Dynamic Capabilities'. EURAM Conference, 2002.

[21] Moore G.A.: *Living on the Fault Line*. Capstone Publishing, London, 2000.

[22] Normann R. and Ramirez R.: 'From Value Chain to Value Constellation: Designing interactive strategy'. *Harvard Business Review*, July 1993.

[23] Optima Media Group: 'Developing and Implementing a CRM Strategy'. 2005. Online: www.business-intelligence.co.uk/reports/crm_strat/contents.asp.
[24] Rappa M.: 'Managing the Digital Enterprise'. 2007. Online: digitalenterprise.org/models/models.html.
[25] Shlyvotsky A.: *Value Migration: How to think several moves ahead of the competition*. Harward Business School Press, Boston, 1996.
[26] Spanos Y.E. and Lioukas S.: 'An Examination into the Caula Logic of Rent Generation: Contrasting Porters's competitive strategy framework and the resource-based perspective'. *Strategic Management Journal*, Vol. 22, 2001, pp. 907–934.
[27] Sultan F. and Rohm A.: 'The Coming Era of "Brand in the Hand" Marketing'. *Harvard Business Review*, September 2005.
[28] Tikkanen H., Lamberg J.-A., Kallunki J.-P., Parviainen P.: 'Business Model: Prospects for a new concept'. Prepared for the European Management Journal, 2003.
[29] von Tunzelmann N.: 'Historical Co-evolution of Governance and Technology in the Industrial Revolution'. *Structural Change and Economics*, Vol. 14, 2003, pp. 365–384.
[30] Van Der Heijden, K.: *Scenarios: the art of strategic conversation*. ISBN 0-471-96639-8, Wiley, 1996.
[31] Venkatraman N.: 'Five Steps to a Dot.com Strategy: How to find your footing on the web'. *Sloan Management Review*, Vol. 41, No. 3, Spring 2000.

10

Legal and Regulatory Framework

Edited by Olli Pitkänen (Helsinki Institute for Information Technology HIIT, Helsinki University of Technology and University of Helsinki)

The earlier chapters have described a comprehensive Mobile Services Architecture, different enabling technologies, reference implementations of applications and business perspectives. This chapter describes the legal and regulatory framework for future mobile applications and services for families and ad-hoc communities in this book's context.

The framework consists of two parts: a general part on competence issues and a specific part on certain legal areas. The framework is based on legal analysis of the scenarios presented in the Appendix, as well as of the service framework and applications presented in the earlier chapters. Additional aspects can be found in [22].

The *competence part* of the chapter (Section 10.2) discusses the basic legal essence of ad-hoc communities and families. Competence, in this context, refers to the ability to accomplish binding legal acts within the sphere of the community's autonomy.

It seems that ad-hoc communities are usually contractual arrangements. The nature of ad-hoc communities, however, suggests that they do not cautiously prepare documents. Therefore, it has to be assumed that they are usually contractual arrangements without written contracts.

Sometimes, the contractual nature of a community can aquire a more specific form because of the activities that the community performs, if they are governed by special legislation. In particular, if the community is involved in gainful employment, labour law can be applicable.

The *scenarios and specific legal areas part* (Section 10.3) describes the most important specific legal areas that are related to future mobile ad-hoc communities. It concentrates especially on privacy and data protection and intellectual property rights. The mobile scenarios (Appendix), the service framework (Chapter 3), and the reference applications (Chapter 7)

Enabling Technologies for Mobile Services: The MobiLife Book Edited by Mika Klemettinen

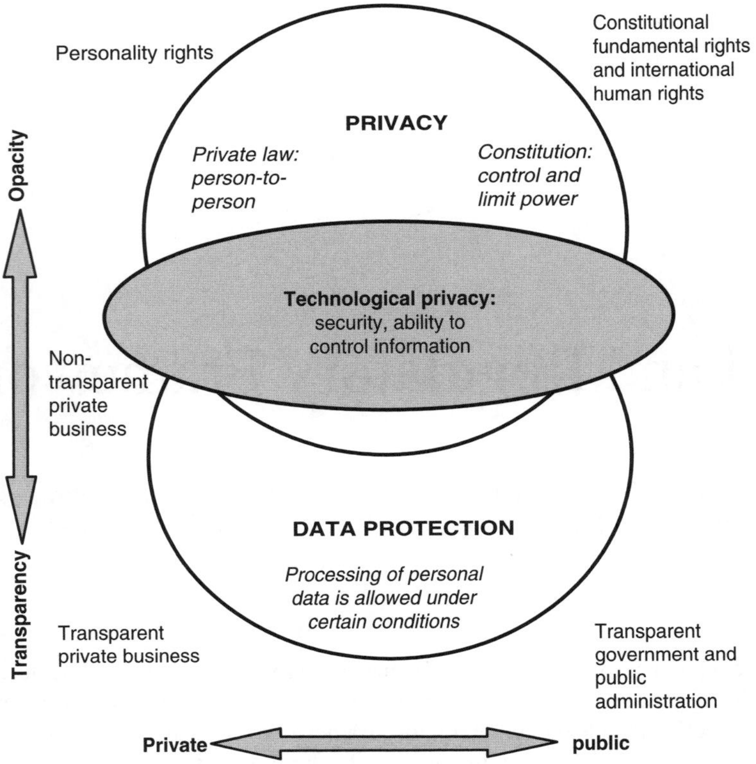

Figure 10.1 Multiple aspects of privacy and data protection.

described in this book are analysed legally. Additionally, a few other scenarios are analysed to complement the illustration of future mobile services and applications from a legal perspective. This part defines a specific legal framework for the scenarios.

Finally, Section 10.4 summarises and concludes the legal discussions.

10.1 Introduction

Developing mobile technologies may cause legal problems. Devices that are able to process data and communicate through networks are spreading everywhere. There is a large consensus that users' privacy may be threatened by that technology. The solution, however, cannot be just technological, the legal and ethical aspects of privacy need to be considered. Also, many traditional legal key concepts are problematic in several existing situations. New business models enabled by new technologies will be even further challenging from this point of view. Technically, it would be attempting to distribute digital content and other information to the end-user adapted in accordance with context, personal preferences and device properties. However, this may violate not only data protection law, but also the intellectual property rights of the content creator. In law, digital phenomena like computer programs or music in digital form are very difficult to handle using the terms and concepts of, for example, copyright and patent law, which have been developed for more traditional objects.

The regulatory and legal problems can be avoided or at least decreased if they are taken into consideration early enough and included in modelling the business. This chapter presents a framework to help service and application providers to prepare for regulatory and legal problems and to overcome them. The focus is on *mobile individuals as members of user communities*. Ad-hoc communities, enabled by mobile technologies, are especially highlighted.

From the legal aspect, an ad-hoc community is a challenging concept, since it does not fit well into any well-known legal form. As discussed below, an ad-hoc community is hardly ever a legal person and to interpret it to be just a contractual arrangement is often unsatisfactory. Usually, this is not a problem, since most ad-hoc communities (say, a group of parents that are gathered in a park to watch over their playing children, or the passengers of a certain bus) do nothing legally significant. However, if an ad-hoc community, for example, processes personal data or creates something valuable, it becomes legally relevant. The scenarios (see the Appendix) suggest that these cases are becoming increasingly common. Therefore, the legal status of ad-hoc communities is turning out to be an ever important subject.

This chapter has two main parts:

- The *competence* part discusses the basic legal essence of ad-hoc communities and families. It defines the general legal framework in relation to communities and presents a list of criteria that helps to decide, in a particular case, how competence is divided.
- The *scenarios and specific legal areas* part describes the most important specific legal areas that are related to the future mobile ad-hoc communities. It concentrates especially on *privacy and data protection*, *intellectual property rights*, as well as *contracts* and *consumer protection* in relation to future mobile services for ad-hoc communities, families and individuals. The scenarios, the service framework and the applications are analysed legally. This part defines a specific legal framework for the scenarios.

The starting point, in this chapter, is the relevant Finnish legislation, extended by European legislation. Comparison with other legal systems, in particular with US laws, has been made when it seems especially fruitful. Additional aspects can be found in [22].

10.2 Competence

This section discusses the basic legal essence of ad-hoc communities and families. Associations, partnerships, work teams and like examples, which perhaps are not ad-hoc but well-known, are the starting point. More general observations on how competence is divided in ad-hoc communities are deduced. The section defines the general legal framework in relation to communities and presents a list of criteria that helps to decide, in a certain case, how competence is divided.

10.2.1 Competence Defined

Imagine an ad-hoc group of parents that are gathered in a park to watch over their playing children. In the future, all the parents are equipped with mobile devices that enable them to communicate, share information, use multiple applications, and access a variety of services. While sitting in the park and discussing, the parents get an idea to share a list of their contact information so that they can for example warn the others if the park is closed some day, and to share the pictures that they have taken in the park. Next day, other parents who happen to be in the park are invited to join their list. One of the parents, who owns a clothing store, notices that the list is actually very useful for sending advertisements on children's clothes. Another parent collects the shared pictures and publishes them. The third parent is insulted and claims that her contact information should not have been given to those who were not present on the first day, the contact information must not be used for commercial purposes, and the pictures she has taken must not be published. The others disagree. They think that the whole group benefits from the list, from the advertisements, and from the possibility to publish the pictures. The question is: Who is allowed to make these kinds of decisions on behalf of an ad-hoc community?

In private law, *competence* refers to the ability to accomplish binding legal acts within the sphere of one's own private autonomy. Only natural or legal persons are capable of bearing subjective rights. The term *natural person* refers to all human beings. *Legal persons,* on the other hand, are certain associations of persons (e.g. companies) or even independently organised assets, as discussed below. The latter two gain legal capacity and competence through certain legal acts, whereas a human being has legal capacity the moment he or she is born.

A central question in the scope of this study is how competence is divided within communities. Each member of the community does have legal capacity. In addition, either they themselves or their guardians also have legal capacity. But what about the community as a whole? Who is able to act on behalf of the community and to what extent?

In general, there are two basic alternatives to arrange competence within communities: competence to act in the name of a community may be based on a *legal person* or on a *contractual arrangement*. The major distinction between these two alternatives is that a legal person is valid, has legal capacity, and can conclude legal acts in its own name both *in relation to third parties* and with its shareholders, partners, or other members while contract is mainly binding between the contractual parties.

According to contractual freedom, the contracting parties may quite freely agree on anything and organise their relationship as they wish, but that does not have much effect on other people's or entities' rights or duties. They may, however, authorise a proxy – or in some cases apply for a trustee – to act on their behalf. Proxies and trustees are legally valid also in relation to third parties as long as they act within the authority. A carefully organised contractual relationship enhanced with a powerful proxy closely resembles a legal person. That is, however, exceptional. The contracts are more flexible, but legal persons provide more legal security since more detailed regulations govern them and public registers have trustworthy information on them.

One of the problems with contractual freedom and purely contractual relationships is that they postulate equal parties. In practice, contractual parties are seldom equal. Usually some of them are more powerful and can dictate terms and conditions. To protect a weaker party, the law sometimes narrows the freedom of contract and requires that certain terms need to be applied even in a contractual relationship. Likewise, the law of legal persons may guarantee certain protection (e.g. for minority shareholders). Yet, it would be too optimistic to believe that, for example, family members were equal and capable of making agreements evenly.

Especially, children's interest may require lawmakers to reduce the contractual freedom of the parents [29].

10.2.2 Legal Persons

A *legal person* is a corporation or a trust. Legal terminology varies in different legal systems, but in general a corporation is formed by a group of people or legal persons and it has members or shareholders. Trust is formed by assets established for a specific purpose. A legal person has a name, separate assets, and a permanent organ that is entitled to represent it [12].

Obligation to a form means that all the legal persons must fulfil a legal form. In most countries, new types of legal persons cannot be formed without new legislation. Therefore, if an entity does not fulfil the legal requirements of a certain form, it is not a legal person [13, 30].

An *association* is a group of several individuals who make a mutual agreement to form a more or less permanent body to accomplish a purpose. The central feature of the association is the right of self-determination; i.e. *autonomy* [13, 30]. An association can be non-profit or for profit. It has an internal organisation and by-laws. The purpose of an association must not be against law and public decency, and it must be rational. Also, the concept of association includes certain membership rights and duties. Not all the countries have any separate association law. In its place, non-profit companies, partnerships or other forms of legal persons can be used instead of associations.

In general, *corporations* are characterised by the following attributes [30]:

- *Partners or shareholders*. A corporation is a legal relationship between two or more entities. Yet, depending on the legal system, the requirement of multiple shareholders can be eased: in many countries it is nowadays possible and also quite common that there is only one shareholder in a company.
- *Corporate purpose*. A corporation needs to have a target that is common to all the partners or shareholders; that is, the activities in the corporation are carried out on partners' or shareholders' common behalf, for their benefit and at their risk. The purpose is usually economic.
- *Co-operation on common behalf*. Co-operation in a corporation is the means through which the corporate purpose is realised. A corporation is always the partners' or shareholders' common enterprise in which investments are mixed to achieve the corporate purpose.
- *Depending on legal acts*. It always requires legal acts to establish a corporation. To establish a partnership, only a contract is required, while to establish a company requires several complex acts. As the types of legal persons are mutually exclusive, a company, for example, cannot be also a partnership although it would seem to fulfil these requirements.

To conclude, the general characteristics of legal persons are:

- They fulfil a legal form (obligation to a form).
- They have the right of self-determination.
- They have a mutual agreement between multiple founders as well as members, partners, or shareholders.
- They are intended to be more or less permanent.

- They are to promote shared interests.
- They have a legal purpose according to public decency: non-profit or economic depending on the form.
- They have by-laws, articles of association, or articles of partnership; as well as internal organisation including an organ that is authorised to represent the legal person.
- Their members have membership rights and duties.

How do ad-hoc communities apply to these characteristics? A truly *ad-hoc* community does not fulfil a legal form since it has not been established and registered in accordance with the laws governing legal entities. However, sometimes an ad-hoc community may want to get organised and form a legal person. This can be well advised if, for example, the community becomes involved in a noticeable business, gains property or earns income. Usually, however, an ad-hoc community does not want to go through the trouble to fulfil any legal form.

Natural and legal persons have certain rights of self-determination or autonomy. Also the members of an ad-hoc community – even if it is not a legal person – have autonomy of their own. However, the community that is not a legal person does not have a right of self-determination irrespective of the autonomy of its members.

If the members of an ad-hoc community accomplish together something that is legally significant, they most likely have some kind of an agreement between themselves. It does not need to be a written formal contract, but at least the members' behaviour suggests that there is a mutual understanding and an implied agreement even if nothing has expressed specifically. Therefore, also ad-hoc communities are usually based on some sort of an agreement though it can be vague.

One of the characteristics of legal persons is that they are intended to be more or less permanent. An ad-hoc community is often quite temporary, and thus it cannot be a legal person. Also, an ad-hoc community that has not seen the trouble to fulfil a legal form does not usually have by-laws, articles of association, or articles of partnership either. Its internal organisation is most likely minimal. However, there can be an organ, especially one of the members, who is somehow authorised to represent the community. In short, ad-hoc communities apply poorly to the general characteristics of legal persons. It is therefore likely that most ad-hoc communities are not legal persons.

10.2.3 Contractual Arrangements and Proxies

It should be noted that usually contracting parties are free to authorise one of the parties or an outsider to act as a *proxy* (or a *legal representative* or an *agent* or a *trustee*) on their behalf. This kind of an agreement and authorisation mostly takes care of the competence-related issues, if it is unambiguous enough. However, most ad-hoc communities and families have not agreed explicitly on authorisation. Then the question is at least: What exactly is the authorisation?

A proxy can be either compulsory or non-compulsory. Minors and legal persons need to have a natural, competent person to act on their behalf. That is called a compulsory proxy. On the other hand, anybody can authorise a natural, competent person to act on his or her behalf. That is called a non-compulsory proxy. Usually, there is no particular form of authorisation, but a non-compulsory proxy can be authorised orally or in writing. Also, it is quite common to have an authorisation on the basis of one's position: for example, a shop assistant is authorised to sell goods.

The following discussion tends to elucidate the situations in which there is no specifically authorised proxy.

10.2.3.1 Employment Relationship

An ad-hoc community can form an employer–employee relation. Let us first consider what an employment relationship is. It has four special characteristics [15]:

- *Contract*. An employment relationship is normally based on an agreement. It does not need to be a written contract, but an oral, electronic or even silent mutual understanding is enough. Furthermore, the contract does not need to be titled as 'Employment Contract', but the content is decisive. If for instance a sales contract is also an employment contract, then both property law and labor law can be applied.
- *Gainful employment for an employer.* Employment is about work: active or passive human behaviour that has economic value. Work is intended to be gainful. An ordinary hobby usually does not include gainful intentions and therefore it is not work. The work is done for an employer. The employment relationship binds the employee personally and he or she may not substitute someone else, like an entrepreneur may. If the employee also owns a significant part of the employer (e.g. a company, a partnership, or a death estate), it is possible that the person in fact works for himself or herself.
- *Remunerativeness*. By default, the employer must compensate the work for the employee even if there is not any specific clause in the contract. The compensation is usually monetary payment, but it can be anything that has economic value.
- *Direction and control*. The employer directs and controls the employee's work. This is often the most important criterion to decide whether it is a question about an employee or an independent entrepreneur that is not under the immediate direction and control of the employer.

An ad-hoc community can be involved in an employment relationship in several ways. First, the community as a whole can work for an external employer. Second, one of the members can be an employer and other members are employees. Third, the community itself can be an employer and the members can be employees. This is the case if there is a contract between the community and each member, the community makes agreements with its customers, and directs, controls and remunerates the employees. It could be a future, more mature phase for a community, but it is hardly called 'ad-hoc' any longer. Speaking of ad-hoc communities, the first alternative – the community as a work team – looks likely.

A *work team* is a group of employees that have jointly committed to complete a job under the direction and control of an employer. The work team's foreman, who makes the agreements with the employer on behalf of the work team, is an *intermediary*. The intermediary is not an employer, but the intermediary's employer is also the others' employer. From the competence viewpoint, it is essential that the foreman as an intermediary has the power to make employment agreements on behalf of the work team [6, 15, 19, 26, 27].

A work team is not a corporation, but workers' consortium in which the community of work arise from the nature of the performance and the wages system. On the work team agreement basis, there is an employment relationship between the employer and each of the members of the team [15, 27].

The legal rules governing work teams originate from issues related to gangs of log floaters, portable sawmills, horse-drawn forest work, and so on. The world has changed a lot since those days. Currently, it is quite rare that people ally to work together in such loose teams. Today, the term work team usually refers to a group of employees within an organisation. The employer directs and controls them and decides who the members of the team are. A team like that is not a work team that is meant here.

If a group of workers gets organised and establishes a legal entity, then it is no longer a work team. There are no clear rules how to distinguish a work team from an enterprise. If the group cannot dismiss the foreman, it is a rather strong circumstantial evidence that the foreman is in fact the employer of the group and an independent entrepreneur. The foreman is probably an independent entrepreneur, if he or she does not participate personally in the accomplishment of the work. However, the most decisive criterion is whether the foreman profits by the work of the others: if the employer pays to the foreman for the employees' work more than the foreman pays to the employees, then the foreman is an independent entrepreneur [6, 15, 19, 26, 27].

If a relationship is considered as an employment relationship, then the freedom of contract is encroached significantly. On one hand, it means that especially the employer does not have the liberty to agree on the terms and conditions of the relationship as it likes, but the employment law and the collective agreements dictate a number of issues. On the other hand, it means that it is not necessary to agree on everything as the law defines the basic defaults.

From the employer's point of view, it can be attempting to try to avoid the applicability of employment law. An enterprise might not want to hire employees, but act as a client and call the employees, for example, 'contractors' or 'independent consultants'. A 'client' obtains several important advantages from hiring an independent contractor rather than an employee. Clients are not required to provide independent contractors with fringe benefits or to pay side costs (social security, unemployed insurance, etc.). The law does not provide independent contractors better protection against dismissal than that specifically agreed on or the rather vague protection provided by contract law. Also, clients do not have vicarious liability for the damages of independent contractors in the same way as they have for their own employees. Obviously, from the employee's viewpoint, it is often more secured to be in an employment relationship with a protection against dismissal, pension benefits, insurances, and so on.

On the other hand, an employment relationship has several advantages also from the employer's viewpoint. Intellectual property laws, as discussed more in detail below, include special provisions that automatically give the employer the rights in the results of employees' work, or at least make it easy to redeem those rights. Data protection is regulated in employment relationships more in detail. Therefore, even from the employer's point of view, employment law has advantages. If an ad-hoc community is a work team, many legal issues become less ambiguous. Employment law, intellectual property laws, tort law, insurance laws, tax laws, and so on, include stipulations how to handle different situations.

10.2.3.2 Civil Law Partnership

Civil law partnership is a corporation that is not covered by statutes. Legally, the concept is quite vague and ambiguous. It includes more or less temporary co-operation that is not clearly distinguished from partners' other activities. With respect to outsiders, it is not a separate, independent entity. Many sorts of phenomena can be considered civil law partnerships: for example, lottery teams, theatre groups, common tours, joint contracted works, pools, and consortiums. It is nevertheless clear that a civil law partnership does not have legal capacity and nobody can conclude legal acts in its name [30].

It seems likely that an ad-hoc community could be a civil law partnership, if it has the typical characteristics of a partnership, but is not formed as a legal person. Especially, an ad-hoc community should have a purpose similar to that of a corporate purpose, is common to all the members of the community, and the activities of the community are carried out on members' behalf, for their benefit and at their risk.

10.2.3.3 Family

The family is the natural and fundamental group unit of society. It is a domestic group affiliated typically by birth, marriage or comparable legal relationships, like adoption. It has the primary function of reproducing society and to locate children socially. Usually there are strong emotional relations between family members. To sum up, families are a very special class of communities.

In legal systems, there exists a field called family law. It covers legal issues related to families, like contracting of a marriage, adopting a child, and getting divorced including related property settlements and alimonies. Parents are usually the legal guardians of an underage child and may legally represent their children and act on their behalf. The law also stipulates how to organise the guardianship in the case of divorce and how to get a guardian *ad litem* if there is a conflict of interest between a minor and the parents.

However, family law does not usually cover issues related to competence and the power to act on behalf of the family in everyday situations. The law does not specifically stipulate how the adult members of a family make their decisions, which of them is competent to represent the family, who is competent to assign something that belongs to the family, and so on. Those issues are left to the family to decide; i.e. the family has a large contractual freedom. That is justified, because the way a family wants to organise its daily matters is very much a private issue and a legal system should not interfere in it.

Therefore, contractual arrangements discussed above are often also applicable to families. Family members can make agreements and they can authorise one of them to act on behalf of the others.

On the other hand, as discussed earlier, it is often too optimistic to believe that family members were equal and capable of making agreements evenly. Especially, children's interest may require lawmakers to significantly reduce the contractual freedom of the parents. Incidents of child abuse and domestic violence may also require the society to step in the private issues of a family. Therefore, considering a family only as a contractual arrangement or an opaque private unit in which the surrounding society may not interfere is all too limited. The lawmakers have needed to find a proper balance [29].

10.2.4 General Legal and Regulatory Framework

Obligation to a form means that *legal persons are mutually exclusive*. An entity cannot simultaneously be of more than one form. For example, an entity cannot be both a company and a partnership.

On the other hand, *contractual forms are not mutually exclusive*. A contractual relationship can be simultaneously, for example, an employment relationship and a co-ownership.

Conversely, the same group of people may have simultaneously several distinguished contracts also.

Therefore, the legal nature of an ad-hoc community can be at most one legal person, typically none, but it may include characteristics of several types of contractual relationships. In the following, some criteria are listed to underline, which sort of a contractual arrangement an ad-hoc community resembles:

- *Contract.* There needs to be some kind of an agreement on the basis of the community in order to call it a contractual arrangement. It means that the members of the community have somehow expressed their intent to co-operate. The expression does not need to be in a written format; it can be oral, but more often it can be mostly deduced from the circumstances and from the behaviour of the members.
- *Work team.* To be interesting from the legal point of view, a community often accomplishes something that is somehow valuable. If the nature of the community is to work for an employer, under the employer's direction and control, and the employer compensates the work, the community may remind of a work team, especially, if the members of the community have jointly committed to complete a job.
- *Civil law partnership* and *unregistered association.* As discussed above, an ad-hoc community could be a civil law partnership, if it has a corporate-like purpose that is common to all the members of the community, and the activities of the community are carried out on members' behalf, for their benefit and at their risk. Especially, a community could be considered as a civil law partnership, if it produces something economically valuable, but it does not have an employer and thus it is not a work team. If an ad-hoc community is not a civil law partnership, it is probably possible to call it an unregistered association as long as it has some sort of a reasonable purpose.

In the following, it is discussed how the criteria above affect competence.

Contract

If the criteria imply that an ad-hoc community is foremost a contractual relationship, the contract should state how the competence is divided. Unfortunately, that is not usually the case. Ad-hoc communities do not often carefully draft written contracts that state who is authorised to represent the community, how the decisions are made, how the disputes are solved, and so on. Instead, as discussed above, the agreements are normally oral or even deduced from the circumstances and from the behaviour of the members. It is typical that there is a spearhead or a leader even in an ad-hoc community. The mutual understanding in the community is that the spearhead is authorised to do something on behalf of them all. However, it can be very challenging to interpret afterwards, what the exact intentions and the understanding of the members in the time when the community emerged were, in case disputes arise. The actual content and the limits of the authorisation remain unclear. Therefore, it is not recommended to organise a community in this way, though it is obvious that this will be the most common way even in the future, since 'ad-hoc' and 'careful contracts' are sort of contradictory in terms. If ad-hoc communities are becoming more significant in the society, the lawmaker is required to set down general rules to reduce uncertainty.

Work team

If an ad-hoc community is considered to be a work team, the labour law becomes applicable. This means first and foremost that the employer has the right to direct and control the team's work. In the relationship between the work team and the employer, the foreman has the authority to negotiate on behalf of the team. It also means that the employer is entitled to benefit from the work, to get profit, and to obtain the rights in the outcome of the work team's efforts. Therefore, it is less important to consider competence issues as the employer is entitled by law to most of the value produced by the team. On the other hand, labour law also stipulates that the employer has important duties towards the employees. In summary, if an ad-hoc community is a work team, the competence issues become less ambiguous and more foreseeable, but also costs will probably rise and the contractual freedom is more limited.

Civil law partnership and unregistered association

As discussed above, a partnership that does not fulfil the legal form can sometimes be called a civil law partnership. On the other hand, if a community has the characteristics of an association, but it has not been registered, then it could be considered as an unregistered association. If a community is considered a civil law partnership or an unregistered association, it probably has some kind of an internal organisation: one of the members is a leader or a chairperson, or the core group of members form a sort of a board. Even if the members have not explicitly agreed on the organisation, it can be deduced that they have authorised the leader to act on their behalf.

10.3 General Legal and Regulatory Framework

This section describes the most important specific legal areas that are related to families and ad-hoc communities that avail of future mobile services and applications. It concentrates especially on privacy and data protection, intellectual property rights, as well as contracts and consumer protection in relation to future mobile services for ad-hoc communities, families and individuals. The mobile scenarios (Appendix), the service framework (Chapter 3) and the reference applications (Chapter 7) described in this book are analysed legally in Section 10.3.2. Additionally, a few other scenarios are analysed to complement the illustration of future mobile services and applications from a legal perspective in Section 10.3.3. This part defines a specific legal framework for the scenarios (Section 10.3.4).

10.3.1 Introducing Special Legal Areas

10.3.1.1 Privacy

There seems to be a large consensus that developing information and communication technologies effect privacy. However, there does not seem to be a consensus on what privacy actually means. Several viewpoints can be taken: at least technological, ethical and legal. Each of them has many definitions of privacy. It is not necessary to present them all here, but a few approaches are described to illustrate the privacy scene.

From the *technological* point of view, privacy is closely related to secrecy:

> Privacy is the requirement of an entity to determine the degree with which it interacts with its environment, including willingness to share information about itself with others. Privacy is related to authorisation. Trust is a firm belief in the competence of an entity to act dependably, securely and reliably within a specified context. Privacy and trust concepts are strictly linked to enable service providers to maintain and manage their customer relationships and proactively promote their services without third-party participation. [16]

However, for analysis of the scenarios from the legal point of view, ethical and legal aspects needs to be considered also.

From the *ethical* point of view, privacy is often divided into several components, like informational, physical and decisional – possibly also dispositional and proprietary [2].

Privacy is often considered as an essential element of democratic societies, because it promotes the ideological variety and political discussions. The claim to privacy finds moral justification in the recognition that people need to have control over some matters that intimately relate to them in order to function as people and be responsible for their own actions. The notion of privacy is essentially dualistic: on one hand there is a boundary between the individual and other individuals, on the other hand a boundary between the individual and the state.

What a person is expected to do in order to respect another's privacy varies with culture. While almost all cultures appear to value privacy, cultures differ in their ways of seeking and obtaining privacy, and differ also in the level they value privacy [5].

According to Gow:

> ... privacy is clearly a value that is important in modern societies and will likely remain so for some time to come. The difficulty lies in establishing a balance between the rights of the community and those of the individual, particularly in the face of new technologies that dramatically increase our ability to collect and use personal information. [9]

According to Allen, this is an age characterised by anxious discourse about 'the end of privacy' being upon us. This end-of-privacy anxiety is peculiar in two ways. The first peculiarity is that the anxiety sometimes seems out of proportion to the threat. The second peculiarity about the anxiety of the age is that all the talk about the involuntary loss of privacy coincides with a good deal of voluntary waiver and alienation of privacy [3].

From the legal point of view, the right to privacy is a human right expressed in several international conventions and a fundamental right stipulated by the constitution of many countries. It is then reflected by national laws.

According to Robertson, the right to privacy is perhaps best understood as combining three related desires or needs [24]:

- The first is the traditional sense of a private *physical space* the state may not enter except in special cases. The typical protection here is found in restrictions against searches of one's person and possessions, or entry into one's home.
- A second major sense has become closely intertwined with *personal morality* that there is a sphere of private activity that the state has no business to regulate.
- The third concept of privacy that gets some legal protection at times is best demonstrated in relation to *religious freedom*, which can sometimes present itself as a freedom not to be bothered by other people's religious concerns, that is, to have a privacy of belief.

Privacy rights are ultimately autonomy rights, the right to act and to develop in one's own way, but there is also a public concern for privacy, the strong sense that it is improper for other people to be nosy. Privacy is typically related to private surroundings, but wherever an individual is – for example at work – there can be a justified need for privacy to some degree [23].

Privacy is restricted by other important rights, like the freedom of speech, the protection of property, the freedom of trade, and the principle of equality. The fundamental and human rights do not have a general priority order, but if two of them are in conflict in a concrete case, it must be considered which of them is more important in those circumstances. National jurisdictions vary somewhat on this issue, but any strong limitation of the media in the interests of privacy runs often against the better defined rules on freedom of speech [24].

10.3.1.2 Data Protection

On the European Union level, data protection is extensively regulated by directives and regulations. For example, Data Protection Directive (95/46/EC) is about the protection of individuals with regard to the processing of personal data and about the free movement of such data, and Directive on Privacy and Electronic Communications (2002/58/EC) applies to the processing of personal data in connection with the provision of publicly available electronic communications services in public communications networks in the Community. On the other hand, numerous national laws include rules that affect data protection. They may stipulate more in detail and more strictly how personal information is to be handled in certain situations, or they may authorise certain usage of private information more freely than general rules would allow. Consequently, the legal construction of data protection rules is quite complex. The rules cannot be found in one law, but they are spread out in numerous statutes.

Data protection law can be applied to a wide area of legal questions. 'Personal data' refers to information relating to an identified or identifiable natural person ('data subject'). 'Processing of personal data' means any operation, which is performed upon personal data, such as collection, organisation, storage, adaptation, retrieval, use, disclosure or destruction. Note that if personal data are anonymised in a way that they cannot be related to an identified or identifiable natural person, data protection law is normally not applicable.

The processing of personal data is not illegal in general. On the contrary, the data protection law tries to enable useful processing of personal data. However, the processing needs to be carried out in accordance with the law. Especially, data protection directive requires that personal data must be:

- processed fairly and lawfully;
- collected for specified, explicit and legitimate purposes and not further processed in a way incompatible with those purposes;
- adequate, relevant and not excessive in relation to the purposes;
- accurate and, where necessary, kept up to date.

Personal data may be processed only if the data subject has given an unambiguous consent or there is another lawful basis for processing. The controller must provide the data subject with certain information, including the purposes of the processing for which the data are intended.

It is also important that disclosing by transmission, disseminating or otherwise making available to others is processing of personal data and thus needs also consent or another lawful basis. Especially, transferring personal data outside the European Union is highly restricted.

There are some important restrictions to the applicability of data protection law. Usually, if a natural person in the course of a purely personal or household activity processes personal data, the data protection law is not applied. Furthermore, the data protection law applies only partially to journalistic and artistic context. Also, the law is not always applied to data processing that is related to, for example, national or public security, criminal investigation, or important national financial interests.

Completely automated individual decisions are restricted. The directive sets strict limitations to decisions which produce legal effects concerning individuals and which are based solely on automated processing of data intended to evaluate the individual's personal aspects, such as performance at work, creditworthiness or reliability

Certain sensitive information should not be processed at all without special lawful reasons. These special categories of data include racial or ethnic origin, political opinions, religious or philosophical beliefs, trade union membership, data concerning health or sex life, and data relating to offences, criminal convictions or security measures.

The European Court of Justice made an important precedent in the *Bodil Lindqvist* case (C-101/01, 2003). The court decided that it constitutes the processing of personal data, if one refers on an Internet page to persons and identifies them by name or by other means, for instance by giving their telephone number or information regarding their working conditions and hobbies. Such processing of personal data is not covered by the exceptions of the Data Protection Directive. Normally, if a natural person in the course of a purely personal or household activity processes personal data, the data protection law is not applied. However, publishing information in a web page and making personal data accessible to anyone who connects to the Internet causes that it cannot be considered purely personal activity. Thus the data protection law may apply to even personal homepages.

10.3.1.3 Intellectual Property Rights, Especially Copyright in Content

Intellectual property rights (IPRs) protect intangible valuables. Some legal systems emphasise the economic aspects of intellectual property rights while others underline moral viewpoints or human rights that limit IPR. According to Jacob *et al.*, 'Intellectual Property' is the umbrella phrase now used to cover all the various rights that may be invoked to prevent imitations of various sorts. It is possible to own physical objects, but one cannot own nor have title to intangible objects like software, multimedia or inventions. Those are objects of intellectual property rights: copyright, patent, trademark, etc. They can be used to prevent some unauthorised gaining of intangible objects, that is, to exclude free-riders [7, 14, 21].

The system of intellectual property rights was developed in quite a different world from the one we live in. Although human creativity and inventiveness have probably not changed a lot, new technologies and business possibilities have remarkably changed the environment in which the intellectual property rights operate [18].

Open source software development is gaining more popularity. It challenges traditional ideas on how strong intellectual property rights promote development. The supporters of open software movement do not want to restrict copying and modifying of their programs. They are still willing to develop software although anybody can copy, change and use it freely. Even if one does not believe that the open source model will dominate in the future, it certainly shows that strong intellectual property rights are not the only way to solve the legal questions about software.

Copyright

Creative works are protected by copyright. National laws, EU directives and international treaties govern it. Anything that is original, expressed and creative is protected by copyright.

The work does not need to be registered or copyright noticed (e.g. © mark). It does not need to be artistic either. The emphasis and the details of the copyright law, however, are on slightly different aspects in different countries. The European Union has strived for harmonising the copyright system in Europe by adapting numerous copyright directives.

The one who has created the work is called an *author*. Normally the author owns copyright originally. Organisations, computers and others cannot create copyright protected works. However, the copyright is in some cases automatically assigned to the employer, if the creative work is a part of the employment. In countries, like the USA, that emphasise publishers' rights, a company may be the author of the work. Thus the employer may be considered the original author of the work created by an employee.

If a work is created in a community, it may have several authors. In that case, copyright in each separate contribution to a collective work is distinct from copyright in the collective work as a whole. Collections of works such as encyclopaedias and anthologies, which may constitute intellectual creations, shall be protected as such, without prejudice to the copyright in each of the works forming part of such collections. On the other hand, the authors of a joint work are co-owners of copyright in the work. A typical example of a collective work is a newspaper in which all the journalists normally have original right in their articles although they have assigned at least some of their rights to the publisher of the paper. A newspaper is thus a collection of copyrighted works that usually have different authors. However, the newspaper as a whole may also be a copyrighted work. In a joint work, the reader is not able to distinguish between the authors. A single newspaper article may be written by a couple of journalists together so that their work is indistinguishable.

Copyright gives some exclusive rights to the author. They include, for example, a right to copy, to sell and to display the work. Those are called economic rights. In many countries the author has also something called the moral rights. Depending on the country, they may include, for example, a right to proclaim or disclaim authorship, and a right to object any modification that would be injurious to the author's reputation. The moral rights cannot be assigned in general.

It is important to realise that there are some significant exceptions to copyright that enable certain usage of copyrighted works even without the consent of the copyright holder. For example, in most countries, a private use clause or a fair use doctrine allows individuals to make copies of works for their private use. That may be very important for some ad-hoc communities and families. As long as they are non-profit and small enough, their usage of copyrighted works can often be considered as private and they can make few copies of the work and distribute them within the community.

A smallish, non-profit community can therefore make copies of a copyrighted work for its members' private use, but a larger community should be more cautious, and a community that benefits members economically may lack that privilege totally. On the other, legal rules vary in different countries. For example, in the USA, fair use doctrine allows the making of copies in certain conditions for a vague set of purposes. Therefore, for example, an ad-hoc community resembling a study circle might be allowed to make a large number of copies of copyrighted material for educational purposes, while in Europe that often requires permission.

One of the very fundamental principles behind copyright is that copyright protection extends only to expressions and not to ideas, procedures, methods of operation or mathematical concepts as such.

As mentioned, to qualify for copyright protection, a work must be original to the author. Especially, no one may claim originality as to facts. Factual compilations, on the other hand, may possess the requisite originality. The compilation author typically chooses which facts to include, in what order to place them, and how to arrange the collected data so that readers may use them effectively. These choices as to selection and arrangement, so long as they are made independently by the compiler and entail creativity, are sufficiently original that such compilations may be protected through the copyright laws. The mere fact that a work is copyrighted does not, however, mean that every element of the work may be protected. No matter how original the format, the facts themselves do not become original through association. Copyright protects only the original compilation, not the compiled facts.

In general, an idea is not copyrightable, but on certain conditions it can be patentable or it may be possible for example to claim it as a trade secret. The expression of an idea may be copyrighted. On the other hand, if the same idea is expressed in different, independent ways, each of those expressions can be a copyrighted work of its own and they do not infringe each other. The physical embodiments or the copies of copyrighted expressions can be for instance sold without assigning copyright [11].

Database protection

More and more information is stored as data in databases. It becomes vital important to understand what kind of rights entities may have in databases. A database can include copyrighted works and even a database as a whole can be copyrighted if it is original enough. However, most databases are not copyrightable and their content is not copyrighted either. Yet, the making of databases requires the investment of considerable human, technical and financial resources while such databases can be copied or accessed at minimal cost. Therefore some kind of protection for databases is needed. The European Union has adopted a directive (96/9/EC) concerning the legal protection of databases. It recognises the possibility of copyrighting a database but also defines a neighbouring right, a specific *sui generis* right.

It should be noted that the word *database* is ambiguous. Especially, a 'database' in information technology and a 'database' in legal context are not necessarily the same. According to the database directive, the term 'database' means a collection of independent works, data or other materials arranged in a systematic or methodical way and individually accessible by electronic or other means. It is further required that in order to get the *sui generis* right in a database, there must have been qualitatively and/or quantitatively a substantial investment in either the obtaining, verification or presentation of the contents [20].

The *sui generis* right provides the maker of a database with the right to prevent extraction and/or re-utilisation of the whole or of a substantial part, evaluated qualitatively and/or quantitatively, of the contents of that database. Although individual data items in a database are not protected by the *sui generis* right, not only the database as a whole is protected but also a substantial part thereof.

Patents

A patent gives an exclusive right to exploit an invention commercially. In principle, patents are granted for any inventions, in all fields of technology, provided that they are new, involve an inventive step and are susceptible of industrial application. An invention can be a product or a process that provides a new way of doing something, or offers a new technical solution to a problem. The invention cannot be commercially made, used, distributed or sold without the patent owner's consent. Patent is limited to a specific period of time – usually for a maximum of around 17 to 20 years – and to a certain territory – usually to a country.

International treaties have harmonised national patent laws worldwide, although some differences still exist. For example, in Europe, patent applies to commercial availing of invention. Therefore, a family or a non-profit ad-hoc community may usually use patented inventions freely for their private purposes. However, in some other countries, like in the USA, patent covers also private usage.

A patent protects information on a higher level of abstraction than for example copyright. In general, a patent protects an idea reduced to practice. That is, it does not protect totally abstract ideas nor only new implementations or expressions of ideas. It does not protect mere data or investments either.

Patents do not appear automatically; they have to be applied for. It is actually quite a laborious and expensive process to get a patent. A patent application normally contains the title of the invention and an indication of the technical field. It also includes the background and a description of the invention, in such a clear and detailed way that others could use or reproduce the invention.

Drawings and other visualisations often help to describe the invention. The application also contains various claims that determine the extent of protection granted by the patent.

The patent rights can be enforced in a court, which holds the authority to stop patent infringement and award damages to the patent owner. On the other hand, a court can also declare a patent invalid if a third party has successfully challenged it. A patent infringement suit in a court can be very expensive. Therefore the threat of trial is often enough to force the parties to negotiate and cases are frequently settled outside the courts.

As mentioned above, to be patentable, the invention must be novel. In other words, there needs to be some new characteristic which is not known in the body of existing knowledge in its technical field. This body of existing knowledge is called 'prior art'. The invention must also show an inventive step. Finally, the subject matter must be patentable. In many countries, discoveries, scientific theories, mathematical methods, aesthetic creations, plant or animal varieties, discoveries of natural substances, schemes, rules, methods for performing mental acts, playing games or doing business, and methods for medical treatment are not patentable.

The patent system is supposed to promote inventions and industrial advances. Arguably, patents provide incentives to individuals by offering them recognition and material reward for their inventions. These incentives should encourage innovation. It is, however, questionable how well the patent system actually achieves that goal. Currently, it is valuable for especially large companies to have an extensive patent portfolio that can be traded with other companies. On the other hand, smaller companies are often required to apply for patents because many venture capitalists and potential acquisitors believe that patents as such add value to a company. Patents are also often used as a marketing and brand-building tool to give a high-tech impression of a company. For many companies, it is enough to file an application as a pending patent brings all the benefits the company is seeking. It seems that the idea of issued patents protecting certain useful technological inventions is giving way to a number of other ways to benefit from the patent system.

For years, there has been a lively discussion about the patentability of computer programs. Previously, programs were likened to mathematical methods, mental acts or games, and thus not patentable. However, case by case, these limitations have crumbled away. The USA has led the development, but the EU and the rest of the world are following. Although the European

Patents Convention (EPC) still states that programs as such are not patentable inventions, in practice computer programs are largely patentable and discussion about patenting business methods is lively. Similar development is ongoing around the world [11, 14, 19].

Other intellectual property rights

A *trademark* is to distinguish goods and services from those of others. Trademarks are typically names and logos, but any kind of mark that can be represented graphically and by means of which goods marketed in business can be distinguished from those of others may be a trademark. A trademark is the legal protection of a brand.

Domain names are distinguished, but in many ways resemble trademarks. Earlier, lots of problems were involved in domain names. Especially, entities were disputing which of them has the right in a certain domain name. In recent years, an international system called Uniform Domain-Name Dispute-Resolution Policy has been successfully introduced, which has reduced the number of problems remarkably.

Lately, IPR protection has strengthened remarkably: the copyright validity period has become longer, criminal sanctions have been made more severe, and so on. New legal means, like database protection, extend the exclusive IPR to cover subject matter that was formerly common property. Anti-circumvention rules – the legal protection of technical protection – or Digital Rights Management (DRM) prevent end-users from exploiting their legal rights. Therefore, the ever-strengthening IPR protection increasingly affects individuals and forms an important legal area in relation to mobile people's everyday life.

10.3.1.4 Contractual Relationships and Consumer Protection

Contracts are the primary legal means to manage rights and duties within bilateral relationships. If two entities know each other and are willing to commit to certain terms and conditions, according to contracting freedom, they are free to agree on issues extensively. On the other hand, however, contracts do not bind outsiders: contracting parties cannot in general give obligations to third parties. Also, especially on computer and telecoms networks, it may sometimes be difficult to identify the contracting parties and be sure what the terms and conditions are. The mandatory laws can limit the contracting freedom furthermore. Therefore, contracts do not always bind the contracting parties either. Instead, laws are required to define the legal framework to control issues that are not governed by contracts.

With respect to individuals and families, one of the major limitations to the freedom of contract is *consumer protection*. Promoting consumers' rights, prosperity and wellbeing are core values of the EU, and this is reflected in its laws. The consumer is considered to be a weak party in contractual relationships and therefore protected by the law. The consumer and the company may not even mutually agree on terms and conditions that may harm the rights of the consumer.

On a mobile network it can be troublesome to decide which is the correct law to govern a certain contract as well as which authorities have jurisdiction over disputes concerning it. EU directives have made these decisions somewhat easier, but yet, in the world of computer and telecoms networks, many severe problems remain. Directives have also introduced new challenges to service providers since they present requirements that are difficult to fulfil with forthcoming mobile and ubiquitous technologies. On the other hand, some scenarios also describe machines that make agreements on behalf of human beings or legal entities. This introduces severe challenges from the contractual point of view.

10.3.1.5 Other Legal Areas

In relation to the context of this book, some other legal and policy areas may also have importance. Especially in connection with telecommunication, there are a number of industry-specific laws and policies. For example, radio frequency allocation may affect the availability and quality of services. If a government decides to free more frequencies for general use, it may promote some useful services.

It is usually preferable that laws are as technology neutral as possible, but it is hardly possible to make them completely independent of any technologies. Thus new technologies may face unnecessary difficulties when old rules are applied to them. Certain functionalities, like emergency calls in telephone services, are required by law, but the regulations do not necessarily apply to new services like peer-to-peer voice-over-IP phone calls. For an end-user, it is essential that the most important functions of current services are also available in the future services, but for a service provider, inappropriate rules may cause extra costs and difficulties.

10.3.2 Scenarios, Service Framework and Applications

In the following, the mobile scenarios (Appendix), the service framework (Chapter 3), and the reference applications (Chapter 7) are briefly analysed from the legal point of view.

It should be noted that the scenarios are to help to define certain technologies. Therefore they intentionally have important limitations. For example, the scenarios do not describe situations in which the technologies fail, but they presume that technology works perfectly. This is completely sensible from the point of view for which the scenarios were developed. On the other hand, legal issues often come up when something goes wrong. Therefore it is necessary to include in these analyses also some discussion what happens if technology fails.

Privacy, trust and security are key topics in this book. That is why the scenarios also highlight privacy issues and the technology takes care of them very well. From the legal point of view, this is somewhat limited. Therefore, to cover also possible technology failures, some issues from scenarios defined elsewhere are added to broaden the legal analysis.

10.3.2.1 Scenario 1: Monday

The Monday scenario describes a community's – namely a family's – ordinary weekday (Figure 10.2). Even a simple example like this illustrates how difficult it is to define the members of the community and how the membership depends on the context and the matters at hand. Also, the scenario shows that the difference between work and leisure time continues to vanish. People are increasingly working at home, but also accomplishing non-work related activities at work.

From the legal point of view, the scenario presents a few interesting issues.

First, *confidentiality* and *privacy* between family members is explicitly highlighted. For example, while working on her presentation, Nicole is using confidential information – possibly protected by trade secrets – that she does not want to share with family members. Also,

Figure 10.2 Monday scenario.

Nicole's private information is not to be disclosed to third parties, like the online CD shop. The confidential information can thus be divided into two groups:

- information protected by privacy and data protection, for example context information, calendars, and personal music play lists;
- business information protected by trade secrets, labour law, and non-disclosure agreements: Nicole's presentation and related information.

These two groups are governed by different laws and the rules are somewhat different, but, nevertheless, they are both legally protected.

The scenario implies that lots of confidentiality issues are taken care of automatically. It does not, however, explicitly describe whether technical solutions are meant to comprehensively take care of confidential information, or if it is just to help users to protect the information themselves. It is possible that the system does not always work as intended. The more complex the policies or rule sets are, the more often errors occur. So, the confidentiality may fail. Who is responsible – the user, the provider or the employer? This depends on circumstances, but it is quite evident that in the near future, human beings are needed to control confidential information and thus they are also finally responsible for it.

Second, the scenario refers to the *fuzzy line between family members and outsiders*. For arrangements involving Elaine and Franco's children, Elaine is considered a trusted family member. It is not unambiguous who is a member of a family. In some situations people like an ex-spouse, a relative or a close friend could be considered family members. Likewise, a person may occasionally belong to several families – or none at all. Considering who is competent to represent a family legally raises difficulties when even the family membership can be unclear.

In the scenario, there is a lot of copyrighted information involved. Yet, the scenario hardly presents any legal issues with respect to copyright. The scenario does not, for example, imply that Nicole is listening to unauthorised copies of music or that she is copying material into her presentation illegally. Therefore the scenario does not present copyright problems.

The disappearing difference between work and free time impose challenges on employment and labour law. Traditionally topics like working hours and right to holidays have been essential in labour law, but their importance is diminishing since the exact amount of work hours or days off are no longer that significant. Instead, legal questions about the employer's right to direct and control the employees become more difficult and central as they define the applicability of labour law. To what extent may the employer direct a person who is not at work, or an employee who is at the office but doing something that is not related to work?

10.3.2.2 Scenario 2: Friday

At least two kinds of communities are presented in the Friday scenario (Figure 10.3). First, the family, but then also those ad-hoc groups that are formed dynamically as needed. It seems that by joining an ad-hoc group a person accepts some rules and delegates the founder or the leader of the group to make some decisions on behalf of the group. However, it is not explicated how this delegation happens and what competence the leader is to have.

The scenario highlights privacy and data protection issues. The system seems to take care of them well. There is a lot of health-related sensitive data, which is specifically protected by data protection law. Liability is strict, if someone fails to keep sensitive data confidential. In the scenario, it seems – although it is not clear – that the system keeps sensitive data secured. That is an issue which should be particularly emphasised.

According to the scenario, insurance policies, liabilities and agreements are managed in electronic form. That is already legal and possible today, but it may be challenging to design devices and software in a way that users are able to understand the terms and conditions so that the agreements bind them legally. Especially, small ubiquitous devices may lack large enough displays to show the user all the necessary information.

According to the scenario, eye-witnesses give their anonymous and provable identities for the insurance company and police. Anonymity is hardly necessary in relation to the police.

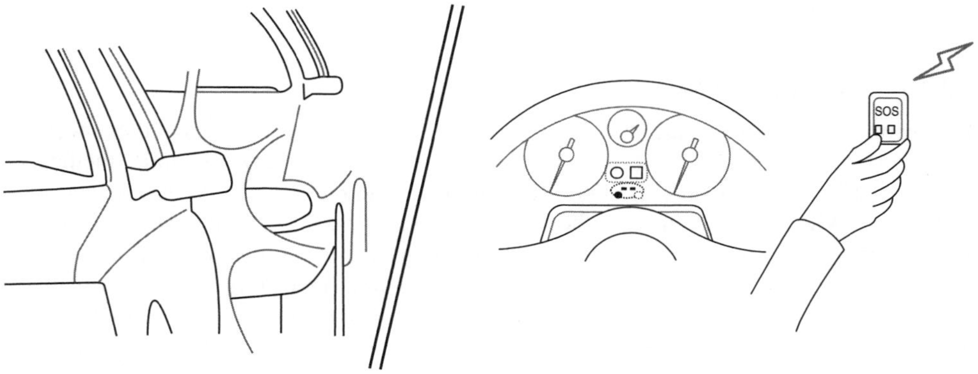

Figure 10.3 Friday scenario.

Actually, the police should know the real identity of witnesses and be able to contact them. Law criminalises the threatening of witnesses and in many countries a court may also forbid a threatening person to approach the witness. It is probably not necessary to withhold the identities of witnesses from the police or the insurance companies, but it may even harm the legal process. During a court trial, it is usually mandatory to disclose the witnesses also to the opposing side to assure a fair trial.

The system appears to be quite complex. The policies and rule sets that govern access to information are compound and dynamic. Are the policies created automatically or are they defined manually? Who may do the configuration? The scenario does not explain how the policies are kept up to date, but it may cause legal problems, if they are outdated or otherwise erroneous.

Interestingly, messages to police etc. are not sent automatically, but to other cars in the proximity. From the data protection point of view that is an issue. The police – at least in the EU member countries – are trustworthy and even personal information can be securely given to the police. It would not be a problem from this viewpoint to send the messages to the police automatically, although people may still want to control themselves when to contact the police.

However, unknown drivers nearby are not trusted by default and therefore it should be more controlled which information is sent to them. Additionally, in some cases, it might be useful to send the message automatically to an ambulance service or to police, if for example the driver is badly injured.

From the copyright viewpoint, it is notable that Maria is able to send further pictures that others have shared to her. It is possible that some of those pictures are copyrightable. Also in some countries, like Finland, a special right exists concerning photographs. Even those photos that are not copyrightable are protected by this somewhat weaker neighbouring right. One might be able to argue that sending the pictures to the insurance company or the police is allowed by copyright exemptions, like private use, but this is hardly the case, since the companies and the officials cannot have private use. Therefore Maria is probably either infringing the rights in photographs or – more likely – the photographers have authorised her to copy and distribute the pictures. The authorisation may have occurred, when people joined the ad-hoc group or when they send the images to Maria. It may have happened explicitly – that is they have told about the authorisation or accepted such an agreement or power of attorney – or implicitly.

10.3.2.3 Scenario 3: Sunday

Once again, the Sunday scenario highlights the fuzzy line between members of the family and outsiders (Figure 10.4). A lot of private information is distributed between family members.

The scenario does not explicitly state who is authorised to make decisions on who gets what information. Will Olivia be able to authorise Maurice to access information? Do parents make this kind of decision in consensus? If one of the family members afterwards thinks that his or her privacy has been violated, it will be difficult to determine who is liable and what went wrong. Typically, these are questions that belong to the autonomy of the family. Any outsiders, like lawmakers, courts or judges, are not wanted to interfere and rule on behalf of the family on those issues. However, if the line between the family and the outsiders gets fuzzier and the new technology makes the questions arise more frequently, then there will be an increasing need to have also some general rules.

Figure 10.4 Sunday scenario.

The scenario describes how advertisements are silently pushed to consumers. The directive on privacy and electronic communications (2002/58/EC) stipulates that advertisements in any text, voice, sound or image message sent over a public communications network are allowed only in respect of subscribers who have given their prior consent. That is, consumers have the right to opt-in, but the advertiser may not send unsolicited messages. Also, marketing law applies to these advertisements: depending on the jurisdiction lotteries, for example, can be illegal or at least strictly restricted and it can be illegal to offer random prises.

The scenario presents interesting examples of privacy and data protection issues. Who is allowed to view others' locations? Who may see the family's calendar? In the game, everybody can get information about players' heart rate, speed and position. That is certainly private information, but possibly also sensitive information, if it discloses anything about one's health. Players (or their guardians if they are under age) should give their consent to publish private data, but sensitive data should not be processed at all.

The data on players is stored in a database. In some cases that database could be valuable and the question on database rights would arise. It does not seem likely that there is substantial investment involved. Thus database *sui generis* right is hardly applicable. However, if there is an investment in either the obtaining, verification or presentation of the contents, it is possible that the creator of the database has an exclusive right in it as described earlier.

The pictures that are taken in the scenario involve important legal questions. A picture can be copyrighted, but it may also be considered as private information. In the scenario, pictures are shared and further distributed quite freely. This may violate both the copyright of photographer and the privacy of the persons that the picture portrays. Also, it is typical that the organisers of large-scale events claim that they have a televising right: no one may broadcast or even record a video on the event without the organiser's consent. That is the legal ground to charge media companies on exclusive rights to televise events. In this scenario, nobody probably claims to own the televising rights, but if it were a commercial event, the technology would allow unauthorised video streaming.

10.3.2.4 Scenario 4: Olympics

The Olympics scenario illustrates how a person may belong to several communities simultaneously, such as family, ad-hoc interest group, and the group of friends (Figure 10.5). This does not make legal issues any easier.

Once again, numerous privacy and data protection issues are presented in the scenario. Logistical information is provided automatically, the system knows their location and directs them to the best transport link. They get notified when they pass locations relevant to their interests, the system recommends services that may be of interest to them based on their previous behaviour, and so on. Therefore, once again, privacy and data protection law is applicable and needs to be obeyed as discussed earlier.

10.3.2.5 Service Framework, Basic Reference Model and Architecture

According to Räisänen *et al.*: 'If privacy is not guaranteed for a user, there will not be sufficient trust towards the system. If users do not trust the system, they will not use it' [25]. This represents well the technology-oriented viewpoint that is characteristic to the scenarios,

Figure 10.5 Olympics scenario.

service framework and applications presented in this book. It implies that the technology alone is often responsible for privacy protection and the end-user's privacy expectations are targeting only the technology. While it provides a reasonable starting point for the technology development, in legal analysis, other aspects need to be considered also. First, it is possible that, in addition to the technology, other factors also support privacy, especially the legal and moral systems in the society. Second, it is plausible that the technology fails, which makes it more important to have alternative means to protect privacy. Third, even today people are using technologies that they do not fully trust.

There seems to be a dichotomy between stated attitudes and actual behaviour of individuals facing decisions affecting their privacy and their personal information security. Surveys report that most individuals are concerned about the security of their personal information and are willing to act to protect it. Experiments reveal that very few individuals actually take any action to protect their personal information, even when doing so involves limited costs.

Technologies as such have little possibilities to directly affect individuals' behaviour with respect to personal information safety, but obviously they have a most important indirect effect. And, of course, even if it is unlikely that the technologies will ever become perfect and flawless, it is still important to try to develop as good technologies as ever possible. Yet, it still pays to study also other means, like regulative and legal aspects, to protect privacy.

Obviously, the technology has also potential to regulate by enabling or disabling behaviour, in comparison with the law that regulates mainly by imposing sanctions. Also, the law has significant limitations. Many undesirable phenomena are out of the reach of the legal system. The law cannot effectively control issues that are hidden or that are not considered to be within the subject matter of the legal system (but, for example, ethical). The technology does not have the same limitations. It is often possible to technologically control issues that are out of the range of the law. Then again, the law can also regulate by influencing the development of the technology. Therefore, it is necessary to consider these approaches simultaneously: Should the law, the technology or the socioeconomic environment and conditions be changed [10, 17]?

As mentioned in Chapter 3, the six essential building blocks of the Mobile Services Architecture are personalisation, adaptation, context awareness, privacy and trust, service management, and service usage. Each of them includes some legal concerns.

Personalisation

The personalisation function provides other functions with profiles and preferences of users and groups. In other words, it processes personal data – unless the profiles and preferences are anonymised in a way that makes it impossible to relate it to an identifiable person. Therefore, data protection law is often applicable.

Context awareness

The context awareness function has additional data protection issues compared to personalisation. First, the automatic fetching of context information (i.e. information about the circumstances of a private person) can make it more difficult to clearly define what information is processed and to inform the person on the processing of that data. Second, the processing of context information involves not only the end-user's personal data, but possibly also information on other people in the proximity. Yet, it is possible to process also context information legally, but a complex set of legal provisions needs to be considered.

Privacy and trust
The privacy and trust function tends to provide technological solutions to privacy and trust problems. It seems, however, as discussed above, that legal and other solutions must not be ignored either. On the other hand, a technology platform like this would be a most interesting target for malicious attackers. For instance, it would be very tempting for many people and companies to install *spyware*, computer programs that covertly gather information on the end-user and delivers it to others. Spyware violates data protection law and may also constitute a criminal offence, but as the information can be extremely valuable, it is likely that there will be lots of attempts to use them.

Adaptation
The user interface adaptation function modifies both users' input modalities as well as services' output. It is also related to other adaptations, such as service quality adaptation. An important legal question here is related to copyright: a lot of content that is to be adapted by this function is copyrighted.

Copyright often enables the copyright owner to object to content modifications. Many content owners (e.g. publishers and media conglomerates, but also individual artists and authors) are concerned about unauthorised adaptations for moral reasons, but also because they are afraid that poor adaptations spoil their valuable brands. A purely technical modification that does not affect the information content, but only data, is typically legal. That is, if changing the file format from one to another or lossless compression has no affect whatsoever to the actual content, then the modification is all right. But if the modification changes the information, and especially if it reduces quality, then it may require the author's permission. If, for example, the resolution of an image of a valuable media character is reduced without permission to fit in a small display of a mobile device, it is likely that the copyright-owner will react. Therefore, it is important to make sure that the adaptation does not violate copyright.

Service usage and service management
The service usage and service management functions provide information about the services, support various forms of service triggering, and support the lifecycle of services. The complex set of services implies that the system as a whole can be distributed to a large extent. There are important legal cross-border issues related to a distributed system like those that implement the architecture. If a system is distributed in several countries, all the applicable laws should be obeyed. For example, transferring personal information even within the system but between organisations and/or countries may violate data protection law. Similar problems arise if the system is connected to other systems. So, both internal and external data processing should be legal. Also, data protection directives are implemented in slightly different ways and they are not applicable outside the EU. Thus there are differences, for example, over which information is to be provided for data subjects (for those whose personal data is processed). These are pretty hard requirements for any system, but especially for systems like those implementing the architecture.

Liability questions in this kind of an environment can be complex. If an ad-hoc group of people is willingly sharing personal information, say context data, but something goes wrong and too much information is shared or some of the information is unwillingly disclosed to outsiders, it can be difficult to find out who is responsible. First, it is difficult to show who was actually doing something wrong, if it was, say, partly a technological failure, partly due to an incorrect configuration that the group members had created together. Second, according to data protection law, the controller is largely liable, but in a dynamic ad-hoc group that distributes data more or less randomly between the members it can be hard to call anybody 'a controller', unless the group as such is the controller. Likewise, if the group violates, for example, copyright law, it can be difficult to show who is responsible unless the whole group can be considered liable.

10.3.2.6 Application 1: Multimedia Infotainer

From the legal point of view, Infotainer (Section 7.5) is a good example of the issues discussed above. Especially, it highlights issues related to personalisation and adaptation. The user profiles are personal data that needs to be processed in accordance with data protection directives and laws (unless they are carefully anonymised). The collector of data needs to specify a purpose for which the data is collected. Personalisation is probably an acceptable purpose, but the collector should be able to specify it. If data is collected for another purpose, they should not be further processed in an incompatible way with that purpose and no inadequate, irrelevant or excessive data in relation to the purpose must be processed. Therefore, one needs to be careful if personalisation avails of data that is collected for other purposes. Also, it must be taken care of that any major decisions are not made based on wrong or partial information.

Content adaptation, on the other hand, might violate copyright, as explained above. A copyright owner has an exclusive right to modify the copyrighted content. If the content is copyrighted, then permission is needed to adapt it. A purely technical modification that does not affect the information content, but only data, is usually legal; but if the modification changes the information, then it requires the consent of the copyright owner.

10.3.2.7 Application 2: Wellness-aware Multimodal Gaming System

From the legal point of view, Wellness-aware Multimodal Gaming System (Section 7.6) is processing personal, even quite sensitive data. It is not clear whether, for example, heart rate as such is sensitive, but refining that information and combining it with other personal data, it may turn into information concerning health. Especially, this kind of system may disclose much more personal data to a bystander than the spectators of a hobby sports event would normally get. As pointed out above, such sensitive data should not be processed at all without explicit consent or other specific lawful reasons, like for example that the processing is necessary for the purposes of carrying out the obligations and specific rights in the field of employment law, or that it is required for medical purposes and conducted by medical professionals. In this case, the controller hardly is the employer of the sports team or a health professional. The other lawful reasons do not apply either. Therefore, the only way to use Wellness-aware Multimodal Gaming System legally is to get an explicit consent in advance from each of the team members. If they are minors, their guardians should give the consent. Still, the controller may be liable in case a third party gets an unauthorised access to the data, and must compensate all the damages even though all care has been taken to ensure that no data leakages would happen.

10.3.2.8 Application 3: TimeGems

TimeGems (Section 7.8) illustrates group formation and related legal challenges. When configuring a group, the questions connected to competence are notable. Who is competent to decide on behalf of the group, how to set up the policies, access rights, and so on? By joining a group and by authorising the group or the application to do something based on the membership, the person relinquishes some of his or her own rights, because the group seems trustworthy enough. For example, the person might relinquish some privacy rights and authorise the group to share some contact information. However, if the group keeps evolving, old members leave the group and new members join in, the person's notion of the trustworthiness of the group may change. The members must be able to cancel authorisations and leave the group if needed. These problems are further discussed below.

Again, some of the user information is personal data that needs to be processed in accordance with data protection directives and laws. The collector of data needs to specify a purpose for which the data is collected. Especially, one needs to be careful if data is used for purposes other than those for which it was collected.

Also, content adaptation, again, can violate copyright, as explained above. A copyright owner has an exclusive right to modify the copyrighted content. If the content is copyrighted, then permission is needed to adapt it. A purely technical modification that does not affect the information content, but only data, is usually legal, but if the modification changes the information, then it requires the consent of the copyright owner.

10.3.2.9 Application 4: MobiCar

From the legal point of view, MobiCar (Section 7.9) once again brings out challenges related to data protection and competences in ad-hoc groups.

First, personal data form an essential part of the system. They need to be processed in accordance with data protection directives and laws. The collector of data needs to specify a purpose for which the data is collected. Especially, one needs to be careful if data is used for purposes other than those for which it was collected.

Second, if a group is authorised to do anything legally significant on behalf of its members, competence issues become important. By joining a group and by authorising the group or the application to do something based on the membership, the person relinquishes some of his or her own rights. As discussed above, the members must be able to cancel authorisations and leave the group if needed. These problems are further discussed below related to the ContextWatcher application.

10.3.2.10 Application 5: FamilyMap

From the legal point of view, FamilyMap (Section 7.7) is another good example of the issues discussed above. FamilyMap highlights primarily issues related to personalisation and adaptation. The user profiles, context data and preferences are personal data that needs to be processed in accordance with data protection directives and laws (unless, again, they are carefully anonymised). The collector of data needs to specify a purpose for which the data is collected. Personalisation is probably an acceptable purpose, but the collector should be able to specify it. If data is collected for another purpose, they should not be further processed in an incompatible way with that purpose and no inadequate, irrelevant or excessive data in relation to the purpose must be processed. Therefore, one needs to be careful if personalisation avails of data that is collected for other purposes. Also, it must be taken care of that any major decisions are not made based on wrong or partial information.

Content adaptation, on the other hand, might violate copyright, as explained above. A copyright owner has an exclusive right to modify the copyrighted content. If the content is copyrighted, then permission is needed to adapt it. A purely technical modification that does not affect the information content, but only data, is usually legal, but if the modification changes the information, then it requires the consent of the copyright owner.

10.3.2.11 Application 6: ContextWatcher

From the legal point of view, ContextWatcher (Section 7.2) is mostly related to privacy and data protection issues. The end-users do have control of their personal information and the system is not supposed to disclose anything more and to any others than what and whom the end-user has specified. Extreme care, however, is needed to ensure that the system works correctly, policies are always up to date, and the dynamical situations are handled correctly. Otherwise, legal liability based on data protection law may take place.

10.3.2.12 Application 7: Personal Context Monitor

Like ContextWatcher, Personal Context Monitor (Section 7.3) is mostly related to privacy and data protection issues – also here the end-users do have control of their personal information. If the information is to be shared to others or within a group, similar legal challenges may appear as described related to, for example, the Wellness-aware Multimodal Gaming.

10.3.3 Comparison with Other Scenarios

As the Mobile Services Architecture in this book emphasises trust and privacy issues, the scenarios highlight them also. Private information is processed a lot. On the other hand, what people do in the scenarios is more or less traditional. They eat together, they prepare presentations, they drive cars and have an accident, and they go to watch a sports event. The scenarios seem to suggest that the everyday life of ordinary people will not change a lot in the forthcoming five or ten years.

To complement the picture that the scenarios above give us on the future, some other scenarios are also briefly discussed and analysed below.

10.3.3.1 SWAMI Scenarios

Summary

The SWAMI project (Safeguards in a World of Ambient Intelligence, funded by the European Commission under contract 006507) looks at the challenges and bottlenecks facing the realisation of the vision of ambient intelligence (*AmI*) as the next stage of the information society in Europe. The project has created an interesting set of dark scenarios and analysed them also from the legal point of view. Unlike most scenarios, SWAMI presents situations in which new technology does not work well and even creates problems [8, 23].

Four SWAMI dark scenarios have been elaborated that encompass both individual and societal concerns, on the one hand, and private and public concerns, on the other hand. These two scenario axes (individual–societal and private–public) have helped to reduce the virtually infinite number of possible futures that could be developed to a manageable number of four [8, 23]:

- *Dark scenario 1*. A typical family in different environments – presents AmI vulnerabilities in the life of a typical family moving through different environments. It introduces dark situations in the smart home, at work and while taking a lunch break in a park.
- *Dark scenario 2*. Seniors on a journey – also references a family but focuses more specifically on senior citizens on a bus tour. An exploited vulnerability in the traffic system causes an accident, raising many different problems related to both travel and health AmI systems.
- *Dark scenario 3*. Corporate boardroom and court case – takes a different stance, involving a data-aggregating company that becomes the victim of theft of the personal data which it has compiled from AmI networks and which fuels its core business. Given its dominant position in the market, the company wants to cover this up but will face the courtroom two years later. The scenario draws attention to the digital divide between developed countries with AmI networks and developing countries without.
- *Dark scenario 4*. Risk society – portrays an AmI risk society from the studios of a morning news programme. It presents an action group against personalised profiling, the digital divide at a global scale and related to environmental concerns, the possible vulnerabilities of AmI-based traffic management systems and crowd control in an AmI environment.

The first two scenarios depict the impact of AmI dark situations on the individual and the family in their everyday life. The impact of the AmI dark situations on the individual is at the micro-level. In scenarios 3 and 4, the impact is on a larger societal scale. The theft of personal data in scenario 3 affects millions of people. Scenario 4 also depicts the societal impact of AmI technologies on privacy, the environment and crowd behaviour. Even though the last two scenarios focus on the macro-level, the problems are still related to individuals' and families' everyday life. Therefore, all the four scenarios are very relevant also from this book's viewpoint, although the technological approach (AmI versus mobile) is slightly different.

Legal analysis

The SWAMI project has accomplished an extensive set of legal analyses on their scenarios. Below, it is mainly referred to their analyses, highlighting issues that are mostly relevant from this book's perspective, but also adding some observations.

The SWAMI project started the analysis of the legal aspects of AmI by describing the existing relevant European law. In the next step, they applied the identified legal framework to the SWAMI 'dark scenarios'. This exercise resulted in the identification of lacunae and problems in the existing European Information Society law. Consequently, in a third step, they moved to the development of legal safeguards addressing key pre-identified threats and vulnerabilities and policy options [10].

SWAMI has highlighted several particularities of the legal regulation of AmI. First, the law is only one of the available sets of tools for regulating behaviour; others include social norms, market rules and the architecture of the technology. On the one hand, the architecture of AmI might well make certain legal rules difficult to enforce (e.g. data protection obligations, copyrights) and it might cause new problems, particularly for the new environment (spam, dataveillance). On the other hand, the architecture has also the potential to regulate by enabling or disabling certain behaviour, while law regulates via the threat of sanction [10].

Another particularity of the legal regulation in cyberspace is the absence of a central legislator. In Europe, the legislative powers are exercised by the Member States, though some powers have been transferred to the European Union, some decision-making competencies have been delegated to the independent advisory organs, and even the technology producers are regulators of the new environments. SWAMI thus recommends allowing all of these actors to play their respective roles, and to involve them in the policy discussion. Development of jurisprudence should also be observed. The legal profession is far from high-level abstract arguments, and tends to solve problems by focusing on concrete situations. Thus, in developing policy options, one should focus on the concrete technologies, and apply opacity and transparency approaches accordingly [10].

10.3.3.2 MC2 Scenarios

Summary

The Mobile Content Communities (MC2) project has created a set of scenarios to study future gaming communities [21]. Two of the themes are briefly introduced and analysed below.

In MC2 scenario 2, a player wants to make a mod (a modification, a change to a published game) but needs help from other players in technical issues. Therefore she gets involved in an on-line community that includes people who are able to solve the technical problems. They start to develop the mod as their hobby, but finally the mod becomes very successful, the development team gets royalties, and some of them are employed by a game development company.

MC2's scenario number 8 is about company-community relations in controlling user-created content. The scenario describes a global online game-play community. When the community matures, there may be a change in power relations between different player groups with different gaming practices, like modding. What is suitable and right for some isn't for others, and one may ask: Whose game is this anyway? The boundaries between fandom and corporate culture change, and there are increasing difficulties in considering different player groups when developing future versions of the game.

Legal analysis

A game can be protected by various intellectual property rights. First, the implementation of the game can be copyrighted. Thus the copyright owners have an exclusive right, for example, to prohibit others from making it available to the public, in either the original or an altered form, in translation or adaptation, in another literary or artistic form or by other technical means. In other words, an adapted game cannot be distributed without the consent of the copyright owners. If a mod can be distributed separately independent on the game itself, the distribution of the mod usually does not infringe the copyright in the game. In some cases, the copyright owner might argue that the mod's sole purpose is to help players to alter the game illegally. At least in some jurisdictions that could make the mod illegal. Of course, many authors of games and computer programs have accepted also the distribution of an altered game in advance by using specific license terms.

The implementation of the mod, on the other hand, can also be copyrightable. Note that a modified game is a derivative work in which both the original authors and those who have modified it have copyright. That is, the players need a licence from both the copyright owner of the game as well as the mod to be able to play the modified game. The copyright in the mod belongs to the modders jointly. That is, it might be difficult to agree on licence terms afterwards, if they have not agreed on them in advance.

Second, in Europe, some games can be partially protected by database *sui generis* right. It provides the maker of a database with the right to prevent extraction and re-utilisation of the whole or of a substantial part of the contents of that database. In accordance with the EU Directive on the legal protection of databases, a database is a collection of independent works, data or other materials arranged in a systematic or methodical way and individually accessible by electronic or other means. To be protected, there needs to be a substantial investment in either the obtaining, verification or presentation of the contents. In principle, the database of the tags within a game could be protected by database *sui generis* right if the game developers have invested substantially in it.

Third, a game may include patentable inventions. Both the original game and a mod may have them. There are limitations on the patentability of games. Schemes, rules and methods for games as such are not patentable. However, an invention related to an implemented game is hardly a scheme, a rule or a method as such.

Therefore a novel invention related to a computer-implemented game, including an inventive step (being non-obvious), and having industrial applicability, could well be patentable. However, the invention is no longer novel, if it has been published. Everything made available to the public in writing, in lectures, by public use or otherwise shall be considered as known and not patentable. On the other hand, patentability also requires that the invention must be technical. As long as an idea is rather abstract and not defined in terms of the technical features of the invention, it is hardly patentable. A game that includes a patented invention cannot be distributed without the consent of patent owners.

Fourth, a game can be trademarked. The names of games are often trademarked. If a trademark protects, for example, the name, then a modified game cannot be distributed using the same name without the consent of the trademark owner. Note, however, that trademark is related only to business: it is usually not applicable to hobby communities that do not make money.

In this scenario, IPR and licence agreements form a tool to control the community. The 'hole' in the licence agreement highlights the difficulties in drafting good agreements. It is unattainable to prepare for all the possible situations.

The contractual framework for a gaming community is complex. It is difficult to build binding contractual relationships between all the members of the community. At least, it is laborious, transaction costs increase rapidly, and the management of contracts gets troublesome. In an unpredictable situation, the existing binding contracts hardly enable the community to sentence a rebelling member to sanctions. Therefore it would be better if the community could form a legal entity, such as a corporation, an association or a cooperative, which has adequate bylaws, reasonably well-defined membership, necessary administration, a clear decision-making process, and an accepted policy to settle disputes among members and between a member and the community.

In MC2 scenario number 8, the community is rapidly spreading all over the world. An internationalising gaming community or company faces mostly the same legal challenges that any growing company heading towards international markets will meet. For example, differences between marketing, IPR and competition laws are difficult to handle. Also, problems related to, for instance, international taxation can be severe. Some of these problems arise faster for a company or a community that operates on computer networks than for a traditional company providing material products. However, in addition to the issues already discussed above, the fundamental nature of those challenges hardly depends on information products and is therefore out of the scope of this analysis.

As long as a game is developed in a hobby community and no one gets paid, the labour and tax laws hardly play any significant role. Compensation could introduce taxation questions, but would hardly bring up labour issues as far as the payer is not controlling the work. When some of the participants become employed by the game development company, the labour law is applied. The company may thereon control their work and is also largely liable for it. Also, during the employment, copyright, database *sui generis* right and patents are usually transferred automatically to the employer. From this chapter's viewpoint, it is interesting to consider whether the community – or some part of it – could form a work team.

10.3.4 Special Legal Framework for the Scenarios

The scenarios, the framework, and the applications as well as the other scenarios described and analysed in this chapter have brought up several legal issues that are concluded below. They form the specific legal framework.

10.3.4.1 Privacy and Data Protection

Computing and communication devices are spreading everywhere in our society. In the future, those devices will become increasingly embedded in everyday objects and places, while communications networks connect the devices together and become available anywhere and anytime. This development is called 'ubiquitous computing' (ubicomp), 'ambient intelligence' (AmI), or 'pervasive computing'. It can be seen partly as a parallel ongoing development with mobile technologies, partly as a successor to them [8, 9].

How are ubicomp or AmI technologies going to affect privacy? It seems obvious that, because devices that are able to exchange information about people are spreading, the *quantity* of privacy problems will arise. The scenarios above illustrate that very well. All of the scenarios include a number of privacy issues. Although privacy problems are not that common today, the scenarios suggest that they will be increasingly ordinary.

But will there be also something else? Will some *qualitative* changes also be likely?

At least two categories of qualitative transforms seem probable. First, people's notions about privacy are changing. People are already getting used to the idea that while they are using, for instance, Internet services, someone can be able to observe their actions. While travelling abroad, one needs to frequently present passports and other documents, even though it makes it possible for authorities to follow people's whereabouts. In the past, that was not possible, but still most people are not concerned about the change. Either they accept the reduction of their privacy, because they think it is necessary or that they get something valuable instead, or they do not care [1]. Anyway, it seems likely that most people will not object to the gradual impairment of their privacy. In the future, people will have a different notion about privacy.

Second, information and communication technologies will no longer affect only informational privacy, but increasingly other sectors of privacy.

One well-known example is Professor Kevin Warwick who carries out research in artificial intelligence, control, robotics and biomedical engineering at the University of Reading, UK. He has shown how the use of implant technology is rapidly diminishing the distance between humans and intelligent networks. In effect, as a human is wired in to the network he/she becomes a part of that ambience. This can have a tremendous impact in the treatment of different neural illnesses. There is a number of areas in which such technology has already had a profound effect, a key element being the need for a clear interface linking the human brain directly with a computer [28].

Professor Warwick's own research has led to him receiving a neural implant which linked his nervous system bi-directionally with the Internet. With this in place, neural signals were transmitted to various technological devices to directly control them, in some cases via the Internet, and feedback to the brain was obtained from such as the fingertips of a robot hand, ultrasonic (extra) sensory input and neural signals directly from another human's nervous system [28].

Professor Warwick's examples show how the technology can be used also to observe and control the human being through computer networks from a distance. It is possible to even affect his brain's decision-making process [28].

Until now, the developing information and communication technology has threatened only informational privacy. Professor Warwick's examples nevertheless clearly show that emerging technologies are not limited to that: they are also capable of jeopardising the other components of privacy. This implies a major qualitative change in privacy problems.

For obvious reasons, especially medical scientists have been interested in ethical and legal questions on privacy in families. For example, if they study a disease that appears to be inherited in some families, they want to collect information not only on research subjects, but also on the whole pedigree. Based on his studies on medical pedigree research, Cook-Deegan has shown that studying a family does not reduce to studying a group of individuals one at a time. This opens the door to legal and moral concepts applied to collectives rather than individuals [4]. Therefore it seems that a family, possibly also any community, may have privacy other than just the sum of its members' privacies.

10.3.4.2 Intellectual Property Rights and Digital Rights Management

Copyright will have a central role in the information society. The right to copy, modify and distribute information products and services – that is, to benefit economically from them – is based on copyright. Therefore, many business models will increasingly depend on copyright. Traditionally, the most important part of copyright has been the exclusive right to make copies. Currently, the situation is changing. The way computers, networks and other digital devices work means that information is all the time copied and copied again. It is no longer essential or even possible to restrict copying, but to try to manage the access to information. Also, as discussed above, the adaptation of content for various devices will become an increasingly important issue. On the other hand, copyright also provides the author with moral rights: for many people, especially amateurs, it is not so vital to make money from the works they have created, but to get credited as an author. Therefore, copyright can be important also for non-profit communities.

Digital Rights Management (DRM) refers to copyright technical protection. Often, it is not enough that the law stipulates the rights of the copyright-owner. Especially, the content industry has required technical tools that give them additional protection. DRM technologies are usually based on encryption: data is encrypted in a way that unauthorised access to information is difficult. A DRM system allows the end-user to access the information – listen to a piece of music, watch a video or play a computer game – only in accordance with the licence terms that are expressed in machine-readable rights expression language (REL). Obviously, the most important licence term is usually that the end-user must pay for the usage in advance. Also, the licence terms may restrict how many copies of the product the end-user may produce, and in how many devices those copies can be used.

DRM technologies cannot protect data completely. It is always possible to circumvent the protection. Sometimes the circumvention is difficult and requires special skills, sometimes it is very easy. Yet, the content industry has lobbied for anti-circumvention rules. In recent years, copyright law has been amended to include this legal protection for DRM.

Digital Rights Management is often considered to be harmful for consumers and other end-users. Yet, there are situations in which an ordinary person may benefit from DRM technologies. As described above, copyright also protects works by common people and non-profit communities. If they want to be sure that their moral rights are respected, they may apply some sort of lightweight DRM technologies that do not necessarily limit accessing the information, but make sure that the work is always attributed to the creators.

On the other hand, DRM technologies may also involve severe privacy problems. Although DRM is meant to ensure copyrights protection, it often also manages information about end-users, their behaviour and their preferences. Therefore, a DRM system should also comply with data protection law.

Other intellectual property rights, including patents, trademarks, database *sui generis* right and domain names, will remain important, but the scenarios do not suggest that their relative importance will grow remarkably.

10.3.4.3 Contractual Relationships and Consumer Protection

Mobile technology introduces new kinds of contractual challenges. For instance, while users are moving, they have many kinds of wireless devices, and their access points keep changing, it can be ever more difficult to identify who the user is. From the contractual viewpoint it is troublesome if the other contracting party is not able to be sure who the other party is. This can be helped using, for example, digital signatures that are certified by a trusted third party. However, that requires technological solutions which will not be available in the near future.

Consumer protection law protects individuals against unfair trade and credit practices. It does not ensure just the safety of goods and services, but also those economic and legal interests that will enable consumers to shop with confidence. The scenarios depict a somewhat wild future world in which various applications and services are provided through networks by numerous providers. It will be challenging for a consumer to know which providers are trustworthy and with whom it is safe to transact. Consumer protection law will have a difficult but increasingly important role to increase consumers' trust and to enable business.

10.3.4.4 Other Legal Areas

A number of other legal areas may be involved in mobile scenarios. For example, the law of unfair business practices may sometimes be applicable. According to the Austrian Supreme

Court, it is an unfair business practice to avail of information, like cell-IDs, in which another business, like an operator, has invested substantially. That would make it illegal to use most of context information in commercial services. It seems that the Austrian High Court is not thinking along the same lines as others in the Western countries: facts should not usually be protected by law and the investment in this case is not directly in the information but in the system as a whole. Therefore, this decision is probably not applicable in other countries. Also, although in Austria there seem to be several cases in line with this one, it is likely that even the Austrian Supreme Court will need to change its line in the future. However, it illustrates how complicated the legal jungle can be even within Europe.

10.4 Conclusions

The central question in this chapter has been the legal essence of mobile ad-hoc communities. Unfortunately, it is impossible to give a precise answer. Instead, the factors and characteristics that affect the legal evaluation of those communities and competence to act on behalf of them have been discussed above. Essentially, mobile ad-hoc communities are only seldom legal entities, but usually they are considered as some sort of contractual arrangements.

Then again, the contracts that the communities are based on are typically implied and their content, terms and conditions are difficult to find out precisely. If disputes arise and the members of the community are not able to resolve them by themselves, ultimately a judge needs to decide what the valid agreement is and how it should be interpreted in the current situation. In default of a written contract, the parties try to show in the court the terms and conditions by hearing witnesses. People often have very different views, impressions and understandings on what their mutual intention had been earlier. Therefore, it is very difficult to construct afterwards a coherent expression on an agreement that has not been articulated properly before. The judge needs to make difficult deductions which may lead to unsatisfying rulings.

Having said that, ad-hoc communities hardly ever have carefully drafted written contracts. The very basic nature of ad-hoc communities suggest that they do not cautiously prepare documents. Therefore, it has to be assumed that ad-hoc communities usually are:

- contractual arrangements;
- without written contracts.

Even if that sounds contradictory, the legal analyses of ad-hoc communities must be based on their vague contractual nature. If ad-hoc communities are becoming more significant in society, the lawmaker is required to set down general rules to reduce uncertainty.

Sometimes, the contractual nature of a community can get a more specific form because of the activities that the community performs, if they are governed by special legislation. Especially, if the community is involved in gainful employment, it is possible that labour and employment law is to be applied.

If a community as a whole has committed to work for an employer, it can be possible to say that the community is a work team and that the labour law applies to it. In that case, the employer has the right to direct and control the team's work. In the relationship between the work team and the employer, the spearhead of the team or the foreman has the authority to negotiate on behalf of the other members. It also means that the employer is entitled to benefit from the work, to get profit, and to obtain the rights in the outcome of the work team's efforts. Therefore, it is less important to consider competence issues as the employer is entitled by law to most of the value produced by the team.

On the other hand, the labour law also stipulates that the employer has important duties towards the employees. The employer must pay significant side costs in addition to the salaries. Also, the employer must obey, in addition to the terms and conditions of the employment contract, the provisions of labour laws and possible collective agreements. In summary, if an ad-hoc community is a work team, the competence issues become less ambiguous and more foreseeable, but the contractual freedom is more limited.

If the community is not a work team and employment and labour law is not applicable, it could be considered as a civil law partnership, an unregistered association or just a contractual arrangement. It means that the competence is not divided by law, but the members are free to authorise one of them or an outsider to act on their behalf. Usually, such an explicit authorisation is not made and therefore it is necessary to reason afterwards, if any disputes arise, what the members intended and understood concerning the authorisation. Typically, there is a spearhead or a leader even in an ad-hoc community who is authorised to do something on behalf of the others by mutual understanding. It can be nevertheless extremely challenging to interpret afterwards the actual content of that understanding

The scenarios, the framework and the applications as well as the other scenarios described and analysed in this chapter have introduced numerous legal issues in several legal areas. They form the specific legal framework.

Privacy and data protection law is highlighted especially in the scenarios and in their analyses. The scenarios, the service framework and the applications underline issues related to personalisation and adaptation. The user profiles are often personal data that needs to be processed in accordance with data protection directives and laws.

Recent developments, especially emerging ubiquitous computing and ambient intelligence technologies, suggest that no longer is it only informational privacy that is threatened by new technology, but also the other components of privacy, like physical, decisional, dispositional and proprietary privacy. Another qualitative change in privacy is that people's notions and expectations about privacy are changing: people are gradually accepting some forms of lessening privacy.

Other important legal areas in the framework are intellectual property rights, especially copyright, and contracts. Content adaptation, especially, might violate copyright. The scenarios, service framework and applications emphasise the adaptation of content. However, a copyright owner may have a right to object to modifications.

Digital Rights Management (DRM) or copyright technical protection poses issues in relation to both intellectual property law and data protection law. DRM systems are meant to prevent unauthorised copying and to control the usage of content in accordance with licence terms and conditions. However, DRM systems also typically process user information. Thus the data protection law may apply to them too.

10.5 Acknowledgements

The following individuals contributed to this chapter: Olli Pitkänen (Helsinki Institute for Information Technology HIIT, Helsinki University of Technology and University of Helsinki).

References

[1] Acquisti A. and Grossklags J.: 'Privacy Attitudes and Privacy Behavior: Losses, Gains, and Hyperbolic Discounting'. In Camp J. and Lewis S. (eds.): *The Economics of Information Security*. Kluwer Academic Publishers, 2004.

[2] Allen A.L.: 'Coercing Privacy'. *William and Mary Law Review*, Vol. 3, No. 1, 1999.

[3] Allen A.L.: 'Is Privacy Now Possible? A brief history of an obsession'. *Social Research*, Vol. 68, No. 1, 2001.

[4] Cook-Deegan R.M.: 'Privacy, Families, and Human Subject Protections: Some lessons from pedigree research'. *Journal of Continuing Education in the Health Professions*, Vol. 21, 2001.

[5] DeCew J.: 'Privacy'. In Zalta E.N. (ed.), *The Stanford Encyclopedia of Philosophy* (Summer 2002 Edition), 2002. Online: plato.stanford.edu/archives/sum2002/entries/privacy/.

[6] Erma R.: 'Työurakka Suomen velvoiteoikeuden mukaisena sopimustyyppinä'. *Suomalaisen lakimiesyhdistyksen julkaisuja*, B-sarja 69, 1955.

[7] Field T.G.: *Introduction to Intellectual Property*. Franklin Pierce Law Center, 2000.

[8] Friedewald M., Vildjiounaite E., Wright D., Maghiros I., Verlinden M., Alahuhta P., Delaitre S., Gutwirth S., Schreurs W. and Punie Y.: 'The Brave New World of Ambient Intelligence: a state-of-the-art review'. Safeguards in a World of Ambient Intelligence (SWAMI). Deliverable D1. A report of the SWAMI consortium to the European Commission under contract 006507, 2006.

[9] Gow G.A.: 'Privacy and Ubiquitous Network Societies'. International Telecommunication Union, ITU Workshop on Ubiquitous Network Societies, Document UNS/05, 2005.

[10] Gutwirth S., De Hert P., Moscibroda A. and Schreurs W.: 'The Legal Aspects of the SWAMI Project'. In Friedewald M. and Wright D. (eds), 'Safeguards in a World of Ambient Intelligence (SWAMI)', Deliverable D5, Report on the Final Conference, Brussels, 21–22 March 2006.

[11] Haarmann P.-L.: *Tekijänoikeus & Lähioikeudet*. Kauppakaari, 1999.

[12] Halila H.: 'Henkilöoikeuden Perusteet', 2005. Online: www.helsinki.fi/oik/tdk/private_law/siviili/oppimateriaalit/Halila2005_Henkilooikeudenperusteet.pdf.

[13] Halila H.: 'Toimivaltajako Yhdistyksissä'. *Suomalaisen lakimiesyhdistyksen julkaisuja*, A-sarja 196, 1993.

[14] Jacob Sir R., Alexander D. and Lane L.: *A Guidebook to Intellectual Property: Patents, trade marks, copyright and designs*. Sweet & Maxwell, 2004.

[15] Kairinen M., Koskinen S., Nieminen K. and Valkonen M.: *Työoikeus*. WSOY Lakitieto, 2002.

[16] Killström U., Galli L., Haaker T., Immonen O., Kijl B., Pitkänen O., Saarinen P.J. and Virola H.: 'Initial Marketplace Dynamics (incl. Business Models) analysis'. IST-MobiLife Deliverable D07b (D1.2b), 2006. Online: www.ist-mobilife.org.

[17] Lessig L.: *'Code and Other Laws of Cyberspace'*. Basic Books, 1999.

[18] Merges R.P., Menell P.S. and Lemley M.A.: *Intellectual Property in the New Technological Age*, 2nd edn. Aspen Law & Business, 2000.

[19] Pekkanen R. 'Sekatyyppinen Työsopimus'. *Suomalaisen lakimiesyhdistyksen julkaisuja*, B-sarja 127, 1966.

[20] Pitkänen O., Virtanen P. and Välimäki M.: 'Legal Protection of Mobile P2P Databases'. International Conference on Law and Technology (LawTech 2002), Cambridge, MA, USA, 2002.

[21] Pitkänen O.: *Legal Challenges to Future Information Businesses*. HIIT Publications 2006-1, Helsinki Institute for Information Technology HIIT, 2006.

[22] Pitkänen O.: 'Legal and Regulation Framework Specification: Competence within mobile families and ad-hoc communities'. IST-MobiLife Deliverable D11 (D1.6), 2006. Online: www.ist-mobilife.org.

[23] Punie Y., Delaitre S., Maghiros I., Wright D., Alahuhta P., De Hert P., Friedewald M., Gutwirth S., Lindner R., Moscibroda A., Schreurs W., Verlinden M. and Vildjiounaite E.: 'Dark Scenarios in Ambient Intelligence: Highlighting risks and vulnerabilities'. Safeguards in a World of Ambient Intelligence (SWAMI). Deliverable D2. A report of the SWAMI consortium to the European Commission under contract 006507, 2006.

[24] Robertson D.: *A Dictionary of Human Rights*, 2nd edn. Europa Publications, Taylor & Francis Group, 2004.

[25] Räisänen V., Karasti O., Steglich S., Mrohs B., Räck C., Del Rosso C., Saridakis T., Kellerer W., Tarlano A., Bataille F., Mamelli A., Boussard M., Andreetto A., Hölttä P., D'Onofrio G., Floreen P. and Przybilski M.: 'Basic Reference Model for Service Provisioning and General Guidelines'. IST-MobiLife Deliverable D34b (D5.1b), 2006. Online: www.ist-mobilife.org.

[26] Vuorio J.: 'Työntekijän käsitteestä työoikeuden eri aloilla'. *Lakimies*, 1952, pp. 690–721.

[27] Vuorio J.: 'Työsuhteen ehtojen määrääminen'. Tutkimus Suomen työoikeuden normi-järjestelmästä. *Suomalaisen lakimiesyhdistyksen julkaisuja*, B-sarja 76, 1955.

[28] Warwick K.: 'Wiring in Humans: Advantages and problems as humans become part of the machine network via implants'. Presentation in SWAMI Final Conference in Brussels, 21–22 March, 2006. Summary of the presentation in Friedewald M. and Wright D.: 'Safeguards in a World of Ambient Intelligence (SWAMI)', Deliverable D5, Report on the Final Conference, Brussels, 21–22 March 2006.

[29] Wilhemsson T.: 'Kertomus parisuhteesta ja sopimuksesta'. *Oikeus*, Vol. 1, 1999, pp. 66–74.

[30] Wilhelmsson T. and Jääskinen N.: *Avoimet yhtiöt ja kommandiittiyhtiöt*. Kauppakaari–Lakimiesliiton kustannus, 2001.

11

Conclusions

Edited by Mika Klemettinen (Nokia, Finland)

11.1 What this Book has Covered

People today take part in varying social contexts and play different roles in their everyday life. This book has addressed the need to manage complex lifestyles by offering facilities and tools to support communication, and share information and time with others.

The user research carried out for this book involved more than 200 end-users, from families to junior football teams and beyond, in qualitative studies with interviews and demonstrations in Finland and Italy. This gave updated information of the users' needs to support their everyday life by means of mobile services and applications, as depicted in Chapters 2 and 8. Based on the needs and requirements, several applications and services were concepted and reference implementations were created (Chapter 7), utilising the enabling technologies and components introduced in this book. The user-centred design (UCD) work was complemented by business and regulatory framework modelling activities (Chapters 9 and 10), thus leading to coverage of new perspectives on the future ecosystem related to mobile applications and services.

The Mobile Services Architecture introduced in Chapter 3 combines multiple enabling technologies together with well-defined interfaces in a novel way following the Service Oriented Architecture (SOA) concept. The architectural solutions can be applied either together or separately, in Internet or IMS-style environments.

In the technological area of *context awareness* (Chapter 4), new domain-independent specifications for context were given and a well-specified API that enables service developers to easily utilise context information in services and applications was presented. These specifications are available in the *Context Management Framework* (CMF) of the *Context Awareness*

Enabling Technologies for Mobile Services: The MobiLife Book Edited by Mika Klemettinen

Function (CAF). Furthermore, ontologies were created that describe generic concepts for users, their context and high-level situations.

In *personalisation* (Chapter 5), this book promoted a unified API enabling personalisation, while allowing exchanging personalisation; i.e. profile learning algorithms. This plug-in mechanism allows swapping currently used algorithms without the need to implement changes in services and applications that use the personalisation API. The personalisation features of the Mobile Services Architecture are offered through the *Personalisation Function* (PF).

For *user interface adaptation*, or *multimodality* more generally (Chapter 5), the *User Interface Adaptation Function* (UIAF) was introduced that is a functional component doing fusion and fission independent of context, supporting multiple modalities. This is the first approach where the full potential of device independence is realised in a complete component, and it is still independent from specific services and service domains. All services and applications can take full benefit of the device independence, so that service developers can offer their applications to a wide audience with different modalities. UIAF is designed in an extensible way so that new device descriptions and content transformers can be added very easily.

In the area of *privacy and trust* (Chapter 6), this book focused on the privacy of users and their trust in the service infrastructure, services and applications. Additionally, the approach that is available through the *Privacy and Trust Function* (PTF) takes context information into account for privacy and trust management.

Regarding *group awareness* (Chapter 6), mainly the OMA Device Management initiative has been trying to specify group issues in a formal way, but it was not primarily intended for the provisioning of application information. In the *Group Awareness Function* (GAF), this book considered static as well as dynamic groups and their trust relations, including visualisation aspects.

After the technologies and architectures presented in this book are commonplace, what will the services landscape look like? Many projects and companies aim at developing enabling technologies, especially re-usable components that can be used like Lego blocks to quickly assemble lots of different types of service. Furthermore, it is envisioned that the services (or some of them) will be dynamically formed from the components based on the available resources and context of the user.

From the end-users' point of view, this design rationale – or rather lack of design rationale – can result either in highly targeted and task-efficient applications, or unprecedented chaos (or something in between the two). Instead of generic and open-ended applications like calendars, email, camera and messaging, will there be single-purpose applications for going out with friends? Will there be location-aware applications for people who are pushing baby prams? How many different types of application can there be, so that collection is still understandable and useful from the end-users' point of view? Will there be interested providers for all of these or even part of these applications? Whose responsibility is it to create and provide both content and context to these applications?

In order to tackle the above questions, user-centred design in a limited sense is not enough; one has to involve other stakeholders, as has been done in this book. This calls for a wider framework (the *Collaborative Service Engineering Framework*) that ties together all the relevant stakeholders to jointly create applications and services that the users want, that are business-model-wise viable, and that follow best practice in service engineering.

11.2 Acknowledgements

The following individuals contributed to this chapter: Andy Aftelak (Motorola Ltd, UK), Mika Klemettinen (Nokia, Finland), and Jukka T Salo (Nokia, Finland). This chapter also presents the joint vision and motivation behind the MobiLife project [www.ist-mobilife.org], which numerous people have contributed to and which has materialised partially in the form of this book.

Appendix

At the beginning of the 'Big Loop' user-centred design process, based on the high-level tasks and behaviours presented in Chapter 2, four scenarios were created: the 'Monday', 'Friday', 'Sunday', and 'Olympics' scenarios. Each story is a sequence of related episodes depicting the life of a hypothetical family. Each scenario focuses on a different aspect of life: planned activities during the work week, dealing with unexpected events during the work week, enjoying leisure activities at the weekend, and taking a special outing as a family. Each story has been divided into episodes, identified by a letter denoting the story and a number denoting the step of the scenario (e.g. M.1 for the first step of the 'Monday' scenario).

A1: The Monday Scenario

The family includes the mother (Nicole), the father (Franco), and their two young children, who live with them. Franco also has another child from a previous marriage, who lives with their mother (Elaine). Two of Nicole's work colleagues, Peter and Ann, participate in the story as well.

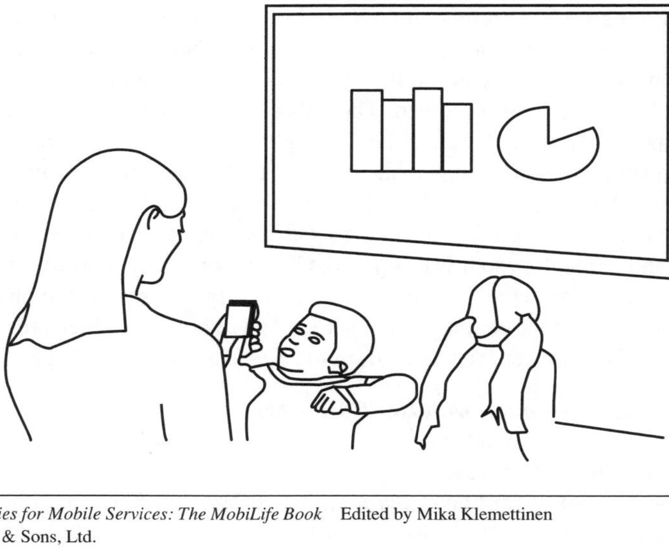

Enabling Technologies for Mobile Services: The MobiLife Book Edited by Mika Klemettinen

M.1: Nicole Wakes up Early to Work on a Presentation

At 6a.m., the alarm goes off. Nicole gets up, while Franco stays in bed.

Nicole has to prepare for an important meeting with a customer later in the day. She needs to collect information for a presentation, partly from colleagues and partly from other sources. Nicole puts into her device a list of keywords, names of collaborating colleagues, and known sources of information. Her company's information management system finds the raw material and delivers it to Nicole's device so that she can construct her presentation more easily. Nicole settles down in the living room with a cup of coffee. She selects the command on her device to switch the information to the large plasma screen in the room. This allows the details of the pictures and graphs in the information to become perfectly visible. She proceeds to assemble the presentation on the plasma screen using her personal mobile device as a controller.

M.2: Franco Gets up and Wakes the Children

At 7a.m. Franco's alarm goes off just in time for him get the children up for school. Because Nicole has been so busy this week, Franco has been helping by doing most of the domestic work associated with the house and their children. While preparing breakfast, he receives an alert/notification from his mobile device. He has received an audio video message from his ex-wife Elaine asking him if he can pick up their son from the school today at 3.30 p.m. as she has to take another of her children for a dental check-up. At the end of the message Franco has the option to 'accept', 'decline' or 'contact'. His schedule for the day has been automatically pulled up underneath with the possible task ghosted into it. Nicole's schedule is shown underneath his. The only things showing in her schedule today is the important client meeting in the afternoon, after which she is scheduled to pick up their own children from the school.

Franco accepts the request and it is fixed into his schedule. He prefers to arrange things with his ex-wife in this way, as when they speak on the phone it can sometimes be difficult. Elaine is automatically sent the acceptance notification.

M.3: Nicole Balances Duties

Nicole is still working on her presentation when the kids come downstairs. They both jump up on the couch next to Nicole and give her a hug to wish her good morning. Her youngest has his jumper on the wrong way around! While she is helping him with it, the attention of her eldest child turns to the plasma screen. The kids usually watch cartoons before breakfast, and she looks expectantly at her mother. Nicole giggles and raises her eyes.

While still adjusting her son's jumper, she uses her mobile device to swap the presentation from the big screen to her laptop, which is sitting on the coffee table. To do this she presses a single button and a list of trusted devices in the local area appear. Nicole tends to flick mostly between the TV and her laptop, so this is top of the list and only a click away. Her presentation appears on her laptop. The screen resolution automatically adjusts for the change in screen size. A brief message appears on the plasma screen 'Switching customer presentation.ppt to Nicole's Laptop' before Tom & Jerry appear on the TV. The kids both cheer.

M.4: Nicole Checks Presentation with Her Co-workers

After breakfast Franco takes the kids to school on his way to work. Nicole sees them off before settling back down to finish her presentation.

Nicole is going to work from home this morning in order to finish the presentation. Unfortunately the data search Nicole carried out earlier has not returned all the information she needs to complete the presentation. Her colleague Peter was working on a couple of slides for her that the search did not find.

Nicole sends a request to have a video conference with Peter who is in the office and also with her boss Ann who is driving into work. Peter and Ann accept the request and this automatically opens Nicole's webcam. Nicole shares the presentation with them both.

The presentation is automatically adjusted so that it is presented in the best way to suit Peter's and Ann's devices and contexts. Peter sees the presentation just as Nicole does, but Ann receives an automatically generated audio dialogue of the presentation content. This allows her to keep her eyes on the road. Peter adds the missing slides and they finalise the presentation together. Ann gives Nicole some advice on how she should pitch to the customer.

M.5: Nicole's Free Time in the City

Nicole has finished the presentation earlier than she expected. As the customer presentation is not until early afternoon, she decides to drive into the city rather than hang around the house. The city has lots of attractions and she is sure she will be able to find something to do. She drives into the centre and parks her car close to the customer's premises.

Nicole looks for suggestions of things to do using her mobile device. The device asks how long she has available. Rather than selecting a time she selects the 'until my next appointment' option. This is useful as the system can then take into account travelling time and unexpected things such as busy queues in the recommendations it provides. Nicole doesn't want to be late for the meeting! She is offered a short list of attractions and events in the local area, which have been specially selected for her based on her personal preferences. She also has access to a much larger list if she doesn't fancy any of the suggestions. Nicole pays a small monthly subscription charge to her mobile operator for this service. She doesn't mind as she is a very active supporter of the arts in the city and by booking through this application she can make good savings on the admission prices for attractions such as the theatre and museum. Nicole even saved money when she took the kids to the zoo by booking the tickets using this service.

As it happens, one of the shortlist recommendations is a new Rodin exhibition at the city art gallery. She selects further details before buying an E-ticket for the exhibition, which saves her 20% compared to paying on the door.

M.6: Rescheduling a Sailing Trip with Friends

The use of mobile devices is not allowed in the gallery, so Nicole's device is automatically set to message divert as she enters. Signs around the building inform her of the gallery's policy. She enjoys walking around the exhibition.

Nicole receives an alert/notification from her mobile device as she leaves the gallery. It is a message from a friend suggesting that a boating trip they had planned with others on Saturday should be rescheduled as the weather over the first part of the weekend is forecasted for heavy showers. Her friend suggests anytime over the following weekend if everyone is free. Nicole visits the boat company's website using her device, they have a booking calendar which shows which boats are available and when. Her schedule has been pulled up underneath. The currently planned boat trip is shown on it (Nicole didn't need to enter this by herself, the trip was automatically added to her schedule when she agreed to her friend's invitation last week). Her friend has shared her calendar with Nicole for the following weekend. From this information Nicole is able to shift the trip over into another slot where the boat they want is available and when they are both free. Nicole provisionally changes the booking with the boat company by altering her entry in their calendar and then shares the company's calendar with her other friends. Almost instantly she gets messages back from two of her other friends that the provisional change is fine. Only one member of the group of friends, Annalise, has yet to confirm the change.

M.7: The Presentation

Nicole receives a reminder about the meeting. The reminder gives her enough warning to ensure she would still have time to make it to the meeting even if she was still at the exhibition, but Nicole is well aware of her afternoon appointment and is already heading there.

Just before she enters the building, Nicole receives a phone call from her friend Annalise. Unfortunately Annalise has arranged to visit relatives the weekend of the rearranged boating trip. Annalise suggests the trip goes ahead without her as she can see from the calendar Nicole shared with her that she is the only one who is unable to attend and the boat they want isn't available at a weekend after the rearranged date for another three months! Nicole reluctantly agrees and wishes Annalise a pleasant visit with her family. They loosely plan to meet up for a coffee when she gets back. Nicole is quite disappointed about this, as Annalise is a good friend but Nicole hasn't had chance to catch up with her for a long while and saw the boating trip as a good opportunity.

Once Nicole arrives at the customer's premises, she is welcomed and shown into the meeting room. Nicole sits down and opens the presentation on her mobile device. With a button press she selects to switch the presentation to a remote device; this is a special option that shows public and unsecured devices outside of her trusted list. The projector in the meeting room is the only device that appears on the list, which she selects. The presentation appears on the projector screen and Nicole is able to start her presentation.

A2: The Friday Scenario

The family is composed of father Giovanni (33), mother Maria (29), and their daughter Elisa, who is aged 20 months.

F.1: Traveling by Car on What Will Turn Out to be a Bad Friday

Maria and her daughter Elisa are in the car. Her husband Giovanni is still working. It's the afternoon, around 4.30p.m.

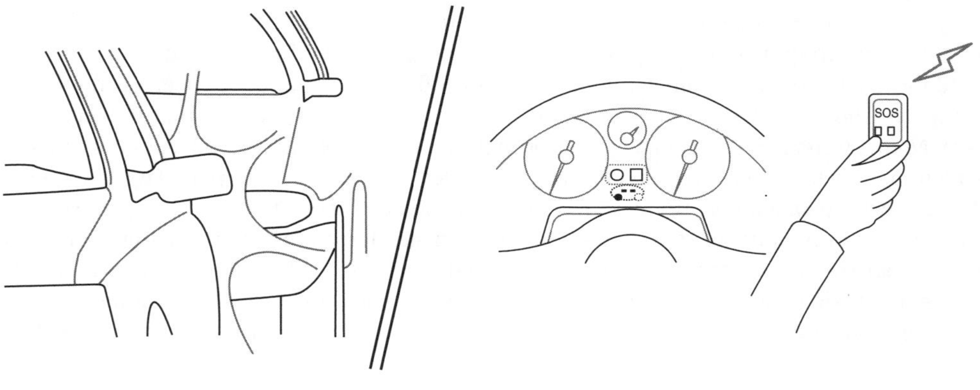

Maria and Elisa have just left Elisa's grandmother's house who they have been visiting. They are on a journey to pick Giovanni up from his work. Giovanni always works from his office on Fridays as he has a weekly meeting with his team. Even though Maria is the main user of the family car, Giovanni would normally drive to work himself on Fridays, but Maria needed the car today. It will take Maria and Elisa about 45 minutes to reach his work.

Giovanni is still working at his office, which is located on the opposite side of the city. Maria is supposed to pick him up around 5.30 p.m.

There is a short trip planned for the weekend (by car). An unplanned strike by one of the major unions impacts the local bus services.

F.2: Car Accident! A Delivery Van Bumps into Maria's Car

A delivery van bumps into Maria's car. The collision is minor, but it has caused damage to the bumper and rear wheel of Maria's car. Maria is obviously concerned about Elisa, but she appears fine and is giggling away in her car seat.

Maria composes herself. Her mobile device starts to bleep. The car has sensed the accident and has sent a message to Maria's device to start emergency procedures to help her. The delivery driver appears at Maria's window apologising, he is just thankful everyone is OK.

Maria looks back at the screen on her device, it shows a message confirming a vehicle accident and asks if anyone has been hurt? She clicks no. This prevents any spurious alerts being sent to the emergency services.

The next prompt asks her if she wishes to start help procedures. Maria clicks yes. Automatically messages are sent to Giovanni and the insurance company. The insurance company will be giving Maria top priority for recovery. The information sent to the insurance company not only includes the car's diagnostic damage report but also confirms that she has a child on board.

Giovanni convinced Maria to opt for this special service from the insurance company when Elisa was born. He was very concerned about Maria and Elisa ever getting stranded with car trouble without him. For this service Maria pays a slightly higher insurance premium.

A tow truck is dispatched straight away from a local garage. Because Maria selected that no one was hurt, the message Giovanni receives is very reassuring regarding Maria and their child's wellbeing, but suggests he should contact them at the earliest opportunity.

F.3: Giovanni Checks on Maria and Elisa after the Accident

Giovanni receives the message and immediately steps out of his meeting to call Maria, he is obviously concerned. Maria reassures him that they are both fine.

To put his mind at ease, Giovanni is also able to check Elisa's basic health status via sensors in her car seat. This feature is used generally by Giovanni and Maria for baby monitoring. Elisa has sensors in her car seat, pushchair and cot. The information Giovanni receives puts his mind at rest that all is well. He's glad he persuaded Maria to go for the special insurance cover after all.

F.4: Car Drivers Close to the Scene Receive an Alert About the Accident

Maria's car and the delivery van's system propagate a message all over the area. The other drivers nearby receive an alert about the incident well in advance of the exact point where the accident happened. This allows approaching drivers to avoid the area, reducing congestion and making the accident area safer for everyone.

A couple of drivers who saw the accident approach; they ask if everything is okay.

F.5: A Short Video Tutorial is Broadcast

A short video tutorial provided by the insurance company on what to do in case of an accident is sent to Maria's mobile device.

Passers-by receive a message from the insurance company propagated through Maria's device asking them to be witnesses. The drivers who saw the accident both answer yes. Their contact details are automatically encoded and sent back to Maria's device. She won't be able to view them but the insurance company will.

The tutorial asks Maria to take some pictures of the damage using the camera on her device. After she does this, an electronic claim document is automatically generated. This includes the photos, witness details and information from the car's systems, all of which is sent to the insurance company in a single package.

At this point the police arrive. They were patrolling in the area and received the broadcast from Maria's car. They just want to make sure the traffic is kept moving. Shortly afterward the tow truck also arrives.

F.6: Giovanni Arranges a Lift Home

With no lift from Maria, Giovanni is wondering how he is going to get home tonight. No one in his office lives in his area of town, so he has a problem. He decides to check the city's car-sharing service from his computer. This is a very popular service because within the inner city there are express traffic lanes that can be used only by cars with two or more people in.

Before he can use the system, Giovanni has to register. He chooses 'Giovanni_Spuri99' as his user ID. He searches for drivers in the area local to his office who commute through his local neighbourhood. The system returns two drivers who match his requirements, 'kelly176' and 'Marco'. Both drivers have been given good ratings for safe driving and friendliness by people who have travelled with them in the past, Giovanni knows that as a male user of the system with no feedback his prospects of getting a lift are not good but he sends a lift request to them both anyway.

Almost immediately 'Kelly176' responds. She notes in her posting that due to personal safety reasons she is willing to offer lifts only to other women, and besides the system must have her categorised as working in the wrong area, as even though she does commute through Giovanni's neighbourhood she actually works on the other side of town! Things aren't looking too good for Giovanni.

About twenty minutes later 'Marco' responds. He asks if Giovanni is the Giovanni Spuri who is friends with Marco's brother Alberto who lives in Giovanni's neighbourhood? As luck would have it Giovanni is Alberto's friend. It turns out that Marco works in the next building and passes Giovanni's area on his drive home out of the city. As Giovanni doesn't have any ratings, lots of people would be put off, but luckily Marco can vaguely remember him from a party at his brother's last Christmas, so agrees to the request. He is always happy to help out a friend of his brother.

F.7: Elisa Needs Changing

Back at the accident scene the police and the delivery van driver have both left. Maria is putting Elisa into her pushchair as her car is about to be towed. Crossed wires at the garage meant the tow truck that turned up had room for the driver and his assistant only, so apologetically they leave Maria by the roadside. So much for the insurance company making her top priority! As

Elisa is removed from her car seat and transferred to her pram, the sensors in the pushchair identify her presence and the chair's systems become active.

Maria sets out towards the nearest bus stop following the directions offered by the system on Elisa's pushchair. Luckily the weather information is showing a sun. Elisa is not happy and is now crying. Maria thinks she needs a nappy change; after all that has happened today this is the last thing she needs!

As they move down the street, the system is receiving special offers and advertisements related to mother and baby items from shops in the area. Maria really isn't interested in these at the moment. She enabled them last week when she was out shopping with Elisa but wants a different sort of information now. She disables these with one touch of the screen. She then selects 'locate' and then 'baby changing facilities' from the drop-down list. There are some public facilities down the road but Nicole notices there is a child-friendly coffee shop around the corner which offers changing facilities; so she sets out to find this intent on a sit down and a calming cup of coffee.

It is indeed a sunny day and it's hard to see the screen on the device. As Maria approaches the coffee shop the handle on the buggy vibrates momentarily. A good job too as the shop is set back off the road and she could easily have missed it.

F.8: Maria and Elisa Reach the Bus Stop

After a quick coffee and a nappy change for Elisa, Maria now has to complete her trip back home. She almost never uses public buses (Maria doesn't like them . . . but now it appears the only option).

As she arrives at the bus stop, she activates the Bus Stop services on her mobile device. She knows that the city bus stops offer travel information to mobile devices based on your location, but as she hardly ever travels on public transport she has never had cause to use it. She clicks on the option and it displays links to relevant travel information and the list of services normally operating from this stop. When Maria views the services, she sees they are nearly all cancelled; a banner running along the electronic sign at the bus stop explains: 'Due to industrial action there are cancellations and delays across the network today. Sorry for the inconvenience'. Maria notices a sign on the bus stop which explains it is possible to broadcast a pickup request to 'hop on' services that would normally operate around the city as well as the scheduled services.

Usually bus drivers tend to ignore pickup requests and leave people to wait for their scheduled services, but today is a little different and the few services that are running are trying their best to help everyone out. A bus driver close by receives the request.

F.9: The Bus Diverts to Pick up Maria and Elisa

Over the bus intercom, the driver asks the people already on the bus if they mind him diverting to pick up Maria. They could easily object by replying using their devices, but everyone is having trouble today so he doesn't receive any complaints. The driver confirms Maria's request from a button on his bus navigation system.

The most efficient route diversion is shown to him and a message alerting Maria that the bus is on its way is automatically sent back to her. As the bus comes into view, Maria's device notifies her that she can enable E-ticketing, which she selects. This makes things much easier boarding the bus as she has enough to deal with, loading Elisa's pushchair on without fiddling with loose change. As Maria boards the bus, the driver receives a simple notification on his navigation system that Maria is using E-ticketing and they start for home. The amount Maria is automatically debited for the bus ride will depend upon where she gets off.

Giovanni is also on his way back. He has met his lift successfully and will arrive home just after Maria and Elisa. The family is grateful to eventually all be heading home, but Maria is a bit annoyed with the mix-up by the garage and plans to complain to the insurance company when she gets home.

A3: The Sunday Scenario

The family is composed of a father, mother, and two kids living at home (Sue and Tim, twins, 13 years old). Their other daughter, Olivia (19), lives in another city nearby and attends the university. She has recently started dating Maurice, who is 21. Also involved are grandfather and grandmother. They live in another part of the country but are visiting for the weekend. This is a big family event as, due to their age and grandmother's poor health, they don't tend to travel much any more.

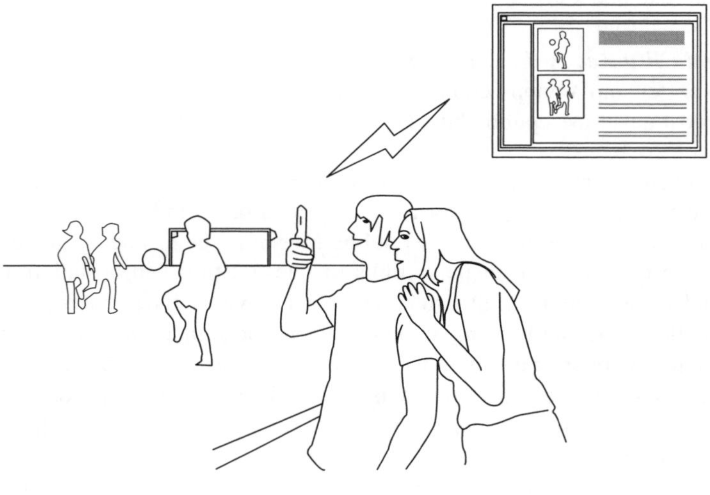

S.1: Sue and Tim are Playing a Video Game
Morning: Olivia contacts the family and an audio connection is opened up in the living room (since the parents are sleeping), where Sue and Tim are playing a video game.

S.2: Olivia Calls About the Family Dinner
Olivia knows that grandfather and grandmother are arriving today and she wants to check what time she should come over. Olivia mentions that she will bring her boyfriend for the family dinner. Sue promises Olivia she will tell dad, but seems so distracted by the video game that Olivia is not sure the message will get through.

S.3: Family Can Use a Shared Notice Board
Olivia decides to leave a message for her mum and dad.

Using her device, Olivia views the family's schedules which are presented to her all together. She blocks in a couple of hours in her dad's schedule in the early evening because she knows he is the one who usually cooks. Because her dad is yet to agree to this, the block is only grey and marked tentative. She links this to a voice message which she leaves posted on the family notice board area. The family notice board area can be viewed by any member of the family from any of their mobile devices, but it is also the default view for the screen on the fridge freezer in the kitchen. Funnily, even though Olivia doesn't live at home any more, she also uses the family's notice board as the default display for her fridge freezer in her flat. It allows her to keep in touch with what's happening at home, and still feel close to the family.

Sue and Tim hear the notification that a message has been left on the notice board by Olivia. They decide to have a bit of fun, and leave their game to go into the kitchen to add some 'extra' content to the message Olivia has left.

When the parents wake up, they are notified that Olivia has left a message. Dad can see Olivia has booked herself and her boyfriend in for dinner and changes the tentative setting to confirmed. He also adds a text message saying they should come earlier so they can help him prepare dinner and spend a bit more time with grandfather and grandmother. Olivia receives the notification and message. When dad opens the voice message left by Olivia, a huge image of a roast chicken flashes up on the screen accompanied by a sound clip of a cockerel crowing! Dad is bemused until he hears Sue and Tim sniggering at the doorway. Obviously they are dropping hints regarding what they would like for dinner when they return from football!

S.4: Olivia Sends a Message to her Boyfriend

Olivia receives the confirmation from her dad that it is OK for her boyfriend and her to come over today. Yesterday Olivia had discussed the possibility of going to her parents with her boyfriend Maurice. Olivia therefore decides to simply send Maurice a reminder.

Maurice receives the message. He replies instantly to Olivia, also he changes his emotional status toward Olivia from friendly to nervous. Olivia knows when Maurice changed his emotional status to nervous, he was just joking about meeting her mum and grandparents. She sets hers to loving and replies to him not to worry. They meet up soon afterwards and take the short bus ride together to Olivia's home town.

They arrive at the family home moments after Sue, Tim and mum have left. Sue's football team has a special match this afternoon, so mum has had to leave to take her to the game before grandmother and grandfather arrive. Tim decided to go along with them as some of his friends would also be at the game.

S.5: Grandfather and Grandmother Arrive

Grandfather and grandmother arrive soon afterwards. Everyone is happy to see them and they are glad to meet Maurice for the first time. Grandmother is disappointed to have missed the twins as she wanted to see Sue in her smart new football strip. Olivia tells her not to worry as if dad is agreeable they could stream video of the game from the club's event space, so she could actually watch the game from the family home.

Grandmother is amazed at what technology we have now, but she knows how things have moved on. She herself has a simple device which she wears around her neck given to her by the hospital. It allows her doctor to monitor her health status remotely and also issues reminders to her when she is due to take her medication. This makes life much easier for grandmother. Before she had the device she used to have to travel to the hospital a number of miles away every few weeks for a check-up, but now her doctor contacts her if he notices anything unusual in the health data sent by her device.

S.6: People Send Photos and Video Clips from the Game
Early afternoon: Sue's football game this afternoon is special. It is a sponsored charity match to help the club buy some extra equipment.

Before the game starts, Sue shows her training log to the coach to get comments on how her personal fitness programme is proceeding. She has been training hard, running almost every day, which is good because the sponsorship for today's game is linked to the distance the team manages to run during the 90 minutes! During the week Sue's device acts as a personal trainer for her, monitoring her training programme and suggesting improvements in both her exercise and dietary plans to help improve her fitness. Sue is very sporty and takes her exercise seriously. When she gets older, she would like to be a sport physiotherapist.

The coach establishes an event-based space for sharing media among the event subscribers. The other coach adds his group. During the game, the parents and other spectators will be able to both add and view captured content and information held within the event space.

S.7: The Family Watches the Game Remotely
Dad has never streamed pictures from Sue's football before. Olivia offers to do this for him, but dad would rather she showed him how to do it, so he can do it on his own next time. They are able to set up his device to receive a live video stream from the event space via mum's mobile device. They then send the pictures to the TV. Mum's mobile services are hosted by a different company from dad's. Her company doesn't charge to stream video like this, but dad's does. Because of this he will be charged for receiving the pictures of the game, but he doesn't mind as it's a special occasion with grandmother and grandfather visiting. As he is preparing dinner, dad watches the game on the TV in the kitchen, while Olivia, grandmother and the others watch the larger TV in the living room. They receive footage from the match as well as live real-time feedback of the amount of money the team are earning in sponsorship as the game is progressing. The teams are doing very well. Dad is happy for Sue but also worried, as this is going to cost him a fortune in sponsor money!

S.8: The Coach Receives Information on the Position and Activity of Each Player
The sponsorship works by using the team's coaching tool. Each team member is wearing a watch. This collects information about heart rate, speed and position, sending it back to the coach's device. He shares the information with all the parents and friends watching around the pitch in real time using the shared event space.

For this special occasion the coach has generated a spreadsheet application with a graph that is continually updated based on the sponsorship amount and distance run by the team. All the mums and dads around the pitch are viewing this. They are discussing how much it will cost them and laughing about it. Special praise is given to Sue by some of the parents as the system is showing she has currently run the furthest. Sue's mum can see how hard she is trying by her raised heart rate information.

Everyone is supporting the team and cheering when they reach the next increment in sponsor money. Some of the parents notice that the system is showing that Eric the goalkeeper appears very active even though the ball is currently at the other end of the field; when they look away from their devices down the pitch at him, they see even he is running up and down the goal line to raise extra money for the team!

S.9: Maurice and Grandfather Walk into Town

Before dinner: Olivia asks her boyfriend Maurice if he would mind walking into town to pick up a few things that her dad needs to finish preparing the dinner. Grandfather asks if he can go along with him, as he fancies stretching his legs before dinner and it would also give him a chance to get to know Maurice. Maurice doesn't know Olivia's town very well so he asks her to just send the list of things she wants him to buy to his mobile device. As they walk into town, he will use it to help him find suitable shops. Maurice requests the device to offer him suggestions. As they walk into town he keeps receiving information regarding nearby locations and shops that sell the items on Olivia's list. The information is provided discreetly, in a kind of 'silent push' manner. They work down the list buying what they need, chatting as they walk around.

Grandfather thinks Maurice's device is marvellous. Grandfather has a mobile device but it is a much older and simpler type, and can't do any of the things Maurice seems to be doing with his. Maurice shows grandfather how the reminder application works; grandfather locates a shop where they can buy olive oil (one of the items on Olivia's shopping list). He comments to Maurice that he is always being told off by grandmother for forgetting things when he goes shopping and this would really be useful for him!

Maurice and grandfather start to walk out of the town towards Olivia's parent's house. He receives an alert on his mobile device. Olivia has added an item to the list, some flowers for her mum (she wants Maurice to make a good impression with her mum). The device knows that Maurice is about to pass the last florist between here and Olivia's mum and dad's. He checks the recommendations on his device and sees the shop in question has a special offer on roses. As Maurice purchases the flowers, the transaction automatically cancels the reminder list and sends a message to Olivia. The message reads 'Maurice has completed this request'.

Maurice and grandfather soon arrive back at Olivia's mum and dad's. Sue and Tim also soon return with mum, who is very grateful for Maurice's gesture with the flowers. Dinner is soon ready and they all sit down to enjoy the afternoon together.

A4: The Olympics Scenario

The main characters are the father (Pekka), the mother (Adriana) and their son (Christian), who is 18 years old. Pekka and Adriana are visiting Turin from Hanover for the Winter Olympics. Christian is based this year at the University in Milan through the Erasmus programme.

O.1 Visiting Turin for the Winter Olympics: The Parents at the Hotel Take a Look at Their Day Plan

Adriana and Pekka start the day at their hotel in central Turin. Christian is not with them because he is studying hard for an exam.

Pekka and Adriana check their schedule for the day. Using her mobile device Adriana pulls up both Pekka's and her own schedules onto the screen in the hotel. This makes it easier to plan the day together, especially as Pekka's eyesight means he finds it difficult to read off the screen on Adriana's tiny device (he prefers a device with a larger screen). All of the events they have bought tickets for are there. Pekka has an E-ticket for the ice hockey tonight but he is planning to swap the ticket for tomorrow night as there is a Germany vs. Brazil football match

he wants to watch in the hotel bar later. Adriana has already made alternative arrangements; she will be going to see the ice skating tonight.

O.2: On the Bus to the Event Location: A Chance to Know More About the City and the Olympic Games

As the first events will be on the mountain pistes, they will need to take public transport there. They will use their virtual bus passes, which were inclusive with the event E-tickets.

Once on the bus, Pekka decides to browse his device for other local attractions they could visit while in Turin. The device pulls up a number of activities. In addition, the system knows Pekka is on the bus and the route, therefore as an Olympics special, one of the activities offered is an on-demand guided tour. He selects it and receives both audio and visual information related to the main sites of the city as they pass them on the way out to the Olympic venue.

As the bus arrives, Pekka is offered the chance to join a 'Ski fans' ad-hoc group. Pekka is a keen skier himself and a member of the German ski federation (DSV). Through its links with other European skiing federations, the DSV is putting on a number of events and activities during the Turin Olympics both virtual and on the pistes. All ski federation members in the crowd today have received the invitation. Pekka declines the opportunity to attend the events on the pistes, but joins the virtual group allowing him to receive special interactive event content filmed at the skiing venue for television, and also discuss the German skiers' performance with other fans at the event. This is really useful as the layout of the course means spectators can only see the skiers for a section of the run.

O.3: Christian Plans to Join his Parents in Turin

Back in Milan, Christian has found out that his exam has been rescheduled for next week – leaving him free to join his parents. He quickly sends them a message to let them know and checks the family's schedules to see where he should try to meet them. He can do this because the system recognises him as a trusted family member. This does not mean Christian can see everything his parents plan to do, but only the things they chose to share with the family group.

His parents are notoriously bad at adding plans to their schedule, but in this case Christian can see the events they have planned to attend because when they bought the E-tickets for events, reminders have been automatically added to their schedules.

He sees his father has a ticket for the ice hockey (Pekka hasn't got around to cancelling it yet!). Christian decides to try to join his father at the event. He is able to get a ticket by contacting the booking office, after which he sets off for the train station. As he boards the train for Turin, he gets a message from his mother asking him to meet them at the skiing event as they don't expect to be back at the hotel until late afternoon.

O.4: The Whole Family Meets in Turin

It is later and Christian has arrived in Turin. As he takes the bus trip up to the ski venue, he calls his mother to tell her he has arrived. She asks him to share his real time location so they can find him when he gets there.

Christian gets off the bus and mingles into the large crowd, but his mum and dad have been keeping an eye on his progress, and leave their seats in the stand to come down and meet him. Using their devices they are easily able to track him down in the crowd and the family meets up.

They watch the medals ceremony together from the public area before deciding to warm up in the mountain café at the venue. Christian tells Pekka he has a ticket for the ice hockey. Oh dear, thinks Pekka, he might not be able to watch the football after all!

O.5: Christian Gets an Invitation

The family are having hot chocolate together in the cafe. Christian receives an alert from his mobile device. Much to his surprise it is an invitation to join a group from his best friend Rolf. Rolf is the only one of Christian's friends who he gives the same level of trust to as his family; therefore when Christian showed his location to his most trusted group, which includes his mum and dad, he also inadvertently showed it to Rolf.

Rolf's invitation has a message attached saying he is also in Turin with friends and should meet up. When Christian selects to join the group, a number of his and Rolf's classmates are shown in Turin including a girl he really likes. Christian can see their names, but only the schedule of Rolf because only his device has been set to share with Christian at this highest level of trust. Christian decides to try to meet up with them.

O.6: Switching Today's Hockey Tickets for Tomorrow's Match

Christian sheepishly tells his parents that he wants to try to meet his classmates (especially this particular girl). Pekka, quite relieved, admits that he wanted to watch the football anyway. They agree to switch their hockey tickets for another match tomorrow, so they post a request to find someone who wants to swap tickets with them.

Pekka and Christian manage to swap their tickets with some other spectators for the following day's hockey game, as the two events had the same ticket value, simplifying the exchange.

O.7: Christian Plans a Night Out With Friends

As they journey back to the hotel, Christian turns his attention to setting up his plans for the evening. Rolf has broadcast a message to everyone saying they should all meet up later to eat. Christian asks for recommendations from his mobile device, but by doing this from within the group, the recommendations generated are based on everyone's pre-learned preferences rather than just his own. Top of the list is a jazz club where they can also get something to eat.

Each classmate gets a message about the meeting and receives directions (based on each person's location) to the place that Christian has chosen for the group.

O.8: Pekka has Everything Under Control, at Least Almost Everything!

Later, back in the hotel bar, Pekka is enjoying a glass of good wine, Adriana has left for the ice skating and Christian has met up with his friends. A large group of Italian guests are watching highlights from the day's Olympic events and celebrating the gold and silver their country won today. Pekka enquires about the football but the barman tells him that unfortunately the Germany vs. Brazil game is not being shown on Italian television tonight.

Pekka decides to retire to his room with his glass of wine to watch it there. He is able to receive the streamed live game broadcast on his mobile device which he then switches to show on the TV in the hotel room.

At half time Pekka decides to check when Adriana will be getting back from the skating. With a smile, he notices that Christian's location settings have been switched back to private.

Glossary

2D	Two-dimensional
3D	Three-dimensional
3GPP	Third-generation Partnership Project
3GPP2	Third-generation Partnership Project 2
AAA	Authentication, Authorisation and Accounting
API	Application Programming Interface
AS	Application Server
AWT	Abstract Window Toolkit
B2B	Business-to-business
BMSD	Best Matching Service Discovery
BSCW	Basic Service for Collaborative Working
BSS	Business Support System
BURP	Basic Usage Record Provider
CAF	Context Awareness Function
CB	Context Broker
CC	Context Consumer
CGI	Common Gateway Interface
CMF	Context Management Framework
CP	Context Provider
CRF	Context Representation Framework
CS	Context Source
CSCW	Computer Supported Cooperative Work
CSV	Comma-separated Values
CW	ContextWatcher
DB	Database
DBMS	Database Management System
DE	Dispatcher Engine
DL	Description Logics
DS	Data Source
ECG	Electrocardiogram
ECP	Enriched Context Provider
EDF	European Data Format

Enabling Technologies for Mobile Services: The MobiLife Book Edited by Mika Klemettinen

eTOM	Enhanced Telecom Operations Map
EU	European Union
FOAF	Friend Of A Friend
FP6	Framework Programme 6
GAF	Group Awareness Function
GCF	Group Context Function
GES	Group Evolution System
GMF	Group Management Function
GPRS	General Packet Radio Service
GPS	Global Positioning System
GSM	Global System for Mobile Communications
GUI	Graphical User Interface
HSS	Home Subscriber Server
HTTP	Hypertext Transfer Protocol
IBM	International Business Machines Corporation
ICT	Information and Communication Technologies
ID	Identity
IETF	Internet Engineering Task Force
IIS	Internet Information Services
IMS	IP Multimedia Sub-system
IMSI	International Mobile Subscriber Identifer
IN	Intelligent Network
IP	Internet Protocol
IrDA	Infrared Data Association
ISBN	International Standard Book Number
ISIM	IP Multimedia Services Identity Module
IST	Information Society Technologies
J2ME	Java 2 Platform, Micro Edition
JVM	Java Virtual Machine
KB	Knowledge Base
KML	Keyhole Markup Language
KR	Knowledge Representation
LAN	Local Area Network
LoD	Law of Demeter
MA	Management Action
ME	Matching Engine
MI	Management Interface
MNO	Mobile Network Operator
MRF	Multimedia Resource Function
MRL	Management and Rules Layer
MSN	Microsoft Network
NFC	Near Field Communication
NoU	Number of Users
OMA	Open Mobile Alliance
OMF	Operational Management Function
OPC	Operational Personal Context

OpenXDF	Open eXchange Data Format
OS	Operating System
OSS	Operation Support System
OWA	Open World Assumption
OWL	Ontology Web Language
PAN	Personal Area Network
PC	Personal Computer
PCF	Personal Context Function
PDA	Personal Digital Assistant
PDP	Policy Decision Point
PEP	Policy Enforcement Point
PF	Personalisation Function
PNO	Public Network Operator
POI	Point of Interest
PSP	Proactive Service Portal
PSP	Proactive Service Provisioning
PSTN	Public Switched Telephone Network
PTF	Privacy and Trust Function
QoC	Quality of Context
QoS	Quality of Service
RAN	Radio Access Network
RDF	Resource Description Framework
REST	REpresentation State Transfer
RF	Reasoning Function
RFID	Radio Frequency Identification
RPID	Rich Presence Information Data format
RSS	Really Simple Syndication
RuBaCRI	Rule-based Context Reasoning Interface
SCM	String Clustering Map
S-CSCF	Serving Call State Control Function
SD	Secure Digital
SDB	Service Database
SDP	Service Discovery Protocol
SER	Services and Applications
SGSN	Serving GPRS Support Node
SHA	Secure Hash Algorithm
SIG	Special Interest Group
SIP	Session Initiation Protocol
SLA	Service Level Agreement
SME	Small and Medium Sized Enterprise
SMS	Short Message Service
SOA	Service Oriented Architecture
SOAP	Simple Object Access Protocol
SOM	Self-organizing Map
SOUPA	Standard Ontology for Ubiquitous and Pervasive Applications
SP	Service Provider

SPF	Service Provisioning Function
SPS	Self-promoting Services
SUF	Service Usage Function
Subs DB	Subscription Database
SUMO	IEEE Suggested Upper Merged Ontology
TAN	Tree Augmented Naïve Bayesian classifier
TCP	Transmission Control Protocol
TE	Trust Engine
TS	Trust Seed
UCD	User-centric Design
UCI	University of California, Irvine
UDDI	Universal Description, Discovery and Integration
UE	User Equipment
UIAF	User Interface Adaptation Function
UI	User Interface
UICC	Universal Integrated Circuit Card
UMA	Unlicensed Mobile Access
UML	Unified Modelling Language
UMTS	Universal Mobile Telecommunications System
UOM	Unit Of Measure
UPnP	Universal Plug and Play
URI	Uniform Resource Identifier
URL	Uniform Resource Locator
WAN	Wide Area Network
Web 2.0	A phrase coined by O'Reilly Media in 2004, refers to a perceived or proposed second generation of web-based services – such as social networking sites, wikis, communication tools, and folksonomies – that emphasise online collaboration and sharing among users. Its exact meaning remains open to debate. (Wikipedia)
WEP	Wired Equivalent Privacy
WiFi	Wireless Fidelity (IEEE 802.11 wireless networking)
WLAN	Wireless Local Area Network
WS	Web Service
WSDL	Web Service Description Language
WWI	Wireless World Initiative
WWRF	Wireless World Research Forum
XCAP	XML Configuration Access Protocol
XDM	XML Document Management Enabler
XML	eXtensible Markup Language
Xpath	XML Path Language

Index

Enabling Technologies for Mobile Services: The MobiLife Book Edited by Mika Klemettinen